Gender, Climate Change and Livelihoods

Vulnerabilities and Adaptations

FSC
www.fsc.org
MIX
Paper from responsible sources
FSC® C022174

Gender, Climate Change and Livelihoods

Vulnerabilities and Adaptations

Edited by

Joshua Eastin

Portland State University, Portland, OR, USA

and

Kendra Dupuy

Chr. Michelsen Institute (CMI), Bergen, Norway

CABI

CABI is a trading name of CAB International

CABI
Nosworthy Way
Wallingford
Oxfordshire OX10 8DE
UK

Tel: +44 (0)1491 832111
Fax: +44 (0)1491 833508
E-mail: info@cabi.org
Website: www.cabi.org

CABI
WeWork
One Lincoln St
24th Floor
Boston, MA 02111
USA

Tel: +1 (617)682-9015
E-mail: cabi-nao@cabi.org

A catalogue record for this book is available from the British Library, London, UK.

Library of Congress Cataloging-in-Publication Data

Names: Eastin, Joshua, editor. | Dupuy, Kendra, editor.
Title: Gender, climate change and livelihoods : vulnerabilities and adaptations / edited by Joshua Eastin and Kendra Dupuy.
Description: Boston, MA : CAB International, [2021] | Includes bibliographical references and index. | Summary: "This book applies a gendered lens to the dynamic linkages between climate change and livelihoods in developing countries. It examines how climate change affects women and men in distinct ways, and the implications for earning income and accessing the natural, social, economic, and political resources required to survive and thrive"-- Provided by publisher.
Identifiers: LCCN 2021006731 (print) | LCCN 2021006732 (ebook) | ISBN 9781789247053 (hardback) | ISBN 9781789247060 (ebook) | ISBN 9781789247077 (epub)
Subjects: LCSH: Gender mainstreaming--Developing countries. | Climatic changes--Developing countries. | Women--Economic conditions--Developing countries.
Classification: LCC HQ1233 .G4524 2021 (print) | LCC HQ1233 (ebook) | DDC 305.309172/4--dc23
LC record available at https://lccn.loc.gov/2021006731
LC ebook record available at https://lccn.loc.gov/2021006732

References to Internet websites (URLs) were accurate at the time of writing.

ISBN-13: 978 1 78924 705 3 (hardback)
978 1 78924 706 0 (ePDF)
978 1 78924 707 7 (ePub)

DOI: 10.1079/9781789247053.0000

Commissioning Editor: David Hemming
Editorial Assistant: Emma McCann
Production Editor: Marta Patiño

Typeset by SPi, Pondicherry, India
Printed and bound in the UK by Severn, Gloucester

Contents

Contributors vii

I. Introduction

1 An introduction to gender, climate change and livelihoods 1
Joshua Eastin and Kendra Dupuy

II. Gendered livelihood vulnerabilities to climate change

2 A gendered approach to understanding climate change impacts: lessons from a coastal region of Bangladesh 11
Salim Momtaz, Muhammad Asaduzzaman and Zobaidul Kabir

3 Understanding the gendered impact of disasters on women, household dynamics and coping strategies: a case study of Bangladesh 26
Nahid Rezwana

4 Gender, weather shocks and food security: empirical evidence from Uganda 38
Francis Mwesigye

5 Managing livelihood in displacement: the politics of land ownership and embodied health and well-being by senior women in Kenya 56
Elizabeth O. Onyango

6 Seeing through water: gender, anxiety and livelihoods in large-scale infrastructural development in the era of climate change 69
Yvonne A. Braun

7 Caring for corn and beans: reassessing subsistence agriculture and climate change 82
Elisabeth Garner

8 Climate change, livelihoods and domestic violence in Indonesia 94
Joshua Eastin

III. Addressing the gendered impacts of climate change on livelihoods

9 Gender and climate-smart agriculture in Africa 107
Agnes A. Babugura

10 Gender differences in awareness and adoption of climate-smart agriculture practices in Bangladesh 123
Elizabeth Bryan, Edward Kato and Quinn Bernier

11 Gender and climate change adaptation in livestock production in Tunisia 143
Dina Najjar and Bipasha Baruah

12 The nexus between climate change, migration and gender 159
Diana Hummel

13 Gendered livelihood adjustments in the context of climate-induced disasters 174
Margaret Alston

14 Climate-induced migration, women and decision making power in the agricultural wage sector in Saiss, Morocco 185
Dina Najjar, Boubaker Dhehibi, Bipasha Baruah, Aden Aw-Hassan, and Abderrahim Bentaibi

15 Bringing women's livelihood and care perspectives into climate decision making 196
Patricia E. Perkins and Balikisu Osman

16 Gender mainstreaming in climate change adaptation strategies in Bangladesh and Nepal 211
Shahreen Shehwar

IV. Conclusion

17 Conclusion: final thoughts and future directions 220
Kendra Dupuy and Joshua Eastin

Index 225

Contributors

Margaret Alston, Professor of Social Work at University of Newcastle and Emeritus Professor at Monash University, 176 Brooks St, Bar Beach, Newcastle NSW 2300, Australia. Email: margaret.alston@newcastle.edu.au

Muhammad Asaduzzaman, School of Environmental and Life Sciences, University of Newcastle, 10 Chittaway Road, Ourimbah NSW 2258, Australia. Email: salim0707@gmail.com

Aden Aw-Hassan, Sustainable Intensification and Resilient Production Systems Program (SIRPSP), International Center for Agricultural Research in the Dry Areas (ICARDA), Amman, Jordan. Email: aawhassan@gmail.com

Agnes Babugura, 144 Peter Road, Ruimsig, Roodepoort 1725, Johannesburg, South Africa. Email: ababugura@gmail.com

Bipasha Baruah, Professor and Canada Research Chair in Global Women's Issues at the Department of Women's Studies and Feminist Research, University of Western Ontario, London, Canada. Email: bbaruah@uwo.ca

Abderrahim Bentaibi, Institut National de la Recherche Agronomique (INRA), Meknes, Morocco. Email: abentaibi@yahoo.fr

Quinn Bernier, Department of Sociology, Institute for Environment and Society, Brown University, Box 1951, 85 Waterman Street, Providence, RI 02912, USA. Email: quinn_bernier@brown.edu

Yvonne Braun, Department of Global Studies, University of Oregon, Eugene, OR 97403-1298, USA. Email: ybraun@uoregon.edu

Elizabeth Bryan, International Food Policy Research Institute (IFPRI/CGIAR), 1201 Eye Street, NW Washington, DC 20005-3915, USA. Email: e.bryan@cgiar.org

Boubaker Dhehibi, Sustainable Intensification and Resilient Production Systems Program (SIRPSP), International Center for Agricultural Research in the Dry Areas (ICARDA), Tunis, Tunisia. Email: b.dhehibi@cgiar.org

Kendra Dupuy, Chr. Michelsen Institute, Norway. Email: kendra.dupuy@cmi.no

Joshua Eastin, Mark O. Hatfield School of Government, Portland State University, 506 SW Mill Street, Suite 650, Portland, OR 97201, USA. Email: jeastin@pdx.edu

Elisabeth Garner, Department of Global Development, B75 Mann Library, Cornell University, NY 14853, USA. Email: eg65@cornell.edu

Diana Hummel, Goethe University, Frankfurt am Main and Institute for Socio-Ecological Research, Germany. Email: hummel@isoe.de

Zobaidul Kabir, School of Environmental and Life Sciences, University of Newcastle, 10 Chittaway Road, Ourimbah, NSW 2258, Australia; Ministry of Fisheries and Livestock, Government of Bangladesh. Email: kabirz85@hotmail.com

Edward Kato, International Food Policy Research Institute (IFPRI/CGIAR), Plot 15m East Naguru Road, Opp. KWESE and SYZ hotel, PO Box 28565, Kampala, Uganda. Email: e.kato@cgiar.org

Salim Momtaz, School of Environmental and Life Sciences, University of Newcastle, 10 Chittaway Road, Ourimbah, NSW 2258, Australia. Email: salim.momtaz@newcastle.edu.au

Francis Mwesigye, Economic Policy Research Centre (EPRC), Plot 51 Pool Road, Makerere University, PO Box 7841, Kampala, Uganda. Email: fmwesigye@eprcug.org

Dina Najjar, Social and Gender Specialist in the Social, Economics and Policy Research Group at the International Center for Agricultural Research in the Dry Areas (ICARDA), Rabat, Morocco. Email: d.najjar@cgiar.org

Balikisu Osman, Faculty of Environmental and Urban Change, York University, 4700 Keele St, North York, ON M3J 1P3, Canada. Email: bosman6@yorku.ca

Elizabeth Opiyo Onyango, Postdoctoral Fellow/Visiting Scholar at University of Waterloo, Balsillie School of International Affairs (BSIA), 67 Erb Street West, Waterloo, ON N2L 6C2, Canada. Email: eopiyoon@uwaterloo.ca

Patricia E. Perkins, Professor in the Faculty of Environmental and Urban Change, York University, 4700 Keele St, North York, ON M3J 1P3, Canada. Email: esperk@yorku.ca

Nahid Rezwana, Associate Professor, Department of Geography and Environment, University of Dhaka, Bangladesh. Email: nahid.rezwana@du.ac.bd

Shahreen Shehwar, Faculty of Environmental and Urban Change, York University, 4700 Keele St, North York, ON M3J 1P3, Canada. Email: sshehwar@yorku.ca

1 An Introduction to Gender, Climate Change and Livelihoods

J. Eastin and K. Dupuy

Climate change is imposing significant and lasting changes to the lives and livelihoods of much of the world's population. The accumulation of greenhouse gasses (GHG) in the atmosphere is increasing global temperatures, altering weather patterns, and magnifying the intensity and duration of climatic disasters (IPCC, 2014). The consequences of these changes – water shortages, food insecurity, social instability, threats to health and safety, and loss of life – are dire, especially for the most marginalized and vulnerable. Agricultural sectors and states that rely on agricultural production face acute risk because of the sensitivity of agriculture to climate disruption and because of the vulnerability of those who rely on it for subsistence and livelihoods (Wiebe *et al.*, 2019). Even if rigorous and comprehensive mitigation initiatives were to suddenly take hold within the world's most profligate GHG emitting states – a far-fetched scenario, at best – climate change has already generated tremendous pressure on the social, cultural, and natural resources necessary to survive and will continue to do so into the foreseeable future.

While the threat of climate change is global, the distribution of threat is both unequal and inequitable: unequal because some states are at greater risk than others, and inequitable because states facing the greatest risk tend to be the least responsible for causing climate change and least capable of mitigating or adapting to its effects (Okereke, 2010). These inequalities and inequities are also reflected within states as well as within vulnerable societal groups. The poor and marginalized are in greater danger than those with superior means because they are more exposed to climate change; more reliant on livelihoods that are sensitive to climate disruption; have fewer resources to invest in livelihood diversification strategies or other adaptations; and often lack the clout necessary to compel robust political action. Disparities persist even within impoverished groups, the most striking and broadly applicable of which are disparities between women and men (Goh, 2012; Eastin, 2018). Indeed, recognition that climate change poses greater risks for women has fostered the view of climate change as a 'double injustice,' a sobriquet signifying both the economic and gendered nature of climate change impacts (Otzelberger, 2014).

By 'gender' we refer to 'the characteristics of women and men... that are socially constructed... [which] include norms, behaviors, and roles associated with being a woman or man' (WHO, 2020, online). Climate change is a gender issue because gender reflects power – most critically, the power of resource access and control, the power to establish and enforce discriminatory cultural norms and legal institutions, and the power to make and implement public policy. The more power an individual has, the more resilient she should be to climate change. Yet, in the vast majority of societies, the social construction of gender leaves women at a

 Gender, Climate Change and Livelihoods: Vulnerabilities and Adaptations (eds J. Eastin and K. Dupuy)
DOI: 10.1079/9781789247053.0001

considerable disadvantage. For evidence, one need not look any further than the fact that women retain only one-quarter of all parliamentary seats in national governments around the world and only one-fifth of ministerial positions; or that women's average annual income is only 53% of that of men; or that women log an estimated 12 billion hours of unpaid care work each year, which are not economically 'valued' and rarely, if ever, reflected in economic policy making (WEF, 2019). Among those for whom climate change poses the greatest threat are rural and Indigenous women, a circumstance which reflects the intersectional nature of societal power differentials, and by extension vulnerability to climate change, as well as the climate sensitivity of agriculture, the predominant rural livelihood for women in developing states. However, while the double injustice persists, acknowledging and acting upon the gendered nature of climate change also illuminates pathways to more sustainable, effective, and just means of bolstering societal resilience and encouraging adaptation.

This book employs gender as a lens to examine the complex linkages between climate change and livelihoods in some of the world's most vulnerable states and regions. The chapters address how climate change affects women and men in distinct ways, and what implications these effects hold for earning income and for accessing the natural, cultural, social, economic, legal, and political resources required to survive and thrive in society. The authors in this volume draw on case research from countries and regions spanning five continents to reflect on how public policies and power dynamics shape gendered climate change vulnerabilities, how gender influences coping and adaptation, and how civil society organizations and other activists incorporate gender into their climate advocacy strategies. The book adopts a livelihoods focus because livelihoods – 'the opportunity set of capabilities, assets, and activities that are required to make a living' (Olsson *et al.*, 2014, p. 728) – are highly gendered, because they are central to ensuring subsistence and development, and because livelihoods, especially those such as agriculture which rely on sensitive natural resources, are under threat from a combination of climate change and population pressures. A livelihoods focus also provides an opportunity to unite and learn from multiple distinct scholarly perspectives and research traditions; it 'provides an institutional bridging function linking people, professions, and practices' in order to facilitate the adoption of a gender-based perspective on climate mitigation and adaptation (Scoones, 2009, p. 172).

Towards Increased Policy Engagement of Gender and Climate Change

It has been nearly 30 years since the United Nations adopted the UN Framework Convention on Climate Change (UNFCCC, 1992). Although the treaty was a major achievement at the time, the role of gender was not considered; the word does not appear in the text even once. The same is true for 'sex,' 'women,' 'woman,' 'female,' or any other word that recognizes, much less emphasizes, gender as a concept relevant to climate change. The exclusion is not surprising; the 1997 Kyoto protocol, the first legally binding accord to emerge under the Framework Convention, is similarly silent (UN, 1997). It is not until the signing of the Paris Climate Agreement (PCA) in 2015, that 'gender' appears in a global climate change treaty. The PCA acknowledges in the Preamble that Parties should 'when taking action to address climate change, respect, promote and consider their respective obligations... [to] gender equality [and the] empowerment of women;' and calls for Parties to be 'gender responsive' when considering adaptation (Article 7) and capacity building (Article 11) measures (UN, 2015, p. 2 and p. 9). Though these references are significant, the treaty falls short in ignoring the role of gender in relation to climate mitigation (Articles 4, 5, and 6), finance (Article 9), and technology (Article 10). Without clear guidance on the importance of gender to all aspects of climate policy, there is a risk that member states will ignore it.

Fortunately, scholars and practitioners have helped to clarify the complex linkages between the environment and development, a process that has increased awareness of how the effects of climate change manifest among different social groups, including women (Brody *et al.*, 2008; Buechler, 2009; Angula, 2010; Dillon and Gill, 2014). Recent years have witnessed greater recognition of the gendered impacts of climate change, and there has been increased concern in policy circles about how to address gender in climate adaptation and mitigation initiatives. This focus is critical; women comprise the majority of the world's poor and are generally more vulnerable to climate change than men (IUCN, 2007; Quan and Dyer, 2008). Women are also vulnerable in different ways. For example, women face greater risk from climate disasters, not only because they are poorer, but also because they are over-represented in climate-sensitive labor sectors, because social customs such as dress codes and greater familial and household burdens can threaten personal safety during escape, and because post-disaster breakdowns in social order and gender discrimination in disaster relief provision can hamper women's recovery capabilities (Neumayer and Plümper, 2007). If policy makers fail to take these considerations into account, then measures aimed at reducing climate vulnerability or enhancing adaptation capacity will fall short.

Despite the PCA's limitations, efforts to place gender and gender-based analysis at the forefront has compelled the UNFCCC to take action. In 2017, member states adopted the first ever comprehensive Gender Action Plan (UN, 2017), and in 2019 the Secretariat produced a synthesis report on gender-differentiated climate change impacts and the integration of gender into national climate policies (UN, 2019). The report stresses the role of climate-sensitive livelihoods, especially agricultural livelihoods, to local and Indigenous communities, as well as the effects of climate change on biomass collection and food production. The livelihoods focus is welcome, as it opens the door to addressing gendered livelihood vulnerabilities and can encourage adaptations that facilitate livelihood diversification and enhance climate resilience. Developing states stand to gain the most from a gendered livelihood focus; they rely more heavily on industries vulnerable to climate disruption and possess fewer resources to invest in adaptation and mitigation efforts (IPCC, 2014). Moreover, increased awareness of how climate change will affect the livelihoods of different groups of people (such as women) is essential to designing and implementing effective policy interventions.

Institutions such as the United Nations that have embraced the role of gender in climate change initiatives have met with some success. In Lesotho, for example, the UN-funded project, 'Women as Agents of Change for Adaptation,' has retained a group of women volunteers to act as 'solar ambassadors' in their villages to help encourage the adoption of solar technologies in order to meet domestic energy requirements and reduce the need to collect scarce vegetation for cooking and heat (UN-Lesotho, undated). In the Mauritian village of Grand Sable, another UN-backed project has sought to reduce climate vulnerability and facilitate livelihood diversification by offering training on the development of mangrove plantations and the cultivation of vetiver, seaweed, and medicinal plants (UN-Mauritius, undated). Project outcomes have enhanced livelihood opportunities for all residents, especially women, through handicraft and herb sales and through increased biological productivity in coastal waters. The plants under cultivation have also offered the village, which is situated between the mountains and the sea, increased protection from hillside (vetiver) and coastal (mangrove) erosion and storm surges, which has helped reduce waterlogging and soil depletion in the fields. In both of these projects, a core component of success has been the recognition that '... women are most likely to suffer from climate change but they are also the most capable of creating change and adaptation within their communities' (UN-Mauritius, undated). While we view these efforts as positive, and the recognition of gender as a step in the right direction, much work remains to be done.

Gender, Livelihoods and Climate Change in Developing Countries

A substantial portion of the global population depends on agriculture and agricultural resources for their livelihoods. This dependence applies not only to the more than 1 billion people who are formally employed in agricultural sectors, but also to individuals who buy and sell crops, farm for subsistence, and who rely on the gathering of biomass to generate energy (World Bank, 2020a). The Food and Agriculture Organization (FAO) estimates that at least 2.5 billion people worldwide draw their livelihoods from agricultural pursuits (FAO, 2012). Climate change poses significant threats to the quantity and quality of resources necessary to maintain these livelihoods, as increasing climate variability alters the predictability of regional weather patterns, exacerbates drought, and elevates the threat of storms, floods, and fires. The world's poor face the greatest risk. Developing countries, which host 96% of the world's estimated 570 million farms, rely disproportionately on agriculture for population subsistence and economic production, and are also among the least capable of adapting to climate change through livelihood diversification (Lowder *et al.*, 2016). Absent effective mitigation or adaptation measures, the impacts of climate change on these states will only reinforce vicious cycles of increasing economic and social inequalities, decreasing recovery capabilities, and greater levels of vulnerability.

As we discuss throughout the book, the burdens climate change places on agriculture and agricultural livelihoods are not borne equally, even among the world's poor. Group-level variations in social and economic status place some at greater disadvantage than others. Gender is an important determinant of these power differentials, and intersects with other identity markers such as age, ethnicity, race, and class in ways that magnify disempowerment. On average, women enjoy less favorable access to credit, land tenure rights, agricultural technologies and inputs, and other assets critical for livelihoods than do their male counterparts (Behrman *et al.*, 2014; Huyer, 2016). Climate change threatens to deepen these gender-based inequalities, especially when it makes natural resources scarcer and more difficult to secure.

The gendered impacts of climate change on livelihoods

Gender, livelihoods, and climate change intersect in multiple and complex ways. A comprehensive review of the literature has identified six areas in which the consequences of climate change – increased rainfall, rising sea levels, flooding, drought, and elevated temperatures – affect women and men differently: agricultural production; food security; health; water and energy resource access; climate-induced migration; and climate-related natural disasters (Goh, 2012). However, the reasons for the disparities often have little to do with gendered variations in biology or capabilities; rather, they tend to reflect 'the persistence of discriminatory laws and social norms that restrict women's rights and opportunities' (Eastin, 2018, p. 290). For example, within the home, common gender inequities in household and familial obligations diminish opportunities for women to participate in the formal labor force, which can increase food insecurity, damage health, reduce social networks, and undermine attempts at livelihood diversification and other adaptation measures. Outside the home, gendered labor force discrimination can reinforce these dynamics, further threatening female health and resource access and increasing disaster vulnerability.

An individual's livelihood is a function of her access to assets, including natural, physical, financial, social, and political. This access informs the nature and content of a person's productive pursuits, as well as her ability to adapt to societal, economic, and environmental changes (Olsson *et al.*, 2014). Livelihoods, especially those that are vulnerable to climate change, are often gendered because ownership and control over the natural resources required to maintain them are gendered (Nelson and Stathers, 2009). Social and political institutions – such as formal laws and cultural norms – dictate the rules of resource access in ways that privilege some and disempower others. Gendered land tenure and inheritance laws, for example, reduce women's access to credit markets, which in turn decreases the availability of agricultural resources, inputs, and technologies necessary to maintain or enhance productivity (Deere and Leon, 2003; Nuryartono, 2005; Scalise, 2009). The overall effect is that women and women's livelihoods are not only more vulnerable to climate change, women also have a decreased capacity to cope with and adapt to changes when they occur. Moreover, because these institutions are often context specific, the gender-differentiated effects of climate change on livelihoods will vary as well. In some places, women will be affected more in certain ways than in others.

A focus on livelihoods provides a conceptual link that enables us to better examine the complex impacts of climate change on both development and gender relations. For instance, agricultural work affects food security by providing subsistence and income. Water and energy access are household assets that directly influence food security, but they are also productive assets in that they are required to maintain agricultural livelihoods. Food security shapes health, and health affects an individual's ability to engage in agricultural production. When climate pressures encourage male outmigration for labor, there can be a shift in the gendered dynamics of a household's livelihoods, if, for example, women must secure new livelihoods or take on additional livelihood burdens to cope (Alston, 2015). Natural disasters such as droughts and floods affect what livelihoods women and men can engage in via their impact on resource quantity and quality, access to which also faces gendered constraints (Schwoebel and Menon, 2004). Cultural barriers to mobility and labor force participation shape gendered livelihood strategies as well, in addition to their impact and effectiveness.

Agriculture is the primary livelihood for a large number of women in developing countries. FAO statistics indicate that 63% of women in the least developed countries are agricultural workers, including 52% of women in sub-Saharan Africa (SSA) and 56% in South Asia, though there is considerable intra-regional variation, especially among the countries of SSA. In Angola, for example, 56% of women are employed in agriculture; in Uganda, the figure is 77%; and in Mozambique, 80%. In Burundi, 96% of women work in the agricultural sector (World Bank, 2020b).

Agriculture is also among the most sensitive sectors to climate change, especially in Africa, a region which contains the largest share of the world's poor, is the least infrastructurally developed, the most reliant on rainfall irrigation, and possesses the lowest access to agricultural inputs such as seeds and fertilizer. The Intergovernmental Panel on Climate Change (IPCC) predicts that if average global temperatures increase by 2°C, the target threshold of the PCA, the results for many African countries will be catastrophic (Niang *et al.*, 2014). At 2°C, southern Africa is expected to lose 20% of its current annual precipitation, which will lead to longer and more powerful droughts and dry spells, and a partial draining of the Zambezi Basin, the home of 32 million people, 80% of whom rely on the water for consumption and irrigation. When the rains do come, the parched earth will be less absorbent and the storms more intense, which will increase the risk of erosion and floods. All regions of the continent will see increases in temperatures, some quite significant, which will generate additional pressure on beleaguered water supplies and crops through evaporation and evapotranspiration. The Sahel region faces substantial threat through a combination of rising populations, declining resource availability and access, and anticipated temperature increases expected to equal 150% of the global average. These factors generate considerable challenges for food security, and are abetted by armed conflict, communal violence, and other forms of civil and political strife that pervade the region. The result has been – and will likely continue to be – unrelenting cycles of hunger, famine, and humanitarian emergency. These projections only reflect 2°C warming. Accounting for current GHG emission rates in 2020, as well as the nationally determined contributions (NDC) already submitted under the PCA, somewhere north of 3°C looks to be more likely (UNEP, 2019).

Given these projections, as well as the significant role women play in agriculture, their greater vulnerability to the effects of climate change, and challenges women face with livelihood diversification, incorporating gender responsiveness into climate change initiatives and addressing gender disparities in agricultural livelihoods are pressing concerns. Doing so can facilitate more just and equitable approaches to addressing climate change, as well as enhancing their effectiveness. The benefits of such a gender-based approach will emerge in the form of greater societal resilience, reduced vulnerability, and improved economic, social, and political conditions for some of the world's most at-risk populations.

Contributions in This Book

This book applies a gender lens to examine the implications of climate change for livelihoods in vulnerable states. The goals are to enhance awareness of climate change as a gender issue, and to highlight the importance of gender in identifying livelihood vulnerabilities and in designing more robust climate adaptation measures, especially in climate-sensitive industries such as agriculture. The contributions in this book examine how the consequences of climate change affect women and men in different ways, and address the implications of climate change for women's livelihoods and resource access. The book's topical coverage is not limited to a single region or time period, though it devotes greater attention to cases such as Bangladesh, which reflect high levels of climate vulnerability, low adaptation capacity, and which are characterized by restrictive gender norms and laws that preserve gender-based livelihood disparities. While one might argue that a country like Bangladesh is atypical in terms of its vulnerability to climate change, it is precisely cases such as this that provide opportunity to elucidate key causal dynamics. Moreover, given that climate change remains an ongoing process, we might expect contemporary Bangladesh to be a precursor to the type of environmental conditions many more states can expect in coming years.

The book is organized into two sections. The first, *Gendered Livelihood Vulnerabilities to Climate Change* (Chapters 2–8) examines disparities in the vulnerability of women's and men's livelihoods to climate change. The chapters in this section address issues such as gender inequalities in the household distribution of labor; differential access to agricultural livelihood inputs and assets; gender-based threats to personal safety and security; and gendered vulnerability to and experiences with climate disasters, food insecurity, and infrastructure development.

In Chapter 2, Salim Momtaz *et al.* employ the sustainable livelihoods framework and disaster crunch model to investigate gender disparities in livelihoods, livelihood vulnerabilities, and coping capacity in coastal areas of Bangladesh. Their analysis demonstrates how exposure to climate disasters influences the gender distribution of livelihood capital in the region, shapes women's efforts at climate adaptation, and threatens their health and well-being. The study also highlights the ways in which women overcome inflexible gender relationships to develop coping mechanisms for the socio-economic and ecological effects of climate change, and recommends a series of policy measures aimed at reducing women's vulnerability.

Nahid Rezwana, also writing on Bangladesh, adds to our understanding by exploring the gendered effects of climate disasters on women's experiences within the household in Chapter 3, especially regarding exposure to gender-based violence. Drawing on a wealth of firsthand interviews with rural women, Rezwana illustrates how women cope with disasters, how they survive, and how they support their families during and after the crisis periods. She also identifies the various ways these processes shape and reinforce women's pre-existing vulnerabilities to climate change. Her recommendations include the adoption of gender-sensitive disaster management plans and the incorporation of local knowledge to combat future disasters more successfully.

Francis Mwesigye's study in Chapter 4 emphasizes important connections between gender inequality, women's role in agriculture, and threats to food security in Uganda. He finds weather shocks – droughts, floods, and irregular rainfall – to be the greatest cause of food insecurity among his survey respondents and for female-headed households to suffer disproportionately higher levels of food insecurity and for longer periods of time than those headed by men. His work indicates the need for policies aimed at enhancing gendered social safety net programs as a means of mitigating the effects of climate change and reducing climate vulnerability.

Elizabeth Onyango's contribution in Chapter 5 takes the reader to Kenya to examine how the combination of climate change and a history of unjust land tenure arrangements have exacerbated gender inequalities, fostered displacement and dispossession, and diminished women's livelihood options in the country. She pays particular attention to how these issues affect the lives, livelihoods, and health and well-being of senior women, who often act as caregivers to children and grandchildren. Her analysis offers a unique perspective on the potential for climate change to influence women's lives and livelihoods beyond traditional accounts of economic and agricultural productivity.

Yvonne Braun's chapter (Chapter 6) evaluates the gendered impacts of infrastructure development – large-scale dam projects – in Lesotho. She takes a narrative approach, drawing on ethnographic research, to document how the push for water resource commodification in the face of changing climatic conditions in the southern African region created unintended consequences for the lives and livelihoods of residents in proximate rural communities in Lesotho, including: a diminution of food security and water access; increases in women's household responsibilities; heightened anxieties and fears about the future; and the loss of connections and relationships between communities distanced by the newly created lakes. Her research also emphasizes the need for incorporating local voices, knowledge, and perspectives as a means of mitigating these consequences, an approach that was not taken prior to the development of the Lesotho Highlands Water Project.

In Chapter 7 Elisabeth Garner draws on her research in western Honduras to enhance our awareness of the links between a woman's familial responsibilities and her productive pursuits. Garner argues that the divides between these spheres are not always meaningful, and that women's role in subsistence agriculture is better understood as 'care work' because of the vital role these efforts play in providing household and community food security. Viewing subsistence agriculture in this way, she argues, is necessary to appreciate why women are more likely to participate in subsistence agriculture than in farming crops of greater economic value, the socio-economic and gendered systems that influence this participation, and to establish a more robust conceptual framework for analyzing women's vulnerability to climate change.

Finally, Joshua Eastin's chapter (Chapter 8) in this section examines the impact of climate change on domestic violence in Indonesia. Through statistical analysis, he demonstrates that high levels of climate variability in December, the core month of the rice-planting season, lead to increases in the incidence of domestic and intimate partner violence the following year. Disruptions to livelihoods, especially agricultural livelihoods, caused by erratic weather patterns exacerbate economic hardship and undermine subsistence, the result of which aggravates psychological pressures commonly associated with domestic violence. At the same time, the economic impact of climate variability on women's lives and livelihoods reduces their ability to flee abusive situations, especially in the context of discriminatory gender norms and laws that emphasize male authority and control.

The second section of the book, *Addressing the gendered impacts of climate change on livelihoods*, takes a gender-based view of various climate adaptation initiatives in areas that rely on agriculture for subsistence and production. The contributions in this section address gender-inclusive participation in climate policy planning and decision making, the role of gender in livelihood adaptation measures, and any successes, failures, or opportunities for improvement that emerge from these efforts.

Two chapters in this section evaluate 'climate-smart agriculture' (CSA) as a means of alleviating the impact of climate change on women's livelihoods. In Chapter 9, Agnes Babugura provides a conceptual overview of CSA, and discusses its implementation in various African states. Drawing lessons from examples of effective CSA practices across the continent, Babugura echoes others in the field when she argues for the necessity of incorporating gender in the planning and execution of CSA initiatives. Elizabeth Bryan *et al.* build on these insights in Chapter 10 to examine gender disparities in the awareness and adoption of CSA practices among residents of rural villages in Bangladesh. Through survey analysis, these authors find that access to information on CSA is low for everyone, but more so for women than men. They also find that when made aware of CSA practices, women are equally, if not more likely to adopt them, and that women are more receptive to adopting CSA practices that correspond to gendered household responsibilities. Among other things, these findings highlight the need for more robust efforts to increase information dissemination, especially to women, in order to expand its use in communities vulnerable to adverse climate change impacts.

In Chapter 11, Dina Najjar and Bipasha Baruah discuss women's participation in another agricultural livelihood strategy: livestock and rangelands management. Their analysis centers on two regions in southern and central Tunisia, where surveys and focus groups reveal individuals' perceptions of rangeland degradation, the gendered impacts of such degradation on pastoral livelihoods, and the adaptation strategies that both men and women adopt to respond to changing environmental conditions. Najjar and Baruah find that while women in these areas are much more involved in rangeland grazing than has been previously understood, the success of their participation is contingent on building awareness of their role, the integration of gender concerns in drought mitigation and other livelihood-related projects, and on the inclusion of women in rangeland-related decision making.

The next three chapters each touch on one of the most prevalent livelihood adaptations to climate change – migration. Diana Hummel provides an expansive analysis of the gendered aspects of climate migration in Chapter 12. She draws from scientific literature on climate change and migration, various conceptual frameworks for integrating gender, and on empirical research on the West African Sahel to demonstrate the complex, multi-causal, and context-dependent relationship between these phenomena. In Chapter 13, Margaret Alston adds to our understanding of gender, climate change, and migration nexus through her discussion of the effects of climate disasters on women in rural agricultural communities. Alston draws on original research from Lao PDR, Bangladesh, and Australia to reveal how livelihood adaptations such as migration reconfigure some aspects of gender relations, while reinforcing others. She finds that while male outmigration compels an increase in women's participation in agricultural labor and food procurement, limitations on women's access to necessary agricultural inputs, training, and technologies remain. However, these cases also reveal that increases in women's contributions to agricultural production have facilitated a relaxation of discriminatory cultural institutions and restrictions on women's involvement in public affairs. In Chapter 14, Dina Najjar and her co-authors draw from an original survey of landless laborers in Morocco, to examine labor migration and the determinants of female migrants' control over decision making in issues related to employment, income expenditures, and assets. Their findings indicate that women are actively engaging in wage-labor migration as a mechanism for livelihood diversification in response to drought, a result which counters the stereotype of women as 'left behind' when husbands migrate for work. However, these authors also find traditional decision making patterns over income and assets to exert sway over migrant communities, suggesting a need for active interventions to overcome discriminatory social norms and institutions.

The final two contributions address the development of climate change adaptation initiatives and policy from both bottom-up and top-down perspectives. In Chapter 15, Ellie Perkins and Balikisu Osman address the role of grassroots community-based organizations, women's groups, and activists in confronting the gendered effects of climate change. These authors pay careful attention to the role of livelihood provisioning and care in these movements' initiatives, as well as their efforts at organizing to enhance women's climate resilience. Overall, this chapter provides a strong case for increasing the availability and sharing of data on women's labor and of encouraging women's participation in climate governance as a means of enhancing the effectiveness of climate policy. Finally, Shahreen Shehwar examines in Chapter 16 the role of gender and gender mainstreaming in the formation of climate policy in Bangladesh and Nepal. She addresses the effects of climate change for rural women in these countries, appraises policies aimed at reducing women's vulnerability, and reflects on the attention to gender concerns in national adaptation initiatives. Overall, she argues that Bangladesh and Nepal have much to learn from one another in developing policy to meet gender-specific threats to women, especially rural women.

We conclude with a chapter (Chapter 17) that connects the book's various ideas, arguments, and findings; discusses the book's themes in light of the global COVID-19 pandemic, which is ongoing as of this writing; and provides some directions and ideas for future research.

References

Alston, M. (2015) *Women and Climate Change in Bangladesh*. Routledge Women in Asia series. Routledge, Abingdon, UK.

Angula, M. (2010) Gender and climate change: Namibia case study. Heinrich Böll Foundation, Southern Africa, Cape Town. Available at: www.boell.de/en/navigation/climate-energy-Namibia-9078.html (accessed 1 February 2021).

Behrman, J.A., Bryan, E. and Goh, A. (2014) Gender, climate change, and group-based approaches to adaptation. International Food Policy Research Institute, Policy Note: 3–6. Available at: www.ifpri.org/publication/gender-climate-change-and-group-based-approaches-adaptation (accessed 1 February 2021).

Brody, A., Demetriades, J. and Esplen, E. (2008) Gender and climate change: mapping the linkages – a scoping study on knowledge and gaps. BRIDGE Institute of Development Studies (IDS), UK. Available at: www.bridge.ids.ac.uk/sites/bridge.ids.ac.uk/files/reports/Climate_Change_DFID.pdf (accessed 1 February 2021).

Buechler, S. (2009) Gender, water, and climate change in Sonora, Mexico: implications for policies and programmes on agricultural income-generation. *Gender and Development* 17, 51–66.

Deere, C.D. and Leon, M. (2003) The gender asset gap: land in Latin America. *World Development* 31, 925–947.

Dillon, A. and Gill, J. (2014) The gender-differentiated impact of climate variability on production possibilities: evidence from cereal production in Mali. In: Ringler, C., Quisumbing, A.R., Bryan, E. and Meinzen-Dick, R. (eds) *Enhancing Women's Assets to Manage Risk Under Climate Change: Potential for Group-Based Approaches*. International Food Policy Research Institute, Washington, DC, pp. 33–35.

Eastin, J. (2018) Climate change and gender equality in developing states. *World Development* 107, 289–305.

FAO (2012) FAO statistical yearbook. Food and Agriculture Organization of the United Nations, Rome. Available at: www.fao.org/3/i3107e/i3107e01.pdf (accessed 1 February 2021).

Goh, A. (2012) A literature review of the gender-differentiated impacts of climate change on women's and men's assets and well-being in developing countries. *CAPRR Working Paper No.106*. International Food Policy Research Institute, Washington, DC. Available at: www.ebrary.ifpri.org/cdm/ref/collection/p15738coll2/id/127247 (accessed 1 February 2021).

Huyer, S. (2016) Closing the gender gap in agriculture. *Gender,Technology and Development* 20, 105–116.

IPCC (2014) *Climate Change 2014: Impacts, Adaptation, and Vulnerability*. Intergovernmental Panel on Climate Change, Cambridge University Press, Cambridge, UK.

IUCN (2007) *Gender and Climate Change*. International Union for the Conservation of Nature, Gland, Switzerland. Available at: www.iucn.org/sites/dev/files/import/downloads/climate_change_gender.pdf (accessed 1 February 2021).

Lowder, S.K., Skoet, J. and Raney, T. (2016) The number, size, and distribution of farms, smallholder farms, and family farms worldwide. *World Development,* 87, 16–29.

Nelson, V. and Stathers, T. (2009) Resilience, power, culture, and climate: a case study from semi-arid Tanzania, and new research directions. *Gender and Development* 17, 81–94.

Neumayer, E. and Plümper, T. (2007) The gendered nature of natural disasters: the impact of catastrophic events on the gender gap in life expectancy, 1981–2002. *Annals of the Association of American Geographers* 97, 551–566.

Niang, I., Ruppel, O.C., Abdrabo, M.A., Essel, A., Lennard, C., Padgham, J. and Urquhart, P. (2014) Africa. In: Barros, V.R., Field, C.B., Dokken, D.J., Mastrandrea, M.D., Mach, K.J. *et al.* (eds) *Climate Change 2014: Impacts, Adaptation, and Vulnerability*. Part B: Regional Aspects. Contribution of Working Group II to the Fifth Assessment Report of the Intergovernmental Panel on Climate Change. Cambridge University Press, Cambridge, UK and New York, pp. 1199–1265.

Nuryartono, N. (2005) Impact of smallholders' access to land and credit markets on technology adoption and land use decisions: the case of tropical forest margins in Central Sulawesi, Indonesia. Cuvillier Verlag, Göttingen, Germany.

Okereke, C. (2010) Climate justice and the international regime. *Wiley Interdisciplinary Reviews: Climate Change* 1, 462–474.

Olsson, L., Opondo, M., Tschakert, P., Agrawal, A., Eriksen, S.H. *et al.* (2014) Livelihoods and poverty. In: Barros, V.R., Field, C.B., Dokken, D.J., Mastrandrea, M.D., Mach, K.J. *et al.* (eds) *Climate Change 2014: Impacts, Adaptation, and Vulnerability*. Part A: Global and Sectoral Aspects. Contribution of Working Group II to the Fifth Assessment Report of the Intergovernmental Panel on Climate Change. Cambridge University Press, Cambridge, UK and New York, pp. 793–832.

Otzelberger, A. (2014) Tackling the double injustice of climate change and gender inequality. CARE International. Available at: www.carefrance.org/ressources/themas/1/4442,CARE_COP20_Tackling-double-injustic.pdf (accessed 1 February 2021).

Quan, J. and Dyer, N. (2008) Climate change and land tenure: the implications of climate change for land tenure and land policy. *IIED/FAO Land Tenure Working Paper No 2*. London and Rome. Available at: www.fao.org/3/a-aj332e.pdf (accessed 1 February 2021).

Scalise, E. (2009) Women's inheritance rights to land and property in South Asia: a study of Afghanistan, Bangladesh, India, Nepal, Pakistan, and Sri Lanka. Report by the Rural Development Institute for the World Justice Project. Available at: www.landesa.org/wp-content/uploads/WJF-Womens-Inheritance-Six-South-Asian-Countries.FINAL_12-15-09.pdf (accessed 1 February 2021).

Schwoebel, M.H. and Menon, G. (2004) Mainstreaming gender in disaster management support projects. Center for Development and Population Activities, Washington, DC. Available at: www.gdnonline.org/resources/DMS%20Report_WID%20India_6%2007%2004.pdf (accessed 1 February 2021).

Scoones, I. (2009) Livelihoods perspectives and rural development. *The Journal of Peasant Studies* 36, 171–196.

UN (1997) Kyoto Protocol to the United Nations Framework Convention on Climate Change. United Nations, New York. Available at: www.unfccc.int/resource/docs/convkp/kpeng.pdf (accessed 1 February 2021).

UN (2015) Paris Agreement. United Nations, Paris. Available at: www.unfccc.int/sites/default/files/english_paris_agreement.pdf (accessed 1 February 2021).

UN (2017) Gender Action Plan, United Nations Framework Convention on Climate Change. United Nations, Bonn, Germany. Available at: www.unfccc.int/resource/docs/2017/sbi/eng/l29.pdf (accessed 1 February 2021).

UN (2019) Differentiated impacts of climate change on women and men; the integration of gender considerations in climate policies, plans and actions; and progress in enhancing gender balance in national climate delegations. Synthesis report. United Nations, New York. Available at: www.unfccc.int/sites/default/files/resource/sbi2019_inf8.pdf (accessed 1 February 2021).

UN-Lesotho (undated) Women as agents of change for adaptation – Lesotho. United Nations, New York. Available at: www.unfccc.int/climate-action/momentum-for-change/activity-database/women-as-agents-of-change-for-adaptation (accessed 1 February 2021).

UN-Mauritius (undated) Enhancing the livelihood of women at grand sable in response to climate change impacts – Mauritius. United Nations, New York. Available at: www.unfccc.int/climate-action/momentum-for-change/activity-database/enhancing-the-livelihood-of-women-at-grand-sable-in-response-to-climate-change-impacts (accessed 1 February 2021).

UNEP (2019) Emissions Gap Report. United Nations Environment Programme, Nairobi, Kenya. Available at: www.unenvironment.org/resources/emissions-gap-report-2019 (accessed 1 February 2021).

UNFCCC (1992) United Nations Framework Convention on Climate Change. United Nations, New York. Available at: www.unfccc.int/resource/docs/convkp/conveng.pdf (accessed 1 February 2021).

WEF (2019) Global Gender Gap Report 2020. World Economic Forum. Available at: www.weforum.org/reports/gender-gap-2020-report-100-years-pay-equality (accessed 1 February 2021).

WHO (2020) Gender and Health. World Health Organization, Geneva, Switzerland. Available at: www.who.int/health-topics/gender#tab=tab_1 (accessed 1 February 2021).

Wiebe, K., Robinson, S. and Cattaneo, A. (2019) Chapter 4 – climate change, agriculture and food security: impacts and the potential for adaptation and mitigation. In: Campanhola, C. and Pandey, S. (eds) *Sustainable Food and Agriculture*. Elsevier, Cambridge, MA, pp. 55–74.

World Bank (2020a) Employment in agriculture (% of total employment). World Bank, Washington, DC. Available at: www.data.worldbank.org/indicator/SL.AGR.EMPL.ZS?end=2018andmost_recent_year_desc=trueandstart=1991andtype=shadedandview=chart (accessed 1 February 2021).

World Bank (2020b) Employment in agriculture, female (% of female employment). World Bank, Washington, DC. Available at: www.data.worldbank.org/indicator/SL.AGR.EMPL.FE.ZS?name_desc=false (accessed 1 February 2021).

2 A Gendered Approach to Understanding Climate Change Impacts: Lessons from a Coastal Region of Bangladesh

S. Momtaz, M. Asaduzzaman and Z. Kabir

Introduction

The purpose of this chapter is to understand the vulnerability of women's livelihoods to climate change impacts in Bangladesh. Although it is generally stated that women are more vulnerable than men to climate change, relatively few studies have investigated this statement in the context of Bangladesh. Underpinned by the concepts of a sustainable livelihood framework and the Disaster Crunch Model (DCM) this study asks the following questions: How are women affected during disasters? How do they respond to and cope with disasters? What level of access do they have to support their coping efforts? This study is informed by the emerging body of literature on vulnerability assessment including that of the Intergovernmental Panel on Climate Change (IPCC).

Bangladesh is one of the countries most affected by climate change. The country experiences climate change in the form of extreme climatic events such as floods, droughts, cyclones, and tidal surges. The coastal zone of Bangladesh consists of 19 districts comprising 147 *upazilas* (administrative sub-districts). A total of 48 *upazilas* in the coastal zone are exposed to the Bay of Bengal. These *upazilas* experience recurring cyclones and associated tidal surges and flooding. In order to understand the structure of women's livelihoods, their vulnerability to natural disasters, and their adaptive capacity we examine two smaller administrative units (unions) in the coastal area of Bangladesh. These unions are located in the area that was badly affected by recent cyclones, including cyclone Aila, which this study uses as a reference point to determine pre- and post-disaster household conditions. Data were collected through a survey of 150 randomly selected women from a sample of households. Focus group discussions, key informant interviews, participant observations, and a transect walk, provided supporting information to substantiate the household surveys.

This chapter first outlines the theoretical foundation on which the research is based. This is followed by examining women's vulnerability in the study area. The chapter then describes women's coping strategies in the face of climate change-induced disasters. The chapter further explores women's adaptive capacity through the examination of their access to various services. It ends with a set of recommendations for policy makers in order to improve the situation of women's vulnerability.

Theoretical Considerations

Understanding and assessing vulnerability

The term 'vulnerability' has its origin in geography and natural disaster studies. The concept

 Gender, Climate Change and Livelihoods: Vulnerabilities and Adaptations (eds J. Eastin and K. Dupuy)
DOI: 10.1079/9781789247053.0002

is now being used in many other contexts including climate change. In climate change research, vulnerability is an integrative measure of the threats to a system (IPCC, 2001). The working definition of vulnerability by the IPCC (2007, p. 883) is 'the degree to which a system is susceptible to, and unable to cope with adverse effects to climate change, including climate variability and extremes.' The IPCC explains vulnerability to climate change as a function of exposure, sensitivity, and adaptive capacity. There are many different definitions of vulnerability, but a common phenomenon across definitions is the link between vulnerability and risk. The literature further explains that vulnerability involves a combination of factors that determine the degree to which someone's life, livelihood, property, and other assets are put at risk by a discrete and identifiable event (or a series or 'cascade' of such events) in nature and in society (Momtaz and Asaduzzaman, 2019). Vulnerability is, therefore, a complex combination of three broad interrelated spheres: physical/material, social/organizational, and motivational/attitudinal (Anderson and Woodrow, 1989) that need an interdisciplinary approach to understand. By 'motivational/attitudinal' the authors mean that the communities with shared values and beliefs tend to help each other at times of disaster and therefore are more resilient than those that do not have these attributes.

Vulnerability assessment

Füssel (2009) has used four fundamental dimensions to describe the context of a vulnerability assessment. This assessment is the process of estimating the susceptibility of elements at risk (people, household, community facilities and services, economic activities, and the natural environment) to various hazards, and analyzing the root causes which place these elements at risk (see also Füssel 2004 and 2005 for further description of these dimensions and elements). Hoddinott and Quisumbing (2003, p. 46) pose five questions that a vulnerability assessment should answer: 'What is the extent of vulnerability? Who is vulnerable? What are the sources of vulnerability? How do households respond to shocks? What gaps exist between risks and risk management mechanisms?' Answering these questions requires multiple data collection methods and analytical techniques. Moreover, a final key feature of vulnerability assessment is the inclusion of community perceptions of vulnerability into the assessment design and definition of vulnerability (Kalibala *et al.*, 2012). Vulnerability assessment incorporates a significant range of parameters in building quantitative and qualitative pictures of the processes and outcomes of vulnerability (Adger, 2006). Therefore, the present study uses both qualitative and quantitative approaches. The following sections feature the vulnerability assessment approaches which have been used to create a comprehensive baseline for analysis of this research.

The Disaster Crunch Model

Oxfam GB (2012) in its report titled 'The Disaster Crunch Model: Guidelines for a Gendered Approach' has provided an elaborate analysis of the DCM, a framework that can be useful to understand vulnerability. This model shows that vulnerability, which is rooted in socio-economic and political processes, has to be addressed first to reduce the risk of disaster. Following the idea of this model, the present study has used a gender-sensitive DCM which underpins the vulnerabilities of women in a certain socio-economic and cultural context. This means that women experience different levels and types of vulnerability to disasters, including those caused by climate change, in a distinctly different way to men. The DCM (Wisner *et al.*, 2004) shows that there are three layers of social processes that cause vulnerability: root causes, dynamic pressures, and unsafe conditions. The basic idea of the model is that disaster risk is the intersection of two opposing forces: those processes generating vulnerability on one side, and the natural hazard event on the other. This model helps researchers to understand people's vulnerability to disasters more fully, and in terms of gender relations, it describes the various levels and types of vulnerability experienced by women in disasters, including those caused by climate change events (Oxfam GB, 2012).

Study Setting and Procedure

The coastal areas of Bangladesh and its millions of inhabitants are exposed to extreme climatic events. The cyclones lashing the coastal zones in

1970, 1985, 1991, 1997, 2007, and 2009 caused significant damage to lives and livelihoods and led to the involuntary displacement of millions of people (Akter, 2009). It is reported that over 35 million people will be forced out of 19 coastal districts of Bangladesh under the scenario of a 1-m sea level rise this century (Rabbani, 2009). The coastal areas' existing salinity problems are expected to be further exacerbated due to climate change and sea level rise. Rising sea level is also causing water levels to rise in the rivers and thereby accelerating the risks of flooding and waterlogging.

The fieldwork for the original study was conducted between May and November 2013. To update information for this chapter the researchers revisited the study sites remotely in the aftermath of cyclone Amphan in late May and early June 2020, using Google Maps and Google Images. A number of local key informants were contacted for information via phone. The local media also provided important information regarding cyclone damages and their effects on women. Two unions in the Shyamnagar *upazila* (approximate population 318,254; BBS 2013a) were purposively selected for this study. The selected *upazila* is part of the coastal Satkhira district (approximate population 2 million; BBS 2013a) in the south-west of Bangladesh near the Sundarbans (the single largest mangrove forest in the world) facing the Bay of Bengal. This district has been one of the areas most affected by natural disasters and climate change-related impacts in recent times. The selected unions were also badly affected by cyclone Aila where many people lost their livelihoods (Kabir *et al.*, 2016). Cyclone Aila hit the south-western part of Bangladesh on 25 May 2009, killing 190 people and destroying crops and properties in its path. It also brought a tidal surge that inundated the study area. The district again suffered severe damages when cyclone Amphan fell on the coastal region of Bangladesh in May 2020. On 23 May 2020, The Daily Star reported:

> The situation of the Amphan-affected people in coastal district Satkhira's Burigoalini union appears to be worse than that of those in Khulna. Swathes of land in the union in Shyamnagar upazila [one of the study sites] have been inundated after an embankment on the Khulpetua River was breached at several points. Many who took shelter at cyclone centers could not return to their homes, which remained under waist-deep water during high tide.
>
> (The Daily Star, 2020a)

It was assumed that by selecting these two locations a general understanding of issues on the ground in this region could be gained as they are representative of similar coastal regions in the country which share the same socio-economic characteristics.

The study utilized a multistage sampling technique to obtain a representative sample of those to be interviewed. First, the two unions, Gabura and Padmapukur (populations of 38,825 and 26,447, respectively), were purposively selected considering their close experience of climate change effects, and then a further seven villages from each of the unions were selected for the current study to provide a wider illustration of the effect of impacts. The villages chosen were: Gabura, Khalishabunia, Dumuria, Kholpetua, Khalisha, Chakbara, and Jelekhali in Gabura union; and Sonakhali, East Patakhali, Pakhimara, Khutikata, Patakhali, Kamalkathi, and Chor Chondipur in Padmapukur union. Having selected the villages, the study then considered a reasonable number of samples to be representative of the population in this region. Through this process, 150 women (over 18 years old) were randomly selected for questionnaire survey. In addition to the researchers, two locally recruited research assistants were employed for conducting the questionnaire survey. The questionnaire included questions about women's livelihoods, their vulnerability, and personal independent adaptation measures. The livelihoods of women were analyzed under the five livelihood capitals: human, natural, financial, social, and physical. Cyclone Aila was used as a reference point to understand the livelihood situations of interviewees, before and after the disaster. The participants were requested to recollect before and after situations. In addition, the study conducted six focus group discussions (FGD) with 8–10 women participating in each group. Women were the main focus of this study although 2 or 3 men attended each FGD. The details of the data collection procedures can be found in Momtaz and Asaduzzaman (2019) – the original work on which this chapter is based. Secondary data regarding climate variables and events, detailed descriptions of the study area, and livelihood status information were obtained from various sources including the Bangladesh Bureau of Statistics (BBS), Bangladesh National Web Portal, and the Upazila Parishad Manual.

General Vulnerability of Women in the Study Area

As stated, women in developing countries are more vulnerable than men to climate change effects. Recent studies have also given credence to that finding by bringing evidence to this effect from developing countries. The Population Reference Bureau (PRB, 2012, p. 1) has identified the reasons as being cultural norms, inequitable distribution of roles, and resources. As an objective, we wanted to test this in our study. Vulnerability of the women in the study area (as in other rural areas of Bangladesh) can be explained as a set of prevailing or consequential conditions, e.g. access to land and security of tenure; dependence on natural resources for livelihood; restricted mobility; limited role in family decision making; and access to services (see section on Women's Access to Major Services, below), which adversely affect their ability to prevent, mitigate, prepare for, or respond to hazard events. These long-term factors, weaknesses, or constraints affect a woman's ability (or inability) to absorb losses after disasters and to recover from the damage. When asked about the vulnerability of women, all participants concurred that women were the most vulnerable group in society. According to the DCM as discussed above, women's vulnerability is progressed due to various root causes, dynamic pressures, and unsafe conditions. This section briefly discusses the causes of women's vulnerabilities in the study area. In order to understand the effects of extreme weather events on women's livelihoods we asked the participants to talk about their situations before and after the tropical cyclone Aila (Table 2.1).

Access to land and security of tenure is often highlighted as an important cause of women's vulnerability (Jacobs, 2002; Agarwal, 2003). The present study also found that women had no ownership of cultivable land. Although according to Muslim Law, women have ownership of the property belonging to their parents, they rarely enjoy the rights. When they become owners of the land of their parents (usually after the death of their parents), often they cannot enjoy it, because in many situations brothers prevent their sisters from establishing ownership of the land they have inherited. Women's limited access to resources was also revealed for other household assets (see Table 2.1). This concurs with the observation of a United Nations study (Osman-Elasha, 2020): 'Women predominate in the world's food production (50–80%), but they own less than 10% of the land.' Although women work on crop lands, they have limited or no control over the land, as they do not own it and therefore cannot make decisions regarding its use. Men, on the other hand, retain greater access to each form of the five livelihood capitals mentioned above. This allows men to diversify their livelihoods and adaptive capacities to climate change to a much greater extent than women.

Women in rural areas in developing countries are especially vulnerable when they are highly dependent on local natural resources for their livelihoods (UN WomenWatch, 2009). As the primary users and managers of natural resources (being typically responsible for fetching water and fuel wood, and bringing it to the house, for example), women depend on the resources most at risk from climate change (UNDP, 2010). High dependency on agriculture, forestry resources, fisheries, and biofuels increases women's vulnerability. This has also been found in the

Table 2.1. Distribution and ownership pattern of major physical assets.

	Percentage of household		Ownership pattern in 2013	
Types of Asset	Before Aila (2009)	After Aila (2013)	Male	Female
Livestock	57 (38%)	45 (30%)	144 (96%)	6 (4%)
Poultry	131 (87%)	120 (80%)	96 (64%)	54 (36%)
Fishing boat and net	65 (43%)	50 (33%)	147 (98%)	3 (2%)
Trees	98 (65%)	29 (19%)	145 (97%)	5 (3%)

case of cyclone Amphan that has damaged all sources of drinking water in the region making it more difficult for women to ensure safe drinking water for the family (The Daily Star, 2020b).

Like in many developing countries of Asia, Africa, and Latin America, rural women in Bangladesh have typically received fewer school years and have grown up as less-educated persons. In the national literacy survey, it was found that only 35% of rural women continued their study after primary level education (BBS, 2013a). For the present study, this figure was even worse, as only 27% of women had continued their study after primary level. Limited education can affect women's ability to understand and act on information concerning climate risks and adaptation measures. Limited education also reduces their opportunities to become involved in income generating activities and, therefore, increases their vulnerability. Women in rural Bangladesh are less mobile due to strict and gendered codes of social behavior and, as such, have fewer chances to escape from affected areas (CCC, 2009). This restricted mobility is one of the reasons that the death rate of women was higher in all cyclones in Bangladesh, as their powerlessness in life was seen to override their survival instincts. This influence of cultural norms is also observed in a recent study by the World Health Organization (WHO, 2020, online): 'A woman cannot receive needed healthcare because norms in her community prevent her from traveling alone to a clinic.' We found in the study area that women could not go to the cyclone shelters during cyclone Aila if they did not have an adult male guardian to accompany them.

Women have a limited role in the decision making processes of households and community level governance in rural Bangladesh, which has implications for their lives, environment, and aspirations (Kabeer, 2005). Despite the fact that the country has a female prime minister, there is limited representation of women in the local level administrations, which attract mostly men through fiercely contested local elections along partisan lines. There are gender inequalities with respect to the enjoyment of human rights; political and economic status; land ownership; housing conditions; exposure to violence; education; and health. All of these inequalities make women more vulnerable before, during, and after climate change-induced disasters. Recent studies by the World Bank and UNDP have found that 103 of 141 developing countries have legalized gender discrimination that decreases women's economic opportunities (World Bank, 2011; UNDP, 2013, p. 3).

Women's Vulnerability in the Study Area in the Context of Climate Change

The researchers asked participants of this study about their experience with climate change events or climatic hazards in the past years. According to the interviewees, the common climatic events or hazards they had observed in their locality were: increased intensity of storms and cyclones; increased soil salinity; increased temperature; waterlogging; and floods following heavy rainfall. The impacts of climate change events are interlinked, and they create vulnerability in women's lives in various interrelated ways. The present study has categorized the degree of vulnerability into four categories: severe, moderate, low, and no vulnerability. This categorization was derived from discussions with the local community during our initial site visits for a pilot survey. It is based on qualitative expressions of affected people about how they felt during disasters and how different aspects of their livelihoods were damaged. The major areas of vulnerability of women's livelihoods are described in Fig. 2.1.

Women's income is more likely to be derived from the informal sector, which is often the worst hit by disasters and the least able to recover (Rahman, 2013). The main livelihood activities of the study area are shrimp farming, fishing, and forest resource collection. The people of the study area largely depend on the natural resources provided by the Sundarbans Forest Reserve for their livelihood, because they live in the closest and most adjacent locality to this reserve. The ecosystem services of the Sundarbans provide an opportunity for the households of this community to continue their livelihoods while they continuously face the severe challenges of climate change. As for the women, the main sources of income are poultry, goat rearing, and kitchen gardening. Some also caught shrimp from the local waterways and

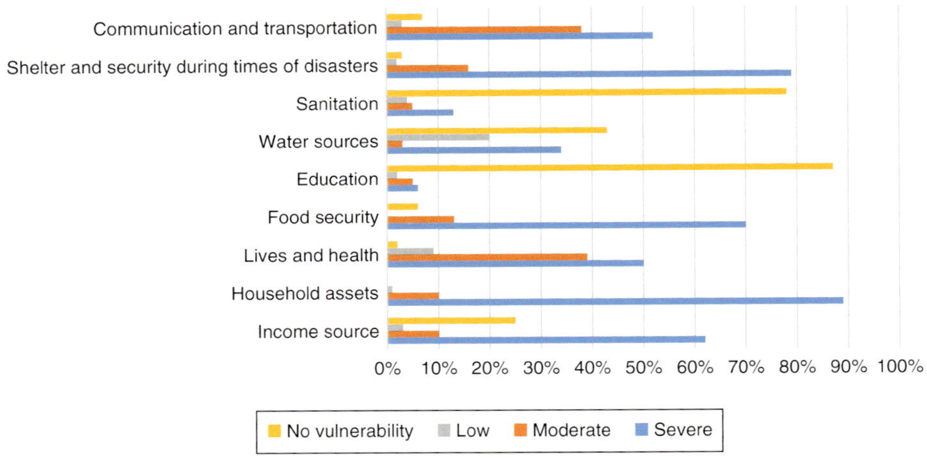

Fig. 2.1. Major areas and extent of women's vulnerability. Based on data from Momtaz and Asaduzzaman, 2019.

collected forest resources from the nearby Sundarbans to sell in the local markets, although this was not seen as a respectable job for women. In the study area, 62% of the women said that their income sources had become 'severely' vulnerable due to recent climate change effects. Another 10% of the women felt that their income sources were 'moderately' vulnerable. One survey participant explained how her income sources were damaged by cyclone Aila:

> I had 20 chickens and 15 ducks from which I earned some money by selling their eggs. I had also a vegetable garden. But Aila came as a curse to our lives which washed away all my poultry, and destroyed my garden. It is 4 years since Aila hit us, but still I could not recover from my losses. I had no ability to buy these poultry again. I tried to grow some fruits and vegetables in my homestead, but because of high salinity after Aila, I failed to do so.
>
> (Fatema Begum, Khalishabunia village, Gabura union)

However, 25% of the women also responded that there was no vulnerability in their income sources, the reason being that these particular women were not engaged in any income generating activities at all (Fig. 2.1).

Discussions with women through personal interviews and FGDs revealed that although women mostly practiced homestead-based livelihoods, the few income generation opportunities they had were lost due to natural disasters, such as cyclone Aila. The deaths of livestock and poultry during disasters increased the vulnerability of women's livelihoods, as these were one of the main income generating sources for women. Many of the respondents used to do homestead gardening from which they earned some money by selling excess fruits and vegetables. The recent climatic events, including increased soil salinity, however, have constrained them in continuing this practice. Women in the study area were in general poor and so had *few assets*. Those they had included a few goats, a few chickens or ducks, a small amount of jewelry, and some pieces of furniture. Whatever they owned, however, was thought of as very valuable in respect to the outcome of various climatic events. During the catastrophe associated with cyclone Aila, most study participants lost the majority of their assets, including livestock and poultry. About 90% of women in the study responded that their assets were 'severely' affected (Fig. 2.1) and that it was very difficult for them to recover from this situation.

Women's lives and health are more vulnerable than those of men in any time of climate variability. More women die during natural disasters compared to men, because they are not adequately warned, cannot swim well, or cannot leave the house alone (CCC, 2009). During cyclone Aila in 2009, the death rate of women was counted at being five times higher than that of men (UN, 2010). The present study could not quantify the number of deaths in the families surveyed, but it was understood that there were more deaths among women than men. The researchers asked the participants if they or any of their female family members had suffered any major health problems from the past climatic events, and 60% of women said that they had faced problems. Moreover, they were continuously suffering from different health-related issues. Among the respondents, 90% of women said they were facing 'severe' to 'moderate' levels of vulnerability with regard to health issues (Fig. 2.1). The FGD participants spoke of suffering from frequent and various diseases including diarrhea, dysentery, cold and flu, and skin diseases. The reasons for these were excessive heat and rain, saline water, waterlogging, and a lack of nutritious food. The recent flood in the northern districts of Bangladesh paints a similar picture for flood-affected women, as mentioned by a female flood victim, Ahela Bewa (65). Her house went under floodwater and she had to take shelter along with her children on high ground. Hundreds of families like her took shelter there. There was an acute shortage of food, pure drinking water, and toilet facilities. Given this dismal situation, women in particular were facing greater difficulty than men (The Daily Star, 2020a).

In the context of climate change, traditional food sources become scarcer and more unpredictable, an outcome which has serious ramifications on the four dimensions of food security: food availability, food accessibility, food utilization, and food systems stability (UN WomenWatch, 2009). It was found that the food situation was very dire for women in the study area, particularly those from the lower economic groups, and 70% of respondents subscribed to this view. Among the respondents, 90% of the women said that their land was degraded due to salinity and had lost its capacity to grow crops. This impact posed serious risks for ongoing food security (Fig. 2.1). The FGD participants and key informants stated that during cyclone Aila, many women took shelter in the river embankments where they were unable to cook and they could not afford to buy food. During that time, families had to live on dry food, often eating only once a day for more than 2 months. As a result of this prolonged food deficiency, most of the women in the study area were suffering from malnutrition. Women's food insecurity has again been exposed in the aftermath of the recent (May 2020) cyclone Amphan. One of the affected single (female) parents interviewed by The Daily Star (2020a) said whatever rice she had in her home was washed away by flooding. There was support from the neighbors but this was not adequate. She did not receive any relief from the government or non-governmental organizations (NGO). One of the FGD participants stated that during cyclone Aila:

> ... Many of us took shelter in the river embankments where there was no facility to cook, and more importantly we did not have money to buy food. We were eagerly waiting and wondering when the speed boat with food from the government and NGOs would come and give us some food, which were mostly dry foods. The amount of food we received in the end was not sufficient for the whole family. We usually fed the other members in the family first and ate what was left over.
>
> (Kulsum Begum, Gabura village)

The term 'vulnerability of education' is used here to represent whether women in the study area faced any problem in providing education for their children. While the respondents themselves did not have much opportunity to be educated, their children were facing a different scenario. They thought that their children were generally not vulnerable to having a lack of access to education due to climate change. Although there were some interruptions in education in post-disaster times, overall, they were happy with the educational services they were receiving for their children (Fig. 2.1). The primary education (up to year 8) in Bangladesh is free for all children. All textbooks and stationery are also provided by the government, creating fairness in the education system. NGOs are also very active in this sector providing support to the education security for children. According to one key informant:

> Now government is emphasizing on education and set target to reduce the dropout rate at primary level. In recent times, many primary schools have been established in this vicinity where children can study without any cost. Therefore, the attendance rate in primary school is very high. In addition, the social and cultural barriers for girls' education also have been reduced by the help of government and NGOs.

> Now, most of the children can go to school at least to primary level.
>
> (Ranjit Mandal, Local School Teacher, Patakhali, Padmapukur, Shyamnagar)

In rural areas all over the developing world, women and girls bear the burden of fetching water for their families and spend significant amounts of time each day hauling water from distant sources, even though the water from such sources is rarely enough to meet the needs of the household (UN WomenWatch, 2009). The FGD participants said that water supply had been a problem in the study area for a long time. The level of vulnerability mainly depends on the availability of fresh water and the distance of water sources from the household. Since women were responsible for most of the water-related works, they were more vulnerable than men due to a lack of local fresh water sources. Overall, 37% of participants mentioned this being an issue of great concern. Conversely, 43% of respondents were of the opinion that they were not vulnerable to water source problems as they were living near water sources, provided by the government or through NGO support (Fig. 2.1).

A large number of rural sanitation facilities are destroyed every time a cyclone or hurricane hits. The FGD participants stated that after Aila they had to live in tents on the riverbank where there were no toilets, and they had to share the same space with men. Many women refrained from using the toilet during the day and consequently suffered from urinary tract infections and other diseases. Moreover, they faced problems in bathing and management of general hygiene while living under the open sky with no private spaces. Overall, during the post-disaster time, the sanitation facilities were disastrous, and women bore the brunt of this. However, various NGOs took the initiative to build toilets in many households, making sanitation facilities more accessible for women. A respondent named Aloka Rani from Sonakhali village said: 'Our house had no structured toilet before. It was very inconvenient for us. But after Aila, NGO Caritas has made toilet in our house. Now, we are happy with our sanitation facility.'

The overall capacity of cyclone centers was found to be very low and decreasing further with population growth; this low capacity occurs despite the fact that climate change events in Bangladesh are predicted to increase exponentially (IPCC, 2007). The situation has improved somewhat in recent years in terms of the rising number of cyclone shelters and their capacity, though problems in location and accessibility remain. There are only five cyclone centers in the vicinity of the villages. Typically, multipurpose cyclone center (MPCS) facilities are multistoried buildings with reinforced concrete that can, on average, accommodate up to 1600 people. In the present study, 97% of women ('low' 2%, 'moderate' 16%, and 'severe' 79%) stated that they felt vulnerable in these disaster shelters (Fig. 2.1). Women often find it risky to go to the shelters if they are not accompanied by men. Due to the locations of these centers (which are often at great distances from the households), limitations on personal security, and inadequate facilities, many people take shelter in neighboring embankments following disasters such as Aila. Yet, sleeping under open skies can be very insecure for women, especially for young women. A UN Assessment Team found that nearly 14,000 families were living on the embankments, even 1 year after cyclone Aila (UN, 2010). The recent flood of Bangladesh has made about 2.5 million people waterlocked. The government has opened 1100 shelters for them but only 60,000 people have taken shelter in these facilities, which is a small percentage of their capacity (Daily Prothom Alo, 2020).

The means of communication is important but can become extremely vulnerable due to climate change as seen in the study area. Women in the study area (52%) thought that they were 'severely' vulnerable to changes in communication and transportation due to climate change effects, while 38% stated they felt 'moderately' vulnerable. Only 7% of the participants registered being 'unaffected' by these events (Fig. 2.1). When Aila hit the study area, most of the roads were destroyed. Many of them were still damaged during our site visits. A key informant described the situation as follows:

> In this area, communication system was not good even before Aila. But during Aila, all village roads were destroyed. At that time people had to depend on boats as a mode of transportation. Afterwards, with the help of government and NGOs, village roads were repaired and still the works are ongoing. However, as the roads are not paved, we suffer a lot in the rainy season.
>
> (Akhtar Hossain, Member of Padmapukur Union Council, Shyamnagar, Satkhira)

Overall, women were found to be vulnerable in various aspects of their livelihood. If proper training programs on disaster awareness and preparedness could be organized for them, their vulnerability could be reduced. However, thus far, women have mostly had to find their own strategies to cope within these stressed situations. The respondents outlined some of these coping strategies as discussed in the following section.

Coping Strategies of Women Against Climatic Events

In the face of disasters over the years, women have developed their own coping strategies to deal with the stressors imposed upon them by unfortunate events. These traditional mechanisms enable them to survive and return to 'normal life' after the disasters are over. Women in the study area, in personal interviews and FGDs, described a number of independent and individual coping strategies (Table 2.2).

All respondents in the study reduced their food intake during and after the times of disaster (Table 2.2). For use during and after disaster times, women preserved firewood, dry foods (such as rice, puffed rice, flattened rice, and molasses), candles, matches, and portable mud stoves in dry places. Despite the limited resources in the coastal areas, women played a significant role in food preservation to combat adverse situations. This study found that the method that most applied in order to recover from the disadvantageous situation was the selling of their resources or assets. In the study area, about 33% of respondents stated that they 'always' sold their assets during times of disaster, whereas about 49% of respondents 'occasionally' sold assets and about 10% of respondents 'very rarely' sold their assets at all (Table 2.2). In order to meet household financial needs during or after disasters, women usually sold any surviving livestock, poultry, stored crops, seeds, ornaments, trees, or whatever they owned. Due to reduced food intake both women and children suffer from malnourishment. This finding can be substantiated by the experiences of women recently affected by floods in Bangladesh where many crop lands have been submerged. In addition, many people living in Dhaka, the capital city, lost their jobs due to the coronavirus outbreak and have left the city to their villages. Also, many people in rural areas have lost their work. In such a situation, this flood has intensified the possibility of malnourishment (The Daily Star, 2020b).

In the years preceding cyclone Aila, a number of microcredit institutions (financial bodies giving small loans to the poor for little or no interest) began their programs. An example of a microcredit institution specifically targeted to disaster-affected people is BURO. Its loans are called 'disaster loans,' initiated in the wake of the disastrous floods of 1998 (Wright and Hossain, 2001). Many women in rural areas are now part of microcredit organizations, using their memberships to access loans. Among the respondents, 35% said that they 'always' took credit on all occasions during disasters, 63% took it 'occasionally,' and another 2% took credit 'very rarely'

Table 2.2. Disaster coping mechanisms of women.

	Frequency of practiced mechanism			
Coping strategies	Always	Occasionally	Very rarely	Never
Reducing food intake	90 (60%)	45 (30%)	15 (10%)	0 (0%)
Selling of assets	49 (33%)	74 (49%)	15 (10%)	12 (8%)
Receiving credit	52 (35%)	95 (63%)	3 (2%)	0 (0%)
Using savings	148 (99%)	2 (1%)	0 (0%)	0 (0%)
Alternative livelihoods	16 (11%)	126 (84%)	6 (4%)	2 (1%)
Vulnerable Group Development (VGD) program card	103 (69%)	5 (3%)	2 (1%)	40 (27%)
Homestead gardening and poultry rearing	135 (90%)	9 (6%)	3 (2%)	3 (2%)

(Table 2.2). Furthermore, Bangladeshi women from past generations have learned to save small amounts from their daily expenses for rainy days, paid either in kind or in cash. The total amount of savings may not be significant, but it is the first option for them to use to survive in times of disaster when their livelihoods have been affected. Almost all of the respondents expressed the same opinion that they used their savings during or after the disaster (Table 2.2). However, this would not be a permanent option for long-term adaptation. Considering their meager amount of savings, they could only utilize them for a short period of time.

Like in all major extreme climatic events, the work opportunities for men were dramatically reduced after Aila. As a result, most of the men left their houses to search for jobs in other districts leaving behind the women with added responsibilities. To cope with this added burden, 84% of respondents said that they tried to find alternative livelihoods 'occasionally' when they could, while 11% of women were forced to do this more often (Table 2.2). Women often caught fish and shrimp fingerlings from the nearby rivers to sell to the local traders, though it was not deemed prestigious work for housewives. Sometimes they worked in wealthy people's houses, and foraged food to feed their family. Some local NGOs gave them support by involving them in some income generating activities, such as goat rearing, sewing, mat preparation, honey culture, crab fattening, etc. These alternative livelihood options, however, are not yet permanent for many and in the short run, while they may help some adapt to their situation, broader spectrum initiatives are needed that focus specifically on women, such that they can bring about long-lasting and sustainable changes in their livelihoods. A key informant working for a local NGO gave support to this statement by saying that their NGO is working with women to make them self-dependent. He said:

> Our NGO is working to improve the livelihood of women of this area. Women of this area were used to raise livestock and poultry, and earned some money. But after Aila they are not able to do this practice anymore. We are particularly working with rural women as they have now few or no income opportunities. They are not interested to work outside of their house. Therefore, we are giving them training on tailoring, making cap, making mat and net, so that they can earn some money from home. We also provide them sewing machine and relevant materials. This will make them self-reliant for the long term.
>
> (Md. Lutfar Rahman, Executive Director, Nowabenki Gonomukhi Foundation, Shyamnagar, Satkhira)

The government of Bangladesh has been running a number of programs for vulnerable people. One of them is the Vulnerable Group Development (VGD) program. The objectives of this program are to overcome prevailing food insecurity, malnourishment, financial insecurity, and social degradation. Women in the study area received some benefits from these programs which helped them to survive during and after disaster times. About 69% of the respondents availed VGD cards 'always' to adapt their livelihoods during scenarios caused by climate change (Table 2.2).

The coping strategies discussed above show that women are capable of developing their own individual and independent measures to adapt to the situations in the aftermath of disasters. While they are encouraging and demonstrate women's resilience to climate shocks, the coping activities are not sustainable unless systematic and continuous support is provided by the government and NGOs to mainstream those measures. There needs to be significant initiatives and targeted policies to improve the present condition of women by removing the barriers for them and helping them adapt to the ongoing changing impacts of climate change.

Women's Access to Major Services

Access to major facilities is another indicator that can be used to assess the adaptation capacity of any target group to climate variability. Women's livelihoods and adaptations can be strengthened by ensuring their access to, control over, and ownership of resources, as well as their access to development resources (credit, information, training) and culturally appropriate technology. These accessibilities can truly make a difference to their lives and personal well-being. After the extreme climatic event of Aila, the devastating condition of the people suddenly came to the forefront of the national and international media's attention, as well as to the different levels of government and NGOs. They started work in the study area immediately after Aila hit there, and they are continuing their work to improve the poor livelihoods of the people in the study areas. Women's access to different welfare facilities are presented in Fig. 2.2. The access levels are categorized as 'adequate access,' 'limited access,' and 'no access,' and evaluated according to perceptions of what this was like before (2009) and after (2013) Aila.

Figure 2.2 reveals that prior to the effects of cyclone Aila, women had limited or no access to medical facilities, including seeing doctors or receiving medicines. After Aila, some NGOs assisted them by providing medical facilities such that they were then able to visit a doctor periodically and acquire some free medicines for general illnesses. The proportion of women who received 'adequate' medical facilities increased to 55% after Aila, from only 7% before Aila. However, the family planning facilities were still not enough to help women, and little has changed since before Aila. Seventeen percent of women had access to adequate family planning facilities in 2013, while it was only 5% in 2009. While there have been slight improvements in the access to these services, for the majority of women, they remain inaccessible and much more needs to be done to ensure that women's needs are provided for on a broader scale. It is evident that women face more challenges than men during cyclones or other natural hazards, and they become more vulnerable when they are the heads of households. In Bangladesh, women become heads of households only when they are divorced or widowed. It is to be noted that there are 12.5% households headed by women in Bangladesh (BBS, 2014). In the absence of husbands, these women face more challenges in earning their livelihood. They are also often excluded from any decision making process for resource management at the community level.

Banking facilities have always been limited or unavailable to women in these areas. In the whole Shyamnagar *upazila*, there are 13 nationalized and private commercial banks (BBS, 2013b). Women in the study area were usually economically so poor that they did not have the capability to run a bank account. After Aila, access to banking facilities increased a little (19% in 2013 from 10% in 2009), but in general the banks were located far enough away to still make them unreachable. The increased access to banking facilities may partially be attributed to the fact that government deregulation of the banking sector has seen a recent increase in the number of branches of private banks in the study area. Women were not used to making contact with local government organizations such as the Union Parishad, and the Upazila Parishad including the Upazila Agricultural Office and the Upazila Fisheries Office, as men usually dealt with these authorities. The social and religious norms and values of the study area do not appreciate or facilitate the need for women to be mobile in respect of doing such jobs. Only 25% of women had adequate access to these authorities when needed.

Communication infrastructure, such as roads and mass transport, improved after Aila due to the direct involvement of government and NGOs. Women (67%) also felt more comfortable in using these improved facilities than before Aila (38%). Overall, however, the transportation options are still limited and in very poor condition. Again, when it comes to services such as electricity, most of the respondents (91%) did not have any access. Only a few people used solar power to produce some electricity in their houses and these technologies were also supported by different NGOs.

Due to extreme and ongoing climate change effects, the most unfortunate thing that happened to the women of the study area was that they had lost many of their income opportunities (e.g. livestock and poultry production, homestead gardening, and working in crop fields as daily laborers). Their sources of income had been destroyed by recurrent cyclones, the impacts of which were compounded by lack of access to credit, low asset ownership, and increased familial burdens due to male outmigration for work. Before Aila, 80% of women had some income, while after Aila this figure had shrunk to only 20% (Fig. 2.2). Although some women were in search of alternative livelihoods, for most of them, the situation had become worse over time. Again, without proper access to a different level of facilities, it is impossible for women to adapt to worsening situations and to grow resilience in all categories of their livelihoods.

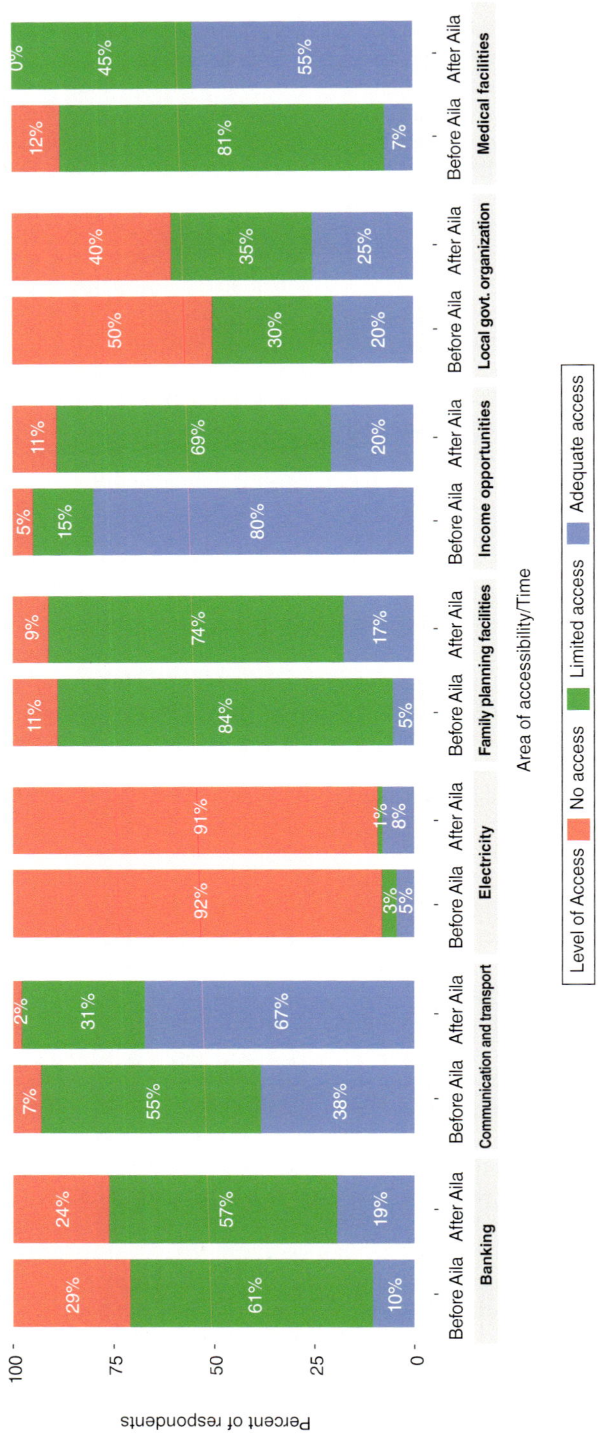

Fig. 2.2. Women's access to major services before (2009) and after (2013) cyclone Aila. Based on data from Momtaz and Asaduzzaman, 2019.

It can be concluded from Fig. 2.2 that huge losses occurred in women's income opportunities, and that over time, medical, communication, and transport facilities improved in positive ways. These findings corroborate with the other similar studies in the context of Bangladesh (see e.g. Garai, 2016). In summary, after cyclone Aila, all areas of accessibility, except income opportunity, have increased. Most notable improvements have happened for medical facilities and communication and transportation. However, income opportunities have reduced dramatically after Aila.

Conclusion

This chapter has investigated the structure of women's livelihoods, livelihood vulnerabilities, and coping mechanisms in a coastal area of Bangladesh vulnerable to disaster. It has explained that, due to their social and culturally constructed roles and responsibilities, and relatively poorer endowments of livelihood capitals, women were more vulnerable during and after climate disasters such as cyclone Aila. Therefore, access to land, greater income opportunities, proper access to information and development facilities, increased independence, and women's empowerment more generally, must be ensured for the rural women of Bangladesh if they are to actively address climate vulnerability and strengthen their adaptive capacity for the future. Government and NGO support are much needed to facilitate these specific requirements, as women still remain the most vulnerable groups to climate change in the country.

Recommendations

In light of the findings of this research, further substantiated by the recent events of floods and cyclone Amphan, the following recommendations are put forward for policy makers. As stated earlier, the study area is representative of the coastal regions in Bangladesh that are particularly exposed to the Bay of Bengal. Therefore, the policy solutions proposed here are generally applicable to these areas. It may be noted that there are many locations in South and South-east Asia that have similar geographical features as the coastal areas of Bangladesh. As such, the recommendations may also be applicable to locations outside of Bangladesh.

- Develop policies to specifically address the vulnerability in the women's livelihood. The policies must be bolstered by adequate legislative framework in order for the policies to be enforced. The bias, if any, should be in favor of the vulnerable women to eliminate inequality.
- Employ adequate resources and workforce for proper implementation of policies.
- Create alternative income generating activities specifically for vulnerable women.
- Create training and awareness education opportunities for vulnerable women.
- Make services accessible for women.

Proper enforcement of the policies and the legal instruments would in the long term eliminate the root causes that make women vulnerable, address the dynamic pressures, and create safer conditions for vulnerable women in the country.

References

Adger, W.N. (2006) Vulnerability. *Global Environmental Change* 16, 268–281.

Agarwal, B. (2003) Gender and land rights revisited: exploring new prospects via the state, family and market. *Journal of Agrarian Change* 3, 184–224.

Akter, T. (2009) Climate change and flow of environmental displacement in Bangladesh. *Scientific Briefing Paper Series on Climate Change and Development.* Unnayan Onneshan – The Innovators, Dhaka, Bangladesh.

Anderson, M.B. and Woodrow, P.J. (1989) *Rising from the Ashes: Development Strategies in Times of Disaster.* United Nations Educational, Scientific and Cultural Organization, Paris. Westview Press, Boulder, CO and San Francisco, CA.

BBS (2013a) District Statistics 2011 Satkhira. Bangladesh Bureau of Statistics, Statistics and Informatics Division, Ministry of Planning, Government of the People's Republic of Bangladesh, Dhaka, Bangladesh.

BBS (2013b) Literacy Assessment Survey 2011. Bangladesh Bureau of Statistics, Statistics and Informatics Division, Ministry of Planning, Government of the People's Republic of Bangladesh, Dhaka, Bangladesh.

BBS (2014) Household Survey in Bangladesh 2014. Bangladesh Bureau of Statistics, Statistics and Informatics Division, Ministry of Planning, Government of the People's Republic of Bangladesh, Dhaka, Bangladesh.

CCC (2009) Climate change, gender and vulnerable groups in Bangladesh. Climate Change Cell, Department of Environment, Ministry of Environment and Forests, Government of the People's Republic of Bangladesh, Dhaka, Bangladesh.

Daily Prothom Alo (2020) Flood and people in Bangladesh. Available at: www.prothomalo.com/bangladesh/article/1665322 (accessed 28 July 2020).

Füssel, H.-M. (2004) Coevolution of the political and conceptual frameworks for climate change vulnerability assessments. In: Biermann, F., Campe, S. and Jacob, K. (eds) *Proceedings of the 2002 Berlin Conference on the Human Dimensions of Global Environmental Change 'Knowledge for the Sustainability Transition. The Challenge for Social Science'*. Global Governance Project, Amsterdam, The Netherlands, pp. 302–320.

Füssel, H.-M. (2005) Vulnerability in climate change research: a comprehensive conceptual framework. *University of California International and Area Studies Breslauer Symposium, Paper 6*. University of California, Oakland, CA.

Füssel, H.-M. (2009) Review and quantitative analysis of indices of climate change exposure, adaptive capacity, sensitivity, and impacts. *Background note to the World Development Report 2010*. Development and Climate Change, Potsdam Institute for Climate Impact Research, Potsdam, Germany.

Garai, J. (2016) Gender specific vulnerability in climate change and possible sustainable livelihoods of coastal people. A case from Bangladesh. *Journal of Integrated Coastal Zone Management* 16, 79–88.

Hoddinott, J. and Quisumbing, A. (2003) Methods for microeconometric risk and vulnerability assessments. *Social Protection Discussion Paper*. Social Protection Unit, Human Development Network, World Bank, Washington, DC.

IPCC (2001) Climate Change 2001: Impacts, Adaptation and Vulnerability. *Contribution of Working Group II to the Fourth Assessment Report of the Intergovernmental Panel on Climate Change*. Cambridge University Press, Cambridge, UK and New York, 1032 pp.

IPCC (2007) Climate Change 2007: Impacts, Adaptation and Vulnerability. *Contribution of Working Group II to the Fourth Assessment Report of the Intergovernmental Panel on Climate Change*. Cambridge University Press, Cambridge, UK and New York, 976 pp.

Jacobs, S. (2002) Land reform: still a goal worth pursuing for rural women? *Journal of International Development* 14, 887–898.

Kabeer, N. (2005) Gender equality and women's empowerment: a critical analysis of the Third Millennium Development Goal 1. *Gender and Development* 13, 13–24.

Kabir, R., Hafiz, T.A., Khan, H.T.A., Ball, E. and Caldwell, K. (2016) Climate change impact: the experience of the coastal areas of Bangladesh affected by cyclones Sidr and Aila. *Journal of Environmental and Public Health* 2016:9654753. Available at: www.dx.doi.org/10.1155/2016/9654753 (accessed 1 February 2021).

Kalibala, S., Schenk, K.D., Weiss, D.C. and Elson, L. (2012) Examining dimensions of vulnerability among children in Uganda. *Psychology Health and Medicine* 17, 295–310.

Momtaz, S. and Asaduzzaman, M. (2019) *Climate Change Impacts and Women's Livelihood: Vulnerability in Developing Countries*. Routledge, New York.

Osman-Elasha, B. (2020) Women in the shadow of climate change. *UN Chronicle*, Magazine of the United Nations XLVI:3 and 4. Available at: www.unchronicle.un.org/article/womenin-shadow-climate-change (accessed 12 May 2020).

Oxfam GB (2012) The Disaster Crunch Model: Guidelines for a Gendered Approach. Oxfam GB, Oxford, UK.

PRB (2012) Population Reference Bureau, Washington, DC.

Rabbani, M.G. (2009) Climate forced migration: a massive threat to coastal people in Bangladesh. Clime Asia, Climate Action Network-South Asia (CANSA) Newsletter, December 2009. Bangladesh Centre for Advanced Studies (BCAS), Dhaka, Bangladesh.

Rahman, S. (2013) Climate change, disaster and gender vulnerability: a study on two divisions of Bangladesh. *American Journal of Human Ecology* 2, 72–82.

The Daily Star (2020a) Cyclone Amphan strikes Bangladesh's coasts. Available at: www.thedailystar.net/cyclone-amphan-strikes-bangladesh-coasts-1904470 (accessed 30 May 2020).

The Daily Star (2020b) Editorial: Massive work needs to be done in Amphan-affected areas. Available at: www.thedailystar.net/editorial/news/massive-work-needs-be-done-amphan-affected-areas-1904806 (accessed 30 May 2020).

UN (2010) Cyclone Aila. Joint UN Multi-sector Assessment and Response Framework. United Nations, New York.

UN WomenWatch (2009) Fact sheet: women, gender equality and climate change. United Nations Inter-Agency Network on Women and Gender Equity (IANWGE), New York. Available at: www.un.org/womenwatch/feature/climate_change/ (accessed 15 June 2020).

UNDP (2010) Gender, climate change and community-based adaptation. United Nations Development Program, New York.

UNDP (2013) Human development report 2013, the rise of the south: human progress in a diverse world. Explanatory note on 2013 HDR composite indices. United Nations Development Program, New York.

WHO (2020) Gender, climate change and health. World Health Organization, Geneva, Switzerland. Available at: www.who.int/phe/en/ (accessed 15 June 2020).

Wisner, B., Blaikie, P., Cannon, T. and Davis, I. (2004) *At Risk: Natural Hazards, People's Vulnerability, and Disasters* (2nd edn). Routledge, London.

World Bank (2011) Gender informed monitoring and evaluation in disaster risk management. *East Asia and the Pacific Region Sustainable Development Guidance Note No. 3*. World Bank, Washington, DC. Available at: www.openknowledge.worldbank.org/handle/10986/26534 (accessed 1 February 2021).

Wright, G.A.N. and Hossain, M. (2001) BURO, Tangail's approach to product development: a case study. Mimeographed, MicroSave Nairobi, Kenya.

3 Understanding the Gendered Impact of Disasters on Women, Household Dynamics and Coping Strategies: A Case Study of Bangladesh

N. Rezwana

Introduction

Bangladesh is a low-lying flat country with 700 rivers and a coastline of 710 km. Natural disasters are common phenomena. The effects of climate change combine with high population density, dependency on agriculture,[1] and developing economic status to increase disaster vulnerability, especially for agricultural sectors and livelihoods (Alam *et al.*, 2017). Disasters also adversely affect poorer inhabitants to a greater degree than wealthier ones, and have varied impacts on women, children, and older citizens. The gendered impacts of disasters alter household dynamics and demand special coping strategies. This chapter discusses the vulnerability of women in Bangladesh, the strategies women adopt to cope and survive in post-disaster periods, and presents firsthand accounts of these dynamics from remote and disaster-prone regions of the country. The data were obtained through household surveys, in-depth interviews, focus group discussions and participant observation in four research sites in Bangladesh during the period 2012–2019. This analysis recommends greater attention to gender mainstreaming in prevailing disaster management plans and policies, and suggests immediate actions to improve women's lives in the disaster-prone regions.

Theoretical Underpinning

Disasters affect people's lives and change their socio-economic conditions. However, the effects are not the same for all: vulnerabilities, and coping and adaptive capacities, are shaped by factors such as geographical location, gender, and socio-cultural, economic, and political structures and processes (Alam *et al.*, 2017). Countries located in disaster-prone regions with large populations and poor economic conditions face the most severe disasters; Bangladesh is one of them.

Floods, cyclones, riverbank erosion, drought, and saline water intrusion are among the most frequent and destructive disasters in Bangladesh. Approximately 12.3% of the population (20.2 million people out of 164.6 million total) live in highly disaster-prone areas (BBS, 2018). Every year the country experiences numerous disaster events, making Bangladesh the third most affected country in the world in terms of loss of lives and damage to the gross domestic product (GDP; INFORM, 2018 in Biswas *et al.*, 2019). In just 1 month (June 2019), at least 11 disasters struck the country, including 59 lightning storms in 32 districts, and flash floods in six districts. Agricultural land, crops, fisheries, and agro-based industries are among the most damaged or destroyed in disasters, an outcome which disproportionately affects poorer populations (Watkins, 2007; Das *et al.*, 2014; Reyes and Lu,

 Gender, Climate Change and Livelihoods: Vulnerabilities and Adaptations (eds J. Eastin and K. Dupuy)
DOI: 10.1079/9781789247053.0003

2017; Rahman and Gain, 2020). Disaster victims must adapt their livelihoods to cope with disaster impacts, as Rahman and Gain explain in their discussion of how victims of the Koyra river erosion in Khulna increased fishing as a means of coping with lost agricultural resources (Rahman and Gain, 2020). When disasters strike, poverty rises, especially at the family level (Das *et al.*, 2014). However, any sudden decrease in family income generates a greater impact on women than on men (IFRC, 2017; Reyes and Lu, 2017).

Household dynamics change (Reyes and Lu, 2017) and coping strategies – 'immediate respons[es] and reaction[s] to hazard events that… draw primarily on capacities within existing institutional frameworks' (Birkmann, 2011; quoted in Rahman *et al.*, 2015, p. 186) – take on gendered dimensions in the post-disaster period (Das *et al.*, 2014). Women and girls are often deprived of food, education, and healthcare, and are forced to marry early, while also relinquishing their source of income and the few resources they possess (agricultural lands, homestead gardens, jewelry) (Rahman *et al.*, 2015; Rezwana, 2018; Akter *et al.*, 2019). Male outmigration is a common coping strategy, especially for poorer families (UN Women, 2015), and when husbands leave, wives take charge of primary familial responsibilities. However, the additional burdens, including collecting relief, drawing loans, earning money, and caring for children and the elderly, can be difficult. While women and girls also migrate in post-disaster periods, migration in conservative patriarchal societies creates many additional difficulties and constraints because of 'unequal gendered power relations that limit women's access to and control over resources' (Bradshaw, 2015, p. 554).

Bradshaw argues that 'for those women who do survive [disasters] there may be longer-term and more intangible "secondary" impacts such as a rise in violence or greater insecurity in employment' (Bradshaw and Fordham, 2013 in Bradshaw, 2015, p. 555). Indeed, gender-based violence (GBV) – 'any harmful act that is perpetrated against a person's will and that is based on socially ascribed (i.e. gender) differences between males and females' – increases in form and severity whether women migrate or stay in disaster-affected regions (Diotima, 2020, online). Victims emerge in a variety of socio-cultural and ecological contexts, from Syrian refugees in Jordan, to drought victims in Zimbabwe, to earthquake victims in Nepal and Ecuador, and examples from countries such as Australia and Canada show that domestic violence increases with disaster stress (HPN, 2014; IFRC, 2017). Women also face discrimination in the distribution of disaster relief aid and the dispensation of loans or other compensation following a disaster (Akter *et al.*, 2019). Female-headed households and small businesses, and women who have lost their husbands to a disaster, are often not included in relief distribution lists in absence of men, and recovery programs also usually pay greater attention to men's jobs rather than those of women (MacDonald, 2005; Dominelli, 2013).

Despite gendered socio-cultural and economic barriers, and persistent discrimination, women adopt distinct strategies to survive and carry on. They become agents of survival in damaged houses; social volunteers in the distribution of relief aid among family members; caretakers for junior and senior citizens; and breadwinners to mitigate losses in familial income. However, while women are active participants or leaders in most coping strategies, they are often presented as victims in the literature and in media, and their survival and savings capacity remains unclear or underrated (Enarson and Chakrabarti, 2009; Terry, 2009). Research on women's contributions to disaster coping strategies is also rare and sporadic, especially in developing countries (Akter *et al.*, 2019).

This chapter draws on women's accounts from different disasters and in different regions of Bangladesh to reveal a truer picture of women's approaches to disaster coping. It is expected that proper attention to women's sensitivity to climate change and disasters, as well as their coping capacities, will enhance understanding of the contextual gender disparities present in society and their complex relationship with disaster impacts and prevailing disaster management plans. These findings will be helpful in taking the steps necessary to make an effective gender-sensitive approach to disaster management.

Research Support: Data, Methods and the Study Sites

This chapter draws on novel data collected in four research sites in disaster-affected regions of Bangladesh from 2012–2019. Two of these sites are in Barguna district, a cyclone-prone district located in the south-western coastal zone of Bangladesh; one site is on Bhola island near the mouth of the Meghna River; and the fourth is in Rangpur, a district in north-western Bangladesh.

Barguna district has experienced eight cyclones within the last 15 years, the latest of which was Amphan on 20 May 2020. Among these eight, cyclone Sidr (2007) was the most severe, killing 1335 inhabitants, and damaging 60–70% of crops and 95,412 houses (NIRAPAD, 2007 in Tamnia, 2009). Every cyclone that strikes Barguna leaves a significant impact on the local economy due to the large population and its dependency on agriculture. Barguna has 950,737 inhabitants living in 1830 km^2. The ratio of men to women is 49:51. Only 11.48% of people in this district live in urban areas (BBS, 2013a). People in this area are mostly poor: approximately 85% of the total population is landless (Byuro, 2006), and households depend on cultivation, fishing, and small business for their livelihoods (Health Bulletin, 2013). The entirety of the coastal community lags behind other parts of the country in various measures of development, especially for women. For example, the literacy rate in Barguna is 57.6%, yet only 56.1% of the female inhabitants are literate, the majority attending school only to primary level. This percentage of women attending school decreases to 4.9% after the age of 19 (BBS, 2013a), mainly because of early marriage, as girls are mostly wedded between the ages of 10 and 16 (BBS, 2006). Approximately 73.2% of women are engaged in household work and only 4.02% are employed in the formal labor force (BBS, 2013a). Employed women mostly work in agriculture (47.69%) or in the service sector (44.89%) while a low number (7.45%) are employed in industry (BBS, 2013a). The patriarchy is strong in Barguna society and women rarely participate in familial decisions. Women's socio-economic conditions, along with prevailing culture and social attitudes to gender roles, have a strong relationship with their access to healthcare, as well as with their capacity to cope with natural disasters (Rezwana, 2018). As a study area, Barguna is useful for evaluating the gendered impacts of disasters, and the researchers applied participant observation, key informant interviews (KII), in-depth interviews, and focus group discussions (FGD) to collect data on the effects of disasters on women's lives, livelihoods, and coping strategies. The author and her team applied snowballing methods among the Barguna inhabitants to recruit interviewees until the saturation point was reached. Local doctors, chief medical officers, government officials, schoolteachers and businesspeople were all interviewed as KII. FGDs were conducted among several groups of men and women.

The second research site is located in the village of Kalikanagar, in Kachia union of Bhola Sadar *upazila*, Bhola. Bhola is the largest island of Bangladesh and is located in the south-eastern part of the country at the mouth of the Meghna river, and on an actively eroding riverbank. According to the government record, there are 120 households in this village of 605 inhabitants. The ratio of males to females is 53:47, and the literacy rate remains low at 12.2%. Only 1.41% of girls ages 15–19 attend school. Early marriage is a common practice. Economic conditions for the inhabitants are very poor and most of them are employed in agriculture (53%) or work as laborers or shopkeepers in the service sector (47%). Women do not work outside the home. About 97.45% of women in this village are unemployed and dependent on family income (BBS, 2013b). Data for this village are drawn from quantitative surveys (93 households) and case study methods. Due to riverbank erosion, Kalikanagar has lost nearly 80% of its land to the river. Some families have moved to the adjacent Bhabanipur village and a few have migrated to cities. Surveys from this village help to reveal the impacts of riverbank erosion on women's lives and their changing livelihoods.

The remaining data used in this study were collected from Kashiganj village, Badarganj *upazila*, in Rangpur, a district in the north-west of Bangladesh. Kashiganj village is located on the bank of the river Jamunashawari. This river usually floods in the rainy seasons; however, in 2017 the floodwater stayed longer and caused damage to crops, cattle, and houses. The village also experiences regular droughts. The village population is 1774, the male/female ratio is 53:47, and the overall literacy rate 41.3%. The literacy rate is lower among the female inhabitants at 38.6%, a number which declines to 2.27% for women between the ages of 20 and 24 years (BBS, 2013c). Women's employment is also low in the village, where 84.84% women are homemakers. Data collection methods include household surveys (204 households); KIIs with schoolteachers, local government members, and other residents; FGDs; and participant observation.

Data from these empirical studies enable us to discern the gendered impacts of natural disasters on affected societies, and especially on women. In the following sections, discussions on women's coping strategies during post-disaster periods are presented, with attention to the effects of these on livelihoods and household dynamics.

Change in Livelihood and Household Dynamics: Women's Coping Strategies Against Gendered Impacts of Disasters

Disasters, natural or man-made, generate severe impacts on people's lives and livelihoods. Casualties, injuries, mental health problems, environmental issues, and resource damages count as direct impacts. However, there are many other indirect effects on economic conditions, education, healthcare access, and well-being (Rahman and Gain, 2020), and the socio-economic development of disaster-affected societies is perennially abated. For example, this chapter reveals that victims of cyclone Sidr (2007) in Barguna still have not recovered from the damages incurred during the disaster. The analysis also finds that the severity of disaster impacts and the capacity of residents to cope with them depend on the socio-economic conditions of affected families as well as gender, age, education, and employment status of the individual. In the following section, the complex situations women face during post-disaster periods and their coping strategies are discussed with reference to the study sites and several disasters occurring in Bangladesh.

'No job for men, women take the responsibility to earn'

As an agriculture-based economy, rural Bangladesh regularly experiences severe disaster impacts. Floods and cyclone storm surges damage crops and saturate fields; fishermen lose their boats and nets; and poultry and dairy sectors lose animals and their shelters. In Bhola, all study participants (100%) lost land (agricultural land or homesteads) to riverbank erosion, 48.39% of them lost crops, and 3.23% lost cattle to the river. In Rangpur, 72.9% of the household survey respondents lost crops in flooding during 2017, and 18.7% lost cattle. Cyclone victims in Barguna lost their crops, and productivity of agricultural lands declined for some time due to excessive saturation. In all these places, the possibility for rapid recovery among small farmers and laborers is minimal; they need time.

Men lose jobs in these disasters, and many remain jobless for months as they must adapt their occupation to find new sources of income. For example, many farmers in Bhola changed their livelihoods from farming to fishing after being affected by riverbank erosion. However, most of the time, this is no easy task. The overall number of jobs declines after disasters, and a lack of requisite skills and experience constrains efforts to find new employment in affected regions. There is also some negligence on the part of jobseekers as they wait for relief and assistance, a point revealed among some male members in the study sites. However, losses in family income also have a direct impact on the basic needs of women and girls, often beyond those faced by men, as women experience gender discrimination in food consumption, education, and healthcare access (Akter *et al.*, 2019; World Bank, 2020). The number of early marriages increases during crises, and many women become involved in generating income to save their families from total ruin. Examples of different experiences during disaster events are presented in the following paragraphs.

Experiences in cyclones

Research from Barguna reveals that when men lose their earnings to cyclones, women are required to supplement family income in order to maintain regular meals, education, and healthcare for their children. Due to low educational status and lack of experience, women often begin work as laborers or part-time housemaids, or they become involved in home-based work if the opportunity is available. There are many cases that show that without women's earnings it would have been very difficult for disaster-affected families to survive during post-disaster periods. However, women's contributions rarely get appropriate recognition and do not bring positive changes to their lives; rather, these contributions become additional gendered responsibilities (IDS, 2008 in UN Women, 2015).

Sathi, a cyclone survivor in Barguna, explains how she earned for her family to cope in post-disaster periods.

> I work as a housemaid in three houses. I did not work outside the home before cyclone Sidr. We had a small amount of land and a house in our village. I took care of the home and he [husband] farmed and worked as a part-time labor. We were in poverty but we could still manage. However, cyclone Sidr changed our life. We lost our crops from that season and the land remained under water for a long time. The house was totally damaged. We did not have money to repair the house. My husband could not find any jobs or any help in the village, so we left and came to the city. However, city life is expensive. His earnings did not meet our ends. I started working as housemaid, so that my children could have meals and went to schools. Now I work full-time. Still sometime, we have to borrow money. He [husband] often becomes sick, he cannot work every day. I take him to the doctors, it's expensive. I also sometime feel sick. I cannot stop working. I need money for the house rent, for food and for healthcare. Beside, after working the whole day outside, I have to cook for the family after coming home. Sometime it is difficult to manage everything, but I have to carry on for my children.
>
> (Sathi, 2013, Barguna, personal communication)

Sathi's experience is very common for poor women in Barguna. Qualitative studies reveal that women must fill the income gap during post-disaster periods. In some cases, they become the main wage earner for the family while also bearing the responsibilities at home. Earning money triples the workload for women, along with their responsibilities as mother and caregiver (Momsen, 2010).

In many cases, and despite contributions to family income, women do not have control over their earnings, nor does their decision making status improve when they generate independent revenue. Husbands control the money (Connell and Pearse, 2015). Maksuda faced this problem when she started earning after her husband lost his boat and fishing net in a cyclone. Maksuda's husband stopped working after the cyclone, and he did not try to find other work. However, Maksuda had small children to feed, so she started working in the other people's fields and houses. Her husband has become addicted to drugs, and forcefully takes most of her money whenever he can. He abuses her if she cannot provide money for his drugs. Maksuda is in 'secondary poverty' (Chant, 2008 in Bradshaw, 2015); domestic violence is everyday terrorism for Maksuda (Pain, 2014).

Again, working outside home at a low-paid job has negative consequences for women in a conservative society like Barguna. Women lose their social status and security. Rina explains that her husband had been injured during the cyclone and could not work for a long time. When the savings were finished, Rina began working outside the home. Her husband was not happy with this decision, but Rina was helpless. She needed money to feed her daughter. Rina explains:

> I knew he was not happy. So, I started working in a house beside the bazaar. But he became suspicious... he started behaving violently, beating me... after some months he left the house and never came back. He abandoned me... I worked and took care of our daughter alone. I married off her... at an early age... what can I do... now I live alone...
>
> (Rina, 2017, Barguna, personal communication)

Rina is not alone; her case is similar to another woman in this study, a victim of riverbank erosion, as well as that of a woman mentioned in a study by Akter *et al.* in Ulania village in Bangladesh (Akter *et al.*, 2019).

Experiences in riverbank erosion and floods

Studies in Bhola and Rangpur also reveal that women's participation in post-disaster income generating work is necessary in order to cope with economic crisis. Women in the study sites take on additional responsibilities, while also having to face further gendered constraints and problems. A family who lost land to riverbank erosion in Bhola became penniless after the main earning member, the 'father,' became ill. In that case, the 'mother' took the responsibility to earn. However, the family left the area because of the shame brought about when the woman became the primary breadwinner. Additionally, following floods in Rangpur, many women began working in the fields to service the increased debt load taken on to repair flood damages, but their contributions do not get proper recognition.

'Damaged kitchen and no money, but food should be prepared'

Women in disaster-hit regions not only participate in family income, they also provide meals. It is very difficult to cook food in a damaged kitchen after a cyclone or in an inundated kitchen during floods. However, the sole responsibility falls on women to collect clean water and manage the cooking and food-gathering. These actions are particularly difficult for women in large households, or for those who are pregnant. Women are often forced to gather the storm-strewn utensils, as most are often submerged in water or thick mud, and clean them to use. They spend a long time collecting clean water for drinking and cooking.

Halima from the flood-prone region describes how she cooks in floods:

> It is very difficult, everything goes underwater, the flood center becomes so dirty that we cannot stay there for long... so many people in one shelter, there is always a long queue for the toilet which remains dirty all the time, there is no privacy... so we try to be at home in inundated rooms... no place to cook... So I cook on my bed... it is made of wood... but collecting water is very difficult... you have to walk in knee-depth water to reach that house which has a tubewell... there are snakes in the water... it is a hard job...
>
> (Halima, 2019, Rangpur, personal communication)

The post-disaster period is a very critical time. People have to deal with injuries and trauma, life and resource loss, and hunger cannot be helped. Following a cyclone, the first thing women in Barguna prepare for is the gathering of food for children and older members of the family. This study reveals that a mother, who was injured herself, ignored her injury and cooked food for the family the day following the cyclone.

In some areas, women preserve dry food items like puffed rice (*muri*), flattened rice (*chira*), and home-made cookies (*pitha*) for emergencies. They preserve dry fuel woods. However, poverty and disaster preparedness are negatively correlated, and while women at all economic levels must prepare for an emergency, it is a matter of concern that their informal household contributions are not counted as work, neither in customary practice nor in official definitions (Mahmud and Tasneem, 2011).

Gender and family displacement

Kalikanagar village in Bhola lost 80% of its land due to erosion of the Meghna riverbank. Most residents lost land and resources, and some lost houses. At present, many of them live in rented houses in the same village or in the adjacent village. Some of the families migrated to the cities or other districts. However, displacement has gendered impacts, especially on women and girls (Sams, 2019). They do not always have access to basic facilities like toilets and bathrooms. Field visits in Bhola reveal that displaced people living on rented land often could not find places to build bathrooms, and the number of existing toilets has been insufficient. Women and girls have had to bathe openly in the river and wait, often until dark, until the toilets became less crowded. The lack of proper sanitation facilities has negative impacts on women's health and increases vulnerability to different types of GBV (MacDonald, 2005). Women, especially those from poor socio-economic backgrounds, also face considerable challenges when collecting water or using common ponds or riverbanks for bathing, including abuse and victimization. Displaced women in this study have reported increases in various types of GBV, including verbal abuse and sexual harassment. Despite their importance, these problems do not receive proper attention by the male family members. A lack of sanitation facilities is considered as a general problem for all, rather than as a gendered problem for women and girls. This perception also holds among the erosion-affected people in Chapai Nawabganj and Rajshahi districts (Islam and Rashid, 2011).

Women lose their previous earning sources and social status when they are displaced by disasters. They lose their social connections, which had served as a social safety net (Arobi *et al.*, 2019) and for social reproduction purposes (Momsen, 2010). Women are highly dependent on 'their surrounding environment to perform a variety of activities' (Akter *et al.*, 2019, p. 76); they lose their homestead gardens, cattle, and income. It is often difficult for them to earn money in new places due to a lack of skills, information, and social connections. There are fewer job options for women in host communities, except those jobs which are low paid (Hossain, 2008). It is also common for displaced people to face social discrimination and negligence in host communities (Islam and Rashid, 2011; Sams, 2019).

Gender and migration

Migration is a common coping strategy for many poor families in disaster-prone regions of Bangladesh (Akter *et al.*, 2019; Rahman and Gain, 2020). People from affected regions migrate with families or alone. When a woman migrates or is displaced with her family, she faces several problems in the new environment (as discussed in the last section). However, problems become more intense and gendered when women migrate alone. In Rangpur, many women and children migrated after severe flooding. According to 18.6% of the inhabitants in Kashiganj village, women and children leave their houses to earn money to cope with disaster impacts. However, these actions increase women's vulnerability to GBV; Bangladesh experienced 15,193 reported cases of violence against women in the year 2017, including rape, acid throwing, dowry violence, serious injury, and others (BBS, 2019). There are also huge negative social consequences. When local culture expects women to be homebound and with their families, a woman migrating alone generally does not receive any appreciation for her efforts to save the family from hunger. The risk of trafficking also increases. Sometimes, parents send their young girls to cities to cope with post-disaster economic crises, but in many cases, these actions lead to trafficking. Again, leaving a conservative village often has a strong impact on the young female migrant's life (Naved *et al.*, 2001); they feel abandoned by their own rural society, which is further aggravated when they cannot cope properly in urban society.

Coping with a new environment becomes a great challenge for female migrants due to continuous social curiosity and pressure. Additionally, they face various legal, cultural, and social barriers and obstacles when accessing the full range of resources, services, and opportunities in cities (IOM, 2015 in Hoque and Kamruzzaman, 2016). Hossain reveals in his research that though poor inhabitants may have been living a long time in cities, most cannot increase their standard of living (Hossain, 2008). They preserve their identity as migrants and keep their rural connections very strong. Especially female migrants, who leave their children at home, regularly travel back. However, despite the contribution to family income while facing these hazards, 'such migration rarely ensures women's position in family.' Their choices remain 'blocked into practical ideologies' (Hoque and Kamruzzaman, 2016, p. v). Male members (mostly husbands or fathers) make decisions about the destinations of migrants, job types, and the sectors where they spend money. The need for money brings these women to the cities but many of them do not want to live here. They wish to return to their village, which also shows their vulnerability to dissatisfaction and poor mental health.

Gender-based violence

As Eastin discusses in Chapter 8 (this volume), there is a strong relationship between disasters and GBV, and the risk of violence increases post-disaster. The severity and number of incidents of GBV might differ depending on the location, disaster preparedness, and socio-economic condition of the society, but the trend is similar for all societies and it is universal (examples of several countries are mentioned in the section on 'Theoretical underpinning,' this chapter). Having analyzed the research findings drawn from the four study sites, this chapter reveals that floods, cyclones, and riverbank erosion increase different types of GBV in Bangladesh. Domestic violence (wife beating, verbal and sexual abuse, dowry violence, wife abandonment), early or forced marriage, rape, trafficking, or forced migration (when migrants are forced to leave home to earn money) are the most common forms of GBV reported. Wife beating is a significant concern and women try to remain patient and cope, but most of the time they do not complain or make reports to legal authorities. In many cases, girls are married off at a very early age to reduce the family cost, lower the dowry price, and to avoid increased risk of sexual harassment in a damaged home and environment (Islam and Rashid, 2011). Child marriage becomes a coping strategy for some families (Goulds, 2013). Girls' education is also halted for the same reasons. However, most of the time early marriage ends with negative impacts on the victim's life. Rumana explains her experiences of early marriage after a cyclone:

> My father married off me in the next month of cyclone... we lost everything... I was only 12 years old... I did not understand anything... He had a wife... but still he married me... she [the other wife] did not take it easy... my father did not know that... this is my fate... I could not finish my school... I cannot leave my husband... cannot change my condition.
>
> (Rumana, 2017, Barguna, personal communication)

Male migration after disasters is common among poor families in all study sites. Women take over all responsibilities for maintaining the family when men leave. It becomes difficult for women to pay for groceries, while continuing their children's education and availing healthcare facilities in a conservative society. Life becomes even more challenging for women when men stop sending money, or abandon/divorce their wives and remarry. This is a very common practice among the male disaster migrants (UN Women, 2015). The women and children who remain fall into acute poverty. In some cases, a woman's rights on her husband's property are diminished, and she is evicted from her husband's house. Moreover, the only available jobs are low paying, do not help to improve economic conditions, and increase exposure to sexual abuse and harassment. A man's absence in a family always attracts unwanted curiosity, raises social pressure, and increases the risk of GBV. A woman's vulnerability increases due to the socio-cultural systems (Hoque and Kamruzzaman, 2016).

Gender and commonalities among disaster types

The discussions above reveal that disasters in different geographical locations and types have commonalities; they affect livelihoods and lives of the victims, change household dynamics, and result in women taking up several strategies to cope with disaster impacts. However, gender is among the strongest factors in increasing women's challenges after disasters; their contributions are not recognized properly and do not always help to improve the conditions of these women or decrease their vulnerability. Table 3.1 shows the commonalities in gendered impacts and women's coping strategies among three disasters.

Women as 'Victim,' 'Survivor' and 'Rescuer' in Disasters: Resilience and Coping Strategies

Disasters affect the livelihoods and economic conditions of disrupted families, but create especially complex consequences for the female members. The previous sections present how women become the primary victims during disasters, often having to confront food and education insecurity, early marriage, forced migration, wife abandonment, and GBV. However, women use their own strategies to cope with these changes, including taking on low-paid jobs; facing all issues that come with settling into a new environment (Akter *et al.*, 2019; Sams, 2019); taking courageous steps to support their families when men migrate; taking care of children alone; and facing harassment and abuse (UN Women, 2015). Women are survivors. It is worth mentioning here that women not only save themselves but their children and, in many cases, save the whole family from post-disaster crisis (Reyes and Lu, 2017). They are the silent rescuers. However, it is a matter of concern that women's efforts in these regards are not recognized properly by the family, by society, or by disaster managers (Enarson and Chakrabarti, 2009). We have seen in the previous sections how these brave women's lives become more difficult and complex. The women become victims of gendered social systems due to the three roles they

Table 3.1. Commonalities among disaster types.

Consequences of disasters	Disaster types: Cyclone	Floods	Riverbank erosion	Coping strategies for women
No job for men; women's responsibility to earn	✓	✓	✓	Doing low-paid jobs: day laborers, housemaids, home-based income generation
Damaged kitchen and no money, but food should be prepared	✓	✓		Fetching water from safe sources, household water management, emergency food preparation
Displacement of family in riverbank erosion			✓	Taking low-paid jobs, managing insufficient sanitation: using open bathrooms, coping with verbal abuse in host communities
Migration of lone woman and gendered problems	✓	✓		Regular travel to visit children, sending money for the family, accommodating new jobs, taking actions to protect from GBV
Double disasters: domestic violence	✓	✓	✓	Taking responsibility of children abandoned by migrant husbands, coping with societal gender discrimination, being patient and tolerating increased wife beatings during post-disaster periods

take on (breadwinner, mother, caregiver). They become victims to the prevailing patriarchal system of gender discrimination. Lack of empowerment and disparity in education, skill, and resources become even larger constraints for these women when attempting to escape these situations, and when attempting to grow and improve their lives. They remain in a vicious cycle of victim, survivor, and rescuer in a disaster-prone region.

Recommendations

To improve these conditions and enhance the coping capacity of women, both structural and non-structural measures should be taken to combat future disasters, which will increase in number and severity due to climate change.

First, proper attention and consideration should be given to women's contributions to disaster coping strategies and their importance in mitigating disaster impacts. Proper recognition of these contributions at the family and planning level will ensure gender-sensitiveness of disaster management plans and improvement of lives, especially those of women and children. Among the non-structural measures, plans should be taken to help women by providing vocational and skill development training, offering small grants/loans, and ensuring security at the workplace. Introducing awareness programs on gender equality and equity, highlighting women's contribution in disaster management, and ensuring the participation of both men and women in these programs will be helpful in making necessary changes. Second, structural measures, such as rehabilitation programs for erosion-hit communities and securing women's land tenure rights (e.g. CDSP program ensures wife's right along with husband on allotted lands, benefits of social forestry, and other facilities provided among the erosion-hit people; CDSP, 2020); proper gender-based sanitation plans for disaster shelters; supplying cooking utensils and stoves as disaster relief; and increasing security at shelters and damaged homes will be helpful for women to cope with post-disaster crises.

Above all, women's contributions as survivors and rescuers should be highlighted in the media, in disaster news, and in reports to inspire others living in disaster-prone regions.

Stories of these brave women deserve proper recognition and appreciation.

Conclusion

Based on empirical research in four disaster-prone study sites in Bangladesh, this chapter reveals the gendered impacts of disasters and the coping strategies of women as 'victims,' 'survivors,' and 'rescuers.' However, the coping strategies of survivors depend on several factors, and most of the time, conditions worsen due to prevailing gender discrimination in the household and in society. Women are left alone to face poverty, humiliation, and harassment without any institutional help. These findings demand immediate action to protect women from poverty, GBV, and disasters. It is essential to recognize women's participation in disaster coping strategies and to take proper actions to include their experiences in disaster management to improve the plans and policies and make them more effective.

Note

[1] Agriculture and the rural sector share 10.67% of national gross domestic product (GDP) and 63% of the nation is farm households (BBS, 2018).

References

Akter, K., Dey, S. and Hasan, S. (2019) Riverbank erosion and its impact on rural women: case study of Ulania village in Bangladesh. *Asian Journal of Women's Studies* 25, 76–95. doi.org/10.1080/12259276.2019.1577343.

Alam, G.M., Alam, K., Mushtaq, S. and Clarke, M.L. (2017) Vulnerability to climatic change in riparian char and river-bank households in Bangladesh: implication for policy, livelihoods and social development. *Ecological Indicators* 72, 23–32.

Arobi, S., Naher, J. and Soron, T.R. (2019) Impact of river bank erosion on mental health and coping capacity in Bangladesh. *Global Psychiatry* 2, 195–200.

BBS (2006) Population Census 2001. *Community Series, Zila: Barguna*. Bangladesh Bureau of Statistics, Statistics and Informatics Division, Ministry of Planning, Government of the People's Republic of Bangladesh, Dhaka, Bangladesh.

BBS (2013a) Population and Housing Census 2011. *Community Report, Zila: Barguna*. Bangladesh Bureau of Statistics, Statistics and Informatics Division, Ministry of Planning, Government of the People's Republic of Bangladesh, Dhaka, Bangladesh.

BBS (2013b) Population and Housing Census 2011. *Community Report, Zila: Bhola*. Bangladesh Bureau of Statistics, Statistics and Informatics Division, Ministry of Planning, Government of the People's Republic of Bangladesh, Dhaka, Bangladesh.

BBS (2013c) Population and Housing Census 2011. *Community Report, Zila: Rangpur*. Bangladesh Bureau of Statistics, Statistics and Informatics Division, Ministry of Planning, Government of the People's Republic of Bangladesh, Dhaka, Bangladesh.

BBS (2018) Statistical Pocketbook 2018. Bangladesh Bureau of Statistics, Statistics and Informatics Division, Ministry of Planning, Government of the People's Republic of Bangladesh, Dhaka, Bangladesh.

BBS (2019) Gender Statistics of Bangladesh 2018. Bangladesh Bureau of Statistics, Statistics and Informatics Division, Ministry of Planning, Government of the People's Republic of Bangladesh, Dhaka, Bangladesh.

Birkmann, J. (2011) First- and second-order adaptation to natural hazards and extreme events in the context of climate change. *Natural Hazards* 58, 811–840.

Biswas, J.C., Haque, M.M., Maniruzzaman, M., Ali, M.H., Kabir, W. and Kalra, N. (2019) Natural hazards and livestock damage in Bangladesh. *Natural Hazards* 99, 705–714. doi.org/10.1007/s11069-019-03768-0.

Bradshaw, S. (2015) Engendering development and disasters. *Disaster* 39(S1), 54–75.

Bradshaw, S. and Fordham, M. (2013) Women and girls in disasters. Report produced for the Department for International Development, London.

Byuro, B.P. (2006) Bangladesh population census, 2001: community series. Bangladesh Bureau of Statistics, Ministry of Planning, Government of the People's Republic of Bangladesh, Dhaka, Bangladesh.

CDSP (2020) Gender development in Char Development and Settlement Project IV. Available at: www.cdsp.org.bd (accessed 12 April 2020).

Chant, S. (2008) Dangerous equations? How female-headed households became the poorest of the poor: causes, consequences and cautions. In: Momsen, J. (ed.) *Gender and Development: Critical Concepts in Development Studies*. Routledge, London, pp. 397–409.

Connell, R. and Pearse, R. (2015) *Gender: In World Perspective*. Polity Press, Cambridge, UK.

Das, T.K., Haldar, S.K., Das Gupta, I. and Sen, S. (2014) River bank erosion: induced human displacement and its consequences. *Living Reviews in Landscape Research* 8. Advance online publication, doi.org/10.12942/lrlr-2014-3.

Diotima (2020) GBV core concepts: gender power violence. Available at: www.diotima.org.gr (accessed 4 May 2020).

Dominelli, L. (2013) Gendering climate change: implications for debates, policies and practices. In: Alston, M. and Whittenbury, K. (eds) *Research, Action and Policy: Addressing the Gendered Impacts of Climate Change.* Springer, London, pp. 77–93.

Enarson, E. and Chakrabarti, P.G.D. (2009) *Women, Gender and Disaster: Global Issues and Initiatives*. Sage, New Delhi.

Goulds, S. (2013) In double jeopardy: adolescent girls and disasters, Commonwealth Education Partnerships. Available at: www.plan-international.org/girls/reports-and-publications (accessed 11 June 2020).

Health Bulletin (2013) Health Bulletin 2013, district: Barguna. Barguna Civil Surgeon Office, Ministry of Health and Family Welfare, Barguna, Bangladesh.

Hoque, M.M. and Kamruzzaman, M. (2016) From kitchen utensil to serving machine: women's internal migration and gender dynamics in rural Bangladesh. *BRAC Institute of Governance and Development (BIGD) Working Paper Series*. Brac University, Dhaka, Bangladesh.

Hossain, S. (2008) Rapid urban growth and poverty in Dhaka City. *Bangladesh e-Journal of Sociology* 5, 1–24.

HPN (2014) Humanitarian exchange Number 60. Available at: www.oldihpn.org. (accessed 20 March 2020).

IDS (2008) Gender and climate change: mapping the linkages. A scoping study on knowledge and gaps. Institute of Development Studies, Brighton, UK. Available at: www.bridge.ids.ac.uk/reports/Climate_Change_DFID.pdf (accessed 10 June 2020).

IFRC (2017) Effective law and policy on gender equality and protection from sexual and gender-based violence in disasters. International Federation of Red Cross and Red Crescent Societies, Geneva, Switzerland. Available at: www.ifrc.org. (accessed 23 March 2020).

INFORM (2018) Index for Risk Management. Available at: www.inform-index.org/ (accessed 25 July 2018).

IOM (2015) *World Migration Report 2015: Migrants and Cities, New Partnership to Manage Mobility*. International Organization for Migration, Geneva, Switzerland.

Islam, M.F. and Rashid, A.B. (2011) Riverbank erosion displaces in Bangladesh: need for institutional response and policy intervention. *Bangladesh Journal of Bioethics* 2, 4–19.

MacDonald, R. (2005) How women were affected by the tsunami: a perspective from Oxfam. *PLoS Medicine* 2:e178.

Mahmud, S. and Tasneem, S. (2011) The under reporting of women's economic activity in Bangladesh: an examination of official statistics. *BDI Working Paper* 1. Brac Development Institute, Brac University, Dhaka, Bangladesh.

Momsen, J. (2010) *Gender and Development*. Routledge, New York.

Naved, R.T., Newby, M. and Amin, S. (2001) The effects of migration and work on marriage of female garment workers in Bangladesh. *International Journal of Population Geography* 7, 91–104.

NIRAPAD (2007) Cyclone SIDR kills hundreds in Barisal and Khulna Division. Network for Information. Response and Preparedness Activities on Disaster, Dhaka, Bangladesh. Available at: www.nirapad.org (accessed 15 July 2008).

Pain, R. (2014) Everyday terrorism: connecting domestic violence and global terrorism. *Progress in Human Geography* 38, 531–550.

Rahman, M.A.T.A., Islam, S. and Rahman, S.H. (2015) Coping with flood and riverbank erosion caused by climate change using livelihood resources: a case study of Bangladesh. *Climate and Development* 7, 185–191.

Rahman, M.S. and Gain, A. (2020) Adaptation to river bank erosion induced displacement in Koyra Upazila of Bangladesh. *Progress in Disaster Science* 5:100055.

Reyes, D.D. and Lu, J.L. (2017) Gender dimensions and women's vulnerability in disaster situations: a case study of flood prone areas impacting women in Malabon City, Metro Manila. *Journal of International Women's Studies* 18, 69–88.

Rezwana, N. (2018) *Disasters, Gender and Access to Healthcare: Women in Coastal Bangladesh*. Routledge, London.

Sams, I.S. (2019) Impacts of climate change induced migration on gender: a qualitative study from the southwest coastal region of Bangladesh. *International Journal of Social Science Studies* 7, 57–68.

Tamnia, U. (2009) Population evacuation need assessment in cyclone affected Barguna district. *Journal of Bangladesh Institute of Planners* 2, 145–157.

Terry, G. (2009) Introduction. In: Terry, G. (ed.) *Climate Change and Gender Justice*. Practical Action Publishing in association with Oxfam GB, Rugby, UK.

UN Women (2015) Climate change and migration in Bangladesh: a gender perspective. United Nations Entity for Gender Equality and the Empowerment of Women, Dhaka, Bangladesh.

Watkins, K. (2007) Fighting climate change: human solidarity in a divided world. UNDP-HDRO Human Development Report 2007/8. United Nations Development Program – Human Development Report Office, New York.

World Bank (2020) Understanding poverty: girls' education. World Bank, Washington, DC. Available at: www.worldbank.org. (accessed 4 May 2020).

4 Gender, Weather Shocks and Food Security: Empirical Evidence from Uganda

F. Mwesigye

Introduction

Climate change-induced weather shocks present a formidable food security and livelihood challenge in many developing countries (Niles and Salerno, 2018). Moreover, weather shocks are deleterious in countries where agriculture continues to play a fundamental role in the rural households' economic portfolio. In sub-Saharan Africa (SSA), more than 80% of rural households depend on agriculture directly or indirectly as a source of income, employment, and livelihood (Davis *et al.*, 2017). SSA's smallholder food producers are viewed as particularly vulnerable to climate change due to their reliance on rain-fed agriculture; limited application of yield enhancing technologies such as fertilizers, improved seed, and pesticides; and limited resources to cope with livelihood losses (Clay and Zimmerer, 2020).

Reducing the food insecurity effects of climate change remains one of the major global challenges (Campbell *et al.*, 2016). In SSA, climate change is already affecting livelihoods through erratic rainfall, droughts, and floods. The devastating locust outbreak across East Africa in 2020 has also been linked to global climate change (Clay and Zimmerer, 2020). The effect of climate change is felt disproportionately depending on vulnerability levels. For instance, the Food and Agriculture Organization (FAO) (2015a, p. 2) emphasizes that 'climate change is already affecting the food security and nutrition of the most vulnerable' and warns that 'if action is not very quickly taken, it will increasingly threaten the achievement of the goal to eradicate hunger.' Studies have shown that climate change-induced shocks negatively affect crop yields (Porter and Semenov, 2005; Challinor *et al.*, 2014), food security (Bandara and Cai, 2014; Akampumuza and Matsuda, 2017), and household welfare (Porter, 2012; Asfaw and Maggio, 2018). The effects of weather shocks are felt disproportionately across gender, and women are severely affected because they play a major role in agricultural production, a critical component of food security, and have fewer options for coping strategies (Mehar *et al.*, 2016; Akampumuza and Matsuda, 2017; Asfaw and Maggio, 2018).

Despite the increase in climate variability and its looming welfare effects, studies examining how weather shocks affect food security by gender are scant. The current study uses a rich household dataset to examine food security effects of weather shocks by gender. The results show that the incidence of food insecurity increases with an increase in all measures of weather shocks (drought, floods, and irregular rains). In addition, the findings reveal that female-headed households are more vulnerable to climate change than male-headed households and hence are most likely to suffer from food insecurity. The results also show that refugees and rural

 Gender, Climate Change and Livelihoods: Vulnerabilities and Adaptations (eds J. Eastin and K. Dupuy)
DOI: 10.1079/9781789247053.0004

households are more prone to food insecurity than national- and urban-based households.

The rest of the chapter is structured as follows. An overview of agriculture sector performance, climate change, and variability in Uganda is provided. The data used in this study and their descriptive statistics are then discussed. The empirical strategy is next presented, followed by presentation and discussion of econometric results. The final section concludes and provides policy implications of the study.

Background

Agriculture, climate change and variability

Agriculture is the main source of livelihoods in Uganda and employs 64% of the population, according to the Uganda Bureau of Statistics (UBoS, 2019). Uganda, like many other countries of SSA, relies heavily on the agriculture sector for income and food security, thus the economic impact of weather and climate change is enormous for small-scale farmers' food security and welfare (Asfaw *et al.*, 2016).

Agriculture is dominated by smallholder farmers producing at the subsistence level with limited use of productivity enhancing technologies such as improved seeds and fertilizers (EPRC, 2018). Agricultural production is largely rainfed, with less than 1% of farmers applying irrigation; it is marked by seasonality and is negatively affected by extreme weather conditions (Mubiru *et al.*, 2018).Thirty-nine percent of the working population (14–64 years) are involved in subsistence agriculture. In addition, there is a disproportionate share of women in subsistence agriculture, with 47% of the working age women employed in subsistence agriculture compared to 31% of men, suggesting that any shock to the agriculture sector has far-reaching implications on women's livelihoods (UBoS, 2019).

Uganda has varying rainfall seasons and planting patterns, especially for annual crops, depending on the region. Seasonal rainfall for Uganda is characterized by a bimodal cycle (two rainy seasons) in the southern, western, and central regions with higher rainfall during the rainy seasons (March–April–May and September–October–November). In the northern region, there is a unimodal cycle with a longer single rainy season that extends across the seasons. The far north-east of Uganda receives little rain during all months of the year. Therefore, the annual rainfall totals for Uganda vary from 500 mm to 2800 mm, with an average of 1180 mm (Irish Aid, 2017). The coasts of Lake Victoria receive the highest precipitation and present the most favorable production conditions in the country.

Climate change is increasingly becoming a threat in Uganda and has induced changes in the frequency and severity of extreme climate events and weather shocks such as droughts, floods, and storms (Lwasa *et al.*, 2009). Accordingly, climate change is a major challenge to food and agricultural development in Uganda (FAO, 2015b). According to the Uganda National Meteorological Authority (UNMA, 2018), there is a general warming trend over Uganda. Since 1950, all the warmest years on record in the country have been observed in the last 18 years. The warming rate over the past 68 years was about 0.23°C per decade compared to 0.25°C per decade over the past 27 years. UNMA also predicts that, with this increasing warming trend, temperatures may reach 2°C above the pre-industrial values (Paris Agreement 2015) in the coming few decades (UNMA, 2018). Mean annual temperature has increased by 1.3°C since 1960, an average rate of 0.28°C per decade, and annual rainfall has decreased at an average rate of 3.4 mm per month (3.5% per decade) (McSweeney *et al.*, 2012). In Uganda, climate change is exacerbated by deforestation, land degradation, and wetland degradation and encroachment. However, like much of the world, the country is victim to forces outside of its control (NPA, 2020).

Climate change and variability present new risks and vulnerabilities in Uganda. Climate-related risks such as prolonged dry seasons are becoming more frequent and intense with negative impacts on agricultural livelihoods and food security (Mubiru *et al.*, 2018). In recent years, repeated dry spells, irregular rains, and floods have characterized the seasons, and lowered crop and livestock productivity. This has negatively affected livelihoods, as households have sold some of their productive assets to acquire food. With the continuing reality of climate change, the food and nutritional security situation is likely to worsen. Many households, especially female-headed ones, lack access to diverse diets, which is indicative of food insecurity (FAO, 2015b). With the change in rainfall patterns, it is expected that the socio-economic impacts of climate change on agriculture, livelihoods, food security, health, and other sectors, will continue to worsen, and perhaps even reverse development gains and jeopardize the government's economic development and poverty eradication goals. Available evidence indicates that climate change in Uganda creates conditions which favor the spread of human, crop, and livestock pests and diseases, including to areas where they were originally non-existent (Environmental Alert, 2010; FAO, 2015b).

To address climate change, Uganda signed and ratified the United Nations Framework Convention on Climate Change (UNFCCC), the Kyoto Protocol, and the Paris Agreement. Through these international frameworks, Uganda has committed to adopt and implement measures to mitigate climate change and adapt to its impacts. At the county level, the third national development plan (NDP III) has identified climate change as one of the 18 programs that the country is prioritizing between 2020–21 and 2024–25. Specifically, NDP III's fifth program on 'Natural Resources, Environment, Climate Change, Land, and Water Management' aims to stop, reduce, and reverse environmental degradation and the adverse effects of climate change as well as improve utilization of natural resources for sustainable economic growth and livelihood security (NPA, 2020).

Women's vulnerabilities and agriculture in Uganda

Agriculture remains a key source of livelihood and the main employer of women in Uganda. According to UBoS (2019), 70% of the working age women (14–64 years) are employed by the agricultural sector. However, women remain trapped in low productive and low paying activities. For instance, 47% of the working age women are employed in subsistence agriculture providing the bulk of 'non-contractible' agricultural outputs, and most of them engage in unpaid agricultural activities such as land preparation, weeding, and harvesting as part of family labor (UBoS, 2019). Despite this, men dominate the majority of decisions related to land use and management, and the security of women's land ownership can be tenuous (Mwesigye *et al.*, 2020).

The concentration of women in low productive activities in Uganda is attributed to the patrilineal society's customs and beliefs which have disadvantaged women in terms of ownership of productive assets such as land, and skill acquisition compared to their male counterparts. For instance, insecurity associated with women's rights to land under customary law is grounded in the assumptions that women are dependent on men and cannot own land in their own right under customary tenure, and hence have what many term 'secondary rights' (Burke and Kobusingye, 2014). Under the patrilineal system, men are by default household heads and women only assume this role when they are either widowed, separated, divorced, or have never married, and according to UBoS (2019), about 75% of the households in Uganda are male-headed. It should be noted that a female becoming a household head does not guarantee that they get full rights over assets; for widows, in most cases, male in-laws assume control of the deceased's assets.

The existing gender norms marginalize women and render them vulnerable. Indeed, Carter *et al.* (2017) note: 'Gender norms and disparities between statutory and customary law that dictate household members' ability to control, retain, and make decisions about investments and assets, and their within-household bargaining power, may result in differential benefit allocation between household members.' For example, the common practice under customary land tenure arrangements is that women are granted land use but not ownership rights, and yet the Constitution recognizes customary tenure systems. The legal pluralism where *de jure* laws coexist and interact with *de facto* institutional arrangements is enshrined in the Constitution, which results in individual's ownership of assets being based on legal and social norms.

To address gender inequities and inequalities associated with resource ownership and use, Uganda's Constitution outlaws practices that discriminate against women and other vulnerable groups. The 1995 Ugandan Constitution guarantees property rights 'without bias to gender or marital status' and 'affirmative action in favor of marginalized groups based on gender or other reasons created by history, tradition, or custom, for the purpose of redressing existing imbalances' (Government of Uganda, 1995; Carter *et al.*, 2017). Further, the 1998 Land Act protects the rights of women over customary land. The Act stipulates that any decision taken in respect of land held under customary tenure, whether regarding land held individually or communally, shall be in accordance with the customs, traditions, and practices of the community concerned, except for a decision which denies women, children, or persons with a disability access to ownership, occupation, or use of any land (Republic of Uganda, 1998).

However, despite the fact that the traditions, customs, and practices which discriminate against women in matters of access and ownership of productive resources have been outlawed, current practices do not acknowledge these changes. For example, *de facto* land tenure arrangements are still not favorable to women in terms of inheritance, tenure security, and land transactions. Indeed, the evidence illustrating gender inequalities in access to land and security of tenure is overwhelming (Mwesigye *et al.*, 2020). Like in other SSA countries, women in Uganda are consistently less likely to own or operate land, and the land they do have access to is often of poor quality and in smaller plots. Where women retain access to land, they do not have secure rights over the land they occupy, and cannot even use the land to obtain credit, thus leaving them poor and vulnerable.

Data and Descriptive Statistics

Data

This study uses data from the 2019 Comprehensive Food Security and Vulnerability Assessment (CFSVA) survey implemented by the UBoS and supported by the World Food Program (WFP). The CFSVA data cover a total sample of 6845 households from all regions in Uganda. CFSVA was designed to produce representative estimates for the main food security and nutrition indicators at national level; urban and rural areas separately; each of the 15 regions; the eight districts of Karamoja sub-region, and the refugee settlements.

The sampling frame used for the 2019 CFSVA is the Uganda Population and Housing Census conducted in August 2014 which was provided by the UBoS. The sampling frame is a complete list of 78,093 census Enumeration Areas (EA) covering the whole country. An EA is a natural village in rural areas or a city block in an urban area. In 2019 Uganda was divided into 128 administrative districts, each district divided into counties, each county into sub-counties, each sub-county into parishes, each parish into villages, and each village further divided into EAs.

The 2019 CFSVA used a two-stage stratified sampling strategy. Stratification was achieved by separating the districts into the various study domains. Samples were selected independently from each stratum according to the sample allocation using 'probability proportional to size' (PPS) selection. With PPS sampling procedure, stratification by residence type was achieved and the sample was proportionally allocated to urban and rural areas. At the second stage, a fixed number of ten households was selected from the up-to-date household listings for each selected EA. Household selection was done centrally at the UBoS headquarters prior to the main survey.

The 2019 CFSVA survey elicited information on household characteristics in terms of household composition, gender, education level of each household member, marital status, and asset ownership. The data also captured the housing materials of the roof, walls, and floor; the quality of latrine; access to clean water; and energy sources such as solar and hydro, among others.

In addition, the CFSVA survey captured data on different indicators of food security experienced by the household over 1 year, and as recent as 1 week preceding the survey. These indicators include the number of meals per day, perceptions about food insecurity incidences such as failure to have enough food to eat, skipping meals, reducing the amount of food eaten by adults, and eating unhealthy food, among others. The data also contain information on different coping strategies adopted by households when faced with food insecurity. The survey captured information on different weather shocks including drought, irregular rains, floods, and landslides.

Household characteristics disaggregated by gender of household head

Figure 4.1 shows that 35% of the households are female-headed and 65% are headed by males. These statistics are consistent with other studies on Uganda. For instance, Milazzo and Walle (2015) found that in 2011, 30% of the households in Uganda were female-headed.

Table 4.1 presents household characteristics disaggregated by the gender of the household head. The results indicate that more than half of female-headed households are based in rural areas (52%) compared to male-headed households (49%), and that a significantly higher percentage of female-headed households claimed to be refugees (8%) compared to male-headed households (5%). The statistics also indicate that female-headed households have smaller families both in numbers and in adult equivalence. Also, female heads are older than male heads. This might be explained by the variables showing marital status of the head. There are more female heads of households who are either divorced or widowed compared to male heads. Specifically, 32% of female heads are widows and 16% are divorced compared to only 2% and 5% male heads, respectively. This statistic makes sense because Uganda follows a patrilineal system whereby a married female becomes a head after the death of her husband.

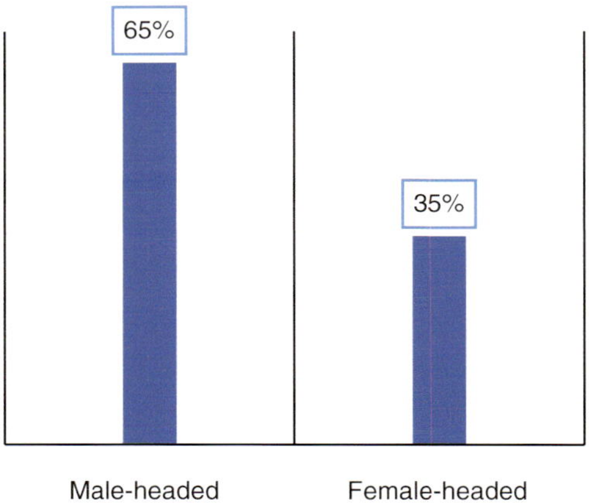

Fig. 4.1. Percentage of female-headed households in Uganda. Author's computation using Comprehensive Food Security and Vulnerability Assessment (CFSVA) survey data, 2019.

More female heads (24%) report to be in a polygamous marriage compared to male heads (19%), and there are more male heads who are monogamously married. The findings are similar to those of Mwesigye *et al.* (2020) who used the nationally representative Uganda National Panel Surveys (UNPS 2015–16) and found that 23% of the female-headed households were polygamously married in 2017 compared to 18% of male-headed households. Milazzo and Walle (2015) found that on average, heads in female-headed households are older (reflecting the many widowed heads) and their households are smaller in size as compared to male-headed households. The same study notes that female heads are largely comprised of widows, and divorced, separated, or abandoned women, married women with a non-resident (polygynous or migrant) husband, single women and mothers.

The female household heads are also less educated compared to the male heads. For example, 46% of the female heads reported that they have never been to school compared to 20% of the male heads. For those that have been to school, a higher percentage of male heads have attained all levels of education than have female household heads. For instance, there is a 9% difference between males who attended primary school and secondary schools than female heads who completed the same levels. There are also more male household heads who have completed diplomas and tertiary degrees than female heads. These findings are similar to national statistics which indicate that males are more educated than females, and that in 2016, 56% of males were enrolled at tertiary level compared to 44% of females (UBoS, 2019). Also, Milazzo and Walle (2015) found that female heads have fewer years of education (4.1) compared to male heads (5.6).

In terms of occupation, 73% of the female heads report that agriculture is their main occupation compared to 68% of the male heads who report the same, and the difference is statistically significant. These findings are consistent with the national figures by the UBoS which show that 70% of working age women (14–64 years) had been employed in agriculture as their main occupation in 2017 compared to 58% of males (UBoS, 2019). The statistics also show that male-headed households have a higher asset value and hold more land than the female-headed ones. The

Table 4.1. Household characteristics disaggregated by the gender of the household head (HH). Author's computation using Comprehensive Food Security and Vulnerability Assessment (CFSVA) survey data, 2019.

	Male-headed (A)		Female-headed (B)		
Variable	Mean	SD	Mean	SD	t-test (A–B)
Rural = 1	0.49	0.50	0.52	0.50	−0.04***
Refugee = 1	0.05	0.21	0.08	0.27	−0.03***
Family size	5.1	2.9	4.6	2.7	0.51***
Family size in Total Adult Equivalence	4.1	2.5	3.5	2.2	0.52***
Age of the HH (years)	41	15	46	17	−4.14***
Marital status of HH					
Monogamously married = 1	0.69	0.46	0.22	0.42	0.46***
Polygamous = 1	0.19	0.39	0.24	0.42	−0.05***
Divorced = 1	0.05	0.21	0.16	0.37	−0.12***
Widow = 1	0.02	0.15	0.32	0.47	−0.30***
Never married = 1	0.05	0.23	0.05	0.23	0.00
Education of HH					
Never been to school = 1	0.20	0.40	0.46	0.50	−0.26***
Some education = 1	0.00	0.04	0.01	0.10	−0.01***
Primary school level = 1	0.45	0.50	0.35	0.48	0.09***
Ordinary level = 1	0.17	0.38	0.08	0.27	0.09***
Advanced level = 1	0.02	0.15	0.01	0.09	0.01**
Diploma = 1	0.07	0.26	0.05	0.21	0.03***
Degree = 1	0.02	0.14	0.01	0.09	0.01*
Agriculture is HH's main occupation = 1	0.68	0.46	0.73	0.44	−0.05***
Value of assets (in US$)	3288	5783	2189	4747	1098.4***
Total HH land holding (ha)	0.98	2.29	0.71	1.46	0.27***
Number of observations	4412		2412		

*p<0.1; **p<0.05; ***p<0.01

assets for male-headed households were valued at US$3288, higher than the US$2189 value for female-headed households, which indicates that male-headed households are wealthier.

Figure 4.2 reveals the wealth status of male- and female-headed households. The wealth quintiles are generated using the asset index, which is constructed using indicators such as the status of toilet facilities (ventilated improved pit, open pit, bush, among others), whether a household is connected to solar power or electricity, quality of roofing for the main house (iron vs grass), the wall and floor material, and education level, among others.

Consistent with the statistics in Table 4.1, 41% of female-headed households fall in the lowest wealth quintile, a figure that is much higher than the 29% of male-headed households who belong to the same quintile. Conversely, more male-headed households belong to the richest wealth quintiles compared to female-headed ones. For example, 43% of male-headed households belong to the highest two quintiles, higher than 35% of female-headed households which belong to the same quintiles. The findings in Table 4.1 and Fig. 4.2 suggest that female-headed households are poorer and more vulnerable than male-headed households, and thus are likely to suffer more from food insecurity due to weather shocks since they may not have robust coping strategies and resilience mechanisms to ameliorate the shock effects.

Household food insecurity disaggregated by gender of household head

Table 4.2 presents indicators of food insecurity disaggregated by gender of the household head. The statistics show that female-headed

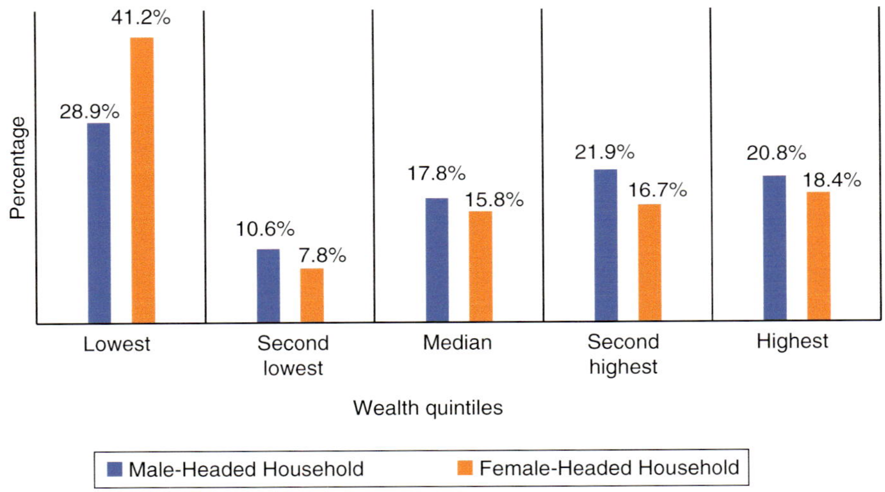

Fig. 4.2. Percentage of male- and female-headed households belonging to different wealth quintiles. Author's computation using Comprehensive Food Security and Vulnerability Assessment (CFSVA) survey data, 2019.

Table 4.2. Household food insecurity disaggregated by the gender of the household head (HH). Author's computation using CFSVA survey data, 2019.

	Male-headed (A)		Female-headed (B)		
Variable	Mean	SD	Mean	SD	t-test (A–B)
Proportion of HHs that were food insecure in any of past 12 months	0.64	0.48	0.71	0.45	−0.07***
Number of months HH experienced food insecurity in past 1 year	2.09	2.35	2.56	2.61	−0.47***
Number of meals per day, including breakfast	2.21	0.79	2.05	0.75	0.16***
Different cases of food insecurity experienced by HH in past 1 year					
HH was worried they would not have enough food = 1	0.71	0.45	0.80	0.40	−0.09***
HH ate unhealthy food = 1	0.72	0.45	0.82	0.39	−0.09***
HH ate few kinds of food = 1	0.72	0.45	0.81	0.39	−0.09***
Any household members skipped meal = 1	0.54	0.50	0.65	0.48	−0.11***
HH ate lower amount of food = 1	0.63	0.48	0.74	0.44	−0.11***
HH ran out of food = 1	0.68	0.46	0.75	0.43	−0.07***
Number of observations	4412		2412		

*p<0.1; **p<0.05; ***p<0.01

households perform poorer in all food security measures than the male-headed households. The survey elicited information on the households that were food insecure in any of the past 12 months preceding data collection. About 71% of female-headed households reported to have felt food insecure at least once compared to 64% of the male-headed households that reported the same, and the difference of 7% is significant.

This study computed the number of months in which the household experienced food insecurity from January to December. The statistics show that in the past 12 months preceding the survey, female-headed households experienced food insecurity for 2.6 months, while male-headed households experienced food insecurity for 2.1 months, and the difference is statistically significant.

The female-headed households also exhibited high levels of food insecurity along many other measures. For instance, more female-headed households (80%) reported that they had been worried they would not get enough food to eat than male-headed households (71%). Also, there were more female-headed households (82%) which reported to have eaten unhealthy food due to food insecurity compared to male-headed households (72%). In addition, 65% of female-headed households reported to have skipped a meal because there was not enough food to eat, compared to 54% of male-headed households. Female-headed households reported that they ate less food (74%) and that they ran out of food (75%) compared to 63% and 68%, respectively. Overall, the statistics indicate that female-headed households are more food insecure than their male-headed counterparts.

Figure 4.3 presents the reported causes of food insecurity. Drought was reported to be the major cause of food insecurity by 55% of the surveyed households, followed by lack of financial resources to buy food which was reported by 41% of the households. The other reported causes included high cost of food, planting fewer crops, floods, pests and diseases, and insecurity and conflict.

The fact that drought is mentioned as the major cause of food insecurity and that floods also affect more than 10% of the households confirms that weather shocks are posing a threat to food security. It should also be noted that drought, through its effects on food scarcity, can cause an increase in food prices and can lead to planting less than required. Therefore, it is highly likely that the other reported causes of food insecurity are driven by weather shocks such as drought and floods.

Weather shocks

Table 4.3 presents weather shocks experienced by households. Households reported to have experiences with three shocks: drought, floods, and irregular rain. The table shows that 56%

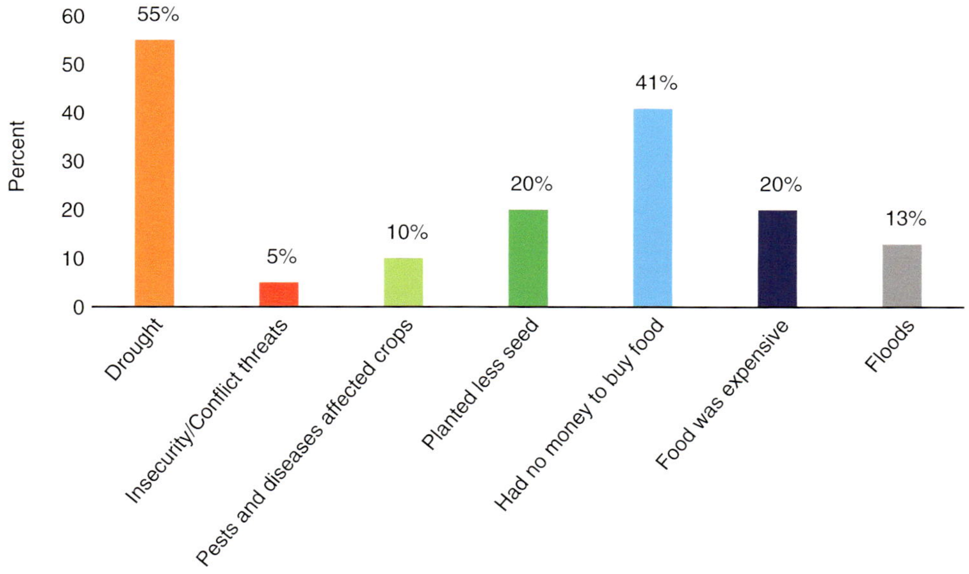

Fig. 4.3. Causes of food insecurity among households. Author's computation using Comprehensive Food Security and Vulnerability Assessment (CFSVA) survey data, 2019.

Table 4.3. Weather shocks experienced by the household in the 12 months preceding the survey. Author's computation using Comprehensive Food Security and Vulnerability Assessment (CFSVA) survey data, 2019.

Variable	Male-headed (A) Mean	Male-headed (A) SD	Female-headed (B) Mean	Female-headed (B) SD	t-test (A–B)
Proportion of households that experienced weather shock in any of past 12 months					
Drought shock	0.56	0.50	0.59	0.49	−0.02
Flood shock	0.20	0.40	0.27	0.44	−0.06***
Irregular rain shock	0.36	0.48	0.29	0.46	0.07***
Total number of months a given shock was experienced in past 12 months					
Number of months with drought	1.91	2.12	2.07	2.18	−0.16*
Number of months with floods	0.44	1.01	0.62	1.19	−0.18***
Number of months with irregular rain	1.09	1.73	0.88	1.55	0.22***
Number of observations	3285		1851		

*p<0.1; **p<0.05; ***p<0.01

and 59% of the male- and female-headed households, respectively, reported to have experienced a drought. The difference between them is not statistically significant. However, there were significantly more female-headed households (27%) who reported to have experienced floods compared to 20% of the male-headed households who experienced the same shock. The difference could be explained by the quality of land occupied by both households, and it might mean that female-headed households are always allocated swampy land which is more susceptible to floods. The results also show that female-headed households experienced floods for a longer period (0.6 months) compared to 0.4 months for male-headed households. However, there were more male-headed households who reported to have experienced irregular rains (36% vs 29%) and for longer periods (1.1 months vs 0.88 months).

Figure 4.4 traces the different weather shocks experienced over different months in 1 year preceding the survey. The survey elicited information about which months each household experienced each of the weather shocks (drought, floods, and irregular rains) during the past 1 year preceding the survey. Panel A shows drought shock, B presents floods, and C is a combination of both drought and floods, which are the most common shocks. The figure shows that households experience shocks in every month of the year, only that these shocks take different forms. For instance, the drought shock is experienced in June and July, after which the trend goes down. It then rises in November and December and starts declining in January and February as shown in Panel A. Panel B shows that when incidence of drought declines in February, a number of households begin to experience floods which continue through May. Panel C shows that a significant proportion of households experience different shocks through the year and only get some breaks when transitioning from one season to another, mainly around March and August at the end of the dry season and beginning of the rainy season.

Empirical Estimation of Gendered Effect of Weather Shocks on Food Security

The statistics above show that there are differences between male- and female-headed households in terms of vulnerabilities. Specifically, the statistics show that female heads are less educated, most of them are either widows or divorced, and most of them report agriculture as a main occupation compared to male heads. In addition, the statistics show that female-headed households are poorer when assessed on value of assets, and on the wealth quintiles they fall into compared to the households headed by males. The statistics also show that female-headed households are more food insecure than those headed by males. Furthermore, the study finds that female-headed households do not only report to have reduced the amount of food consumed, they have eaten less healthy food, skipped meals, had

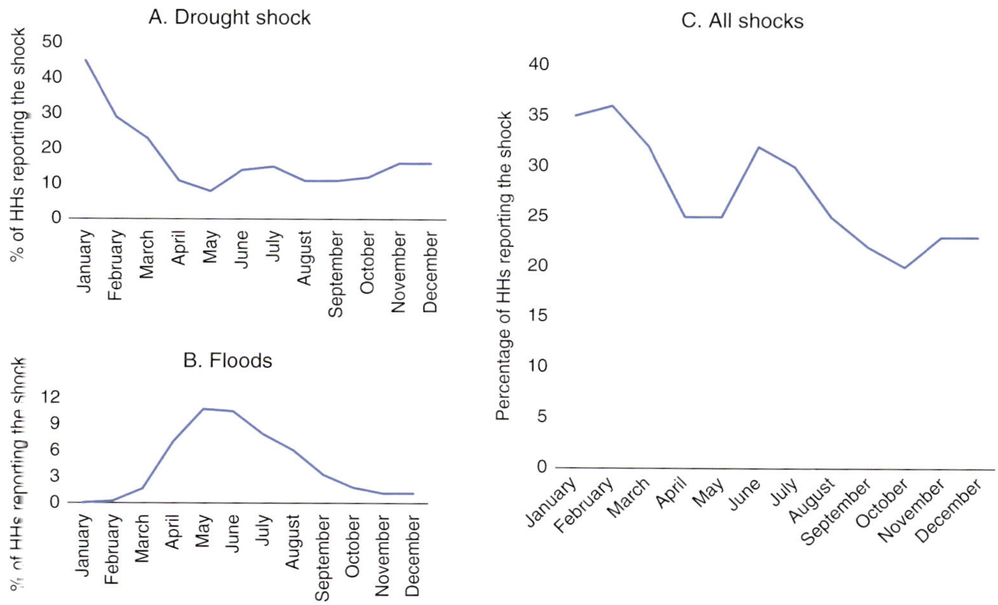

Fig. 4.4. Major weather shocks reported by household heads (HH) during the 12 months preceding the survey. Author's computation using Comprehensive Food Security and Vulnerability Assessment (CFSVA) survey data, 2019.

fewer meals in a day, and experienced food insecurity for a longer period of time than male-headed households.

In this section, the study runs simple ordinary least squares (OLS) regression and controls for all observed characteristics that might influence both food consumption and the extent to which the households are affected by shocks, so as to examine whether gender of the household head and weather shocks are associated with food insecurity. The study follows the following specification:

$$FS_i = \alpha + \beta \sum_{1}^{3} W\_Shocks_i + \vartheta\, FemaleHead_i + \partial X_i + \varepsilon_i \qquad (4.1)$$

where FS_i represents different measures of food insecurity such as the number of months the household experienced food insecurity during the year preceding the survey, and the number of meals eaten per day. In other specifications the outcome variable FS_i is a dummy which takes 1 if the household worried they would not have enough food to eat, ate less food, or ran out of food, and 0 otherwise. When the dummy outcome variables are used, the results are estimated using a linear probability model.

W_Shocks_i, is a vector of three main weather shocks: the number of months the household experienced drought, the number of months with irregular rains, and the number of months the household experienced floods. The study utilizes the self-reported number of months that a household experienced different forms of weather shocks. In other specifications, dummy variables were used to measure whether the household experienced any of the three shocks or not during the past 1 year.

$FemaleHead_i$ is a dummy variable which takes 1 if the household is female-headed and 0 otherwise. A number of studies have used gender of household head to examine the association between gender and food security, and also to examine if weather shocks have a differential effect on food security by gender (see Akampumuza and Matsuda, 2017). The hypothesis is that female-headed households are more vulnerable in terms of access to resources and have limited income generating activities. Indeed, the statistics show that female-headed households have small land holdings and fewer assets, and the majority fall in the lowest wealth quintiles compared to male-headed households. This statistic suggests that female-headed households are less resilient to shocks and are more likely to suffer from food insecurity. The author thus expects that when vulnerability indicators (e.g. assets, land size, education levels, and other household characteristics) are not controlled for, the incidence of food insecurity is higher among the female-headed households compared to their male-headed counterparts.

X_i is a vector of other household characteristics such as location (rural or urban); whether the household is a refugee or national; age of the household head; household head's education level; family size in adult equivalence; the value of household assets (in US$); size of land holding; marital status of the head (e.g. polygamously married, monogamously married, divorced, widowed, or single); and the dummy variable which takes 1 if agriculture is the household's main occupation. Marital status is included because it affects the level of vulnerability. The statistics show that many female heads were either divorced or widowed. In Uganda widows are denied land ownership rights, and even when they get to share land, they are allocated poor quality land in swamps which are prone to floods (Mwesigye *et al.*, 2020). Thus, marital status also appears to bear on the vulnerability of female-headed households. ε_i is the idiosyncratic error term that is assumed to be normally distributed.

Results from Regression Analysis and Discussion

Table 4.4 presents the results on the association between weather shocks, gender of the household head, and food insecurity. The author does not control for other household characteristics in this analysis. This follows the argument, in the section titled 'Data and descriptive statistics,' that male-headed and female-headed households are systematically different along many attributes and that female-headed households are more vulnerable. The author, therefore, expects that without controlling for other characteristics such as assets, marital status, family size, and education level, among others, food insecurity should be higher among the female-headed households.

In specification (1) of Table 4.4, the outcome variable is the number of months the household was food insecure during the year

Table 4.4. Association between weather shocks, gender of household head (HH) and food insecurity.

	# months the HH was food insecure	# meals per day	1 if HH worried they would not have enough food	1 if HH ate less food	1 if HH ran out of food
	(1)	(2)	(3)	(4)	(5)
# months drought	0.267***	−0.043***	0.0121***	0.0143***	0.0297***
	(17.86)	(−8.757)	(5.333)	(5.359)	(9.919)
# months irregular rains	0.0770***	−0.00681	0.00805***	0.0138***	0.0113***
	(3.943)	(−1.067)	(2.715)	(3.968)	(2.899)
# months floods	0.383***	−0.155***	0.0440***	0.0662***	0.0890***
	(12.70)	(−15.69)	(9.612)	(12.27)	(14.73)
1 if female HH	0.337***	−0.127***	0.0614***	0.0796***	0.0825***
	(5.043)	(−5.827)	(6.052)	(6.663)	(6.154)
Constant	1.771***	2.298***	0.777***	0.669***	0.510***
	(31.72)	(126.0)	(91.81)	(67.07)	(45.63)
Observations	5136	5136	5136	5136	5136
R-squared	0.095	0.070	0.032	0.045	0.069

t-statistics in parentheses. *p<0.1; **p<0.05; ***p<0.01

preceding the survey. In specification (2), the outcome variable is the average number of meals consumed per day by the household. In specifications (3), (4), and (5), the outcome variables are dummy variables that take on a value of 1 if a household was worried they would not have enough food, if they ate less food, and if they ran out of food, respectively, and 0 otherwise. In all the specifications, the outcome variables represent different measures of food security and are strongly correlated.

The results show that food insecurity increases with an increase in weather shocks: drought, irregular rains, and floods. For instance, an increase in the number of months a household experienced drought by 1 is associated with a 0.27-month increase of food insecurity and a reduction in the number of meals per day by 0.04, all statistically significant at 1%. The results also show that other measures of food insecurity increased with an increase in the drought period. The likelihood that a household was worried they would not have enough food to eat and that they actually ate less food increased by 1% with a 1-month increase in the drought period. In addition, the likelihood that a household ran out of food increased by 3% with a 1-month increase in the drought period.

The results also show that besides drought, other weather shocks that affect food security are floods and irregular rains. For instance, when floods last for an extra month, there is a 0.4-month increase of food insecurity and a 0.2 reduction in the number of meals per day. An increase in the number of months a household was faced with floods by 1 month was also associated with 4% increase in the likelihood that the household would worry that they would not have enough food to eat, 7% increase that they would eat less food, and 9% increase that they would actually run out of food. Irregular rains present a similar relationship with food insecurity as the other weather shocks. The results show that an increase in the number of months with irregular rains is associated with an increase in months with food insecurity. Also, an increase in the number of months with irregular rain is associated with a higher likelihood of other measures of food insecurity, such as worrying that the household will not have enough food to eat, eating less food, and skipping a meal. The results, thus, indicate that all kinds of weather shocks affect food security.

The results show that food insecurity is higher among female-headed households compared to their male-headed counterparts. For instance, the number of months a household was food insecure were 0.34 (equivalent to 10 days) higher, and the number of meals were 0.13 lower among the female-headed households compared to male-headed households. In addition, the probability that a household was worried they would not have enough food, they ate less food, and that they ran out of food was 6.1%, 8% and 8.3%, respectively; higher for female-headed households compared to the male-headed households.

Table 4.5 examines whether there is a differential effect of weather shocks by gender. Different weather shocks are interacted with the gender dummy to establish whether food insecurity effects of weather shock differ by gender. However, the results do not show that drought, irregular rains, and floods have a differential effect on food security between female- and male-headed households.

Table 4.6 presents the results of determinants of food insecurity, with all variables that proxy vulnerability, in addition to weather shocks and the gender of the household head presented in Table 4.4. The author argues that controlling for other measures of vulnerability attenuates the relationship between gender of the household head and food security. Indeed, the findings are consistent with this expectation. The results show that the female-headed dummy is not significantly associated with most measures of food insecurity when the controls are included. It is only weakly associated with the likelihood that the household was worried they would not have enough food, and that they actually ate less food.

The other control variables, especially those that proxy vulnerability such as value of assets, whether agriculture is the household's main occupation, education level, and family size, are strongly associated with food insecurity. The results indicate that an increase in the value of assets is associated with fewer months of food insecurity, more meals per day, and a lower likelihood of worrying about food insecurity, eating less food, and even not having food at all. Wealth

Table 4.5. Weather shocks, gender and food insecurity: interacting different shocks with the gender of the household head (HH).

	# months the HH was food insecure	# meals per day	1 if HH worried they would not have enough food	1 if HH ate less food	1 if HH ran out of food
	(1)	(2)	(3)	(4)	(5)
# months drought	0.286***	−0.042***	0.0138***	0.0139***	0.0296***
	(15.15)	(−6.754)	(4.827)	(4.112)	(7.832)
1 if female HH	0.473***	−0.15***	0.0946***	0.0945***	0.105***
	(4.448)	(−4.297)	(5.864)	(4.973)	(4.943)
# months drought*female HH	−0.0507	−0.00317	−0.00428	0.00136	0.000545
	(−1.642)	(−0.314)	(−0.914)	(0.246)	(0.0882)
# months irregular rain	0.0815***	−0.0119	0.0115***	0.0168***	0.0145***
	(3.466)	(−1.550)	(3.220)	(4.001)	(3.073)
# months irregular rain*female HH	−0.0138	0.0162	−0.0107*	−0.00935	−0.009
	(−0.327)	(1.172)	(−1.674)	(−1.239)	(−1.149)
# months floods	0.401***	−0.165***	0.0554***	0.0729***	0.101***
	(9.995)	(−12.57)	(9.106)	(10.17)	(12.52)
# months floods*female HH	−0.0383	0.0240	−0.0264***	−0.0161	−0.0270**
	(−0.627)	(1.203)	(−2.858)	(−1.475)	(−2.213)
Constant	1.721***	2.306***	0.765***	0.663***	0.502***
	(27.20)	(111.5)	(79.78)	(58.68)	(39.60)
Observations	5136	5136	5136	5136	5136
R-squared	0.095	0.071	0.034	0.046	0.070

t-statistics in parentheses. *p<0.1; **p<0.05; ***p<0.01

Table 4.6. Weather shocks, gender and food insecurity with controls.

	# months the HH was food insecure	# meals per day	1 if HH worried they would not have enough food	1 if HH ate less food	1 if HH ran out of food
	(1)	(2)	(3)	(4)	(5)
# months drought	0.256***	−0.0296***	0.00784***	0.00737***	0.0213***
	(17.30)	(−6.428)	(3.546)	(2.900)	(7.510)
# months irregular rain	0.0887***	−0.0141**	0.00900***	0.0150***	0.016***
	(4.623)	(−2.360)	(3.136)	(4.554)	(4.301)
# months floods	0.320***	−0.103***	0.0264***	0.0393***	0.055***
	(10.49)	(−10.81)	(5.780)	(7.485)	(9.337)
1 if female HH	0.0393	0.0238	0.0225*	0.0262*	0.0133
	(0.492)	(0.954)	(1.883)	(1.911)	(0.866)
1 if refugee	0.547***	0.0722	0.00754	0.0468*	0.00811
	(3.689)	(1.561)	(0.340)	(1.838)	(0.285)
1 if rural	0.255***	−0.212***	0.0741***	0.0899***	0.086***
	(3.858)	(−10.29)	(7.500)	(7.921)	(6.782)
Age of HH	0.00751***	−0.000740	−0.001***	−0.0013***	−0.0006
	(3.196)	(−1.009)	(−2.732)	(−3.131)	(−1.307)
Family size	0.0535***	−0.0272***	0.0197***	0.0288***	0.019***
	(3.436)	(−5.591)	(8.461)	(10.77)	(6.376)
Marital status of HH[a]					
1 if monogamous	0.128	0.0709	0.0570**	0.0840***	0.0740**
	(0.775)	(1.371)	(2.301)	(2.950)	(2.328)
1 if polygamous	0.309*	−0.0115	0.105***	0.135***	0.159***
	(1.761)	(−0.210)	(3.995)	(4.465)	(4.720)
1 if divorced	0.154	0.110*	0.0626**	0.0611*	0.00213
	(0.792)	(1.806)	(2.147)	(1.824)	(0.0570)
1 if widowed	0.225	−0.0135	0.0860***	0.109***	0.0717*
	(1.147)	(−0.221)	(2.933)	(3.225)	(1.907)

Continued

Table 4.6. Continued.

	# months the HH was food insecure	# meals per day	1 if HH worried they would not have enough food	1 if HH ate less food	1 if HH ran out of food
	(1)	(2)	(3)	(4)	(5)
Education level of HH[b]					
1 if some education	−0.335	0.0865	0.0318	−0.00355	0.00930
	(−0.805)	(0.666)	(0.511)	(−0.0496)	(0.116)
1 if completed primary	−0.188**	0.204***	−0.0370***	−0.0798***	−0.125***
	(−2.385)	(8.274)	(−3.139)	(−5.884)	(−8.239)
1 if completed O-level	−0.297***	0.272***	−0.0453***	−0.140***	−0.203***
	(−2.671)	(7.823)	(−2.721)	(−7.301)	(−9.510)
1 if completed A-level	−0.950***	0.441***	−0.215***	−0.328***	−0.326***
	(−3.743)	(5.573)	(−5.674)	(−7.523)	(−6.699)
1 if completed diploma	−0.697***	0.439***	−0.124***	−0.229***	−0.253***
	(−4.450)	(8.967)	(−5.292)	(−8.511)	(−8.413)
1 if completed degree	−1.227***	0.437***	−0.208***	−0.292***	−0.350***
	(−4.286)	(4.885)	(−4.867)	(−5.928)	(−6.375)
1 if agriculture is main occupation	0.104	−0.132***	0.0264**	0.0536***	0.061***
	(1.360)	(−5.541)	(2.306)	(4.071)	(4.179)
Log of assets in US$	−0.257***	0.108***	−0.0403***	−0.0498***	−0.063***
	(−11.91)	(16.05)	(−12.50)	(−13.43)	(−15.27)
Total land size	−0.00595	−0.000869	−0.00501**	−0.00415	−0.00556*
	(−0.392)	(−0.183)	(−2.208)	(−1.592)	(−1.910)
Constant	2.917***	1.605***	0.954***	0.886***	0.881***
	(13.43)	(23.67)	(29.38)	(23.74)	(21.13)
Observations	5120	5120	5120	5120	5120
R-squared	0.153	0.207	0.123	0.175	0.202

[a]Reference category = 'single'
[b]Reference category = 'never been to school'
t-statistics in parentheses. *p<0.1; **p<0.05; ***p<0.01

as measured by the value of assets, is a measure of vulnerability. The poor remain more vulnerable to different causes of food insecurity. The results also show that food insecurity is higher for households whose main occupation is agriculture, as they had more months of food insecurity, fewer meals per day, and a higher likelihood of experiencing different measures of food insecurity. Agriculture in Uganda remains subsistence and is highly vulnerable to weather shocks. Households who only depend on the sector as a source of livelihood are susceptible to shocks and are also more likely to face food insecurity compared to the formally employed families of those who run businesses. The rural dummy is negatively and significantly associated with all measures of food insecurity. For instance, being in a rural area is associated with a further 0.3 months of food insecurity, 0.2 fewer meals per day, 7% higher likelihood of being worried of not having enough food, and 9% higher likelihood of running out of food. This might be explained by the fact that rural households have limited access to other income generating activities, or because of limited access to produce markets.

The findings also show that education level is associated with food insecurity. Food insecurity is lower for households where the head has had a minimum of primary school education compared to households where the heads had never been to school. In addition, family size is associated with food insecurity. A one-person increase, in adult equivalence, is associated with 0.06 months of further food insecurity, 0.03 fewer meals per day, and a higher likelihood of being worried the household would not get enough food, and running out of food. Marital status is weakly associated with food security. These results indicate that weather shocks and other measures of vulnerability such as wealth status and other household characteristics play a key role in influencing food insecurity. In addition, female-headed households present higher levels of vulnerability and hence are more likely to be food insecure compared to male-headed households.

Conclusion and Policy Recommendations

This study uses nationally representative household level data to examine how weather shocks and the gender of the household head influence food security. The findings reveal that female-headed households are more vulnerable and hence experienced more food insecurity than male-headed households. A larger percentage of female-headed households belong to the lowest wealth quintiles, are less educated, divorced, widowed or in a polygamous marriage, and agriculture is their main occupation compared to the male-headed households. The results show that when we do not control for other measures of vulnerability, female-headed households have a higher level of food insecurity compared to male-headed ones. However, the relationship between gender and food security disappears when we control for household characteristics and wealth indicators. This suggests that having a female head does not in itself lead to food insecurity; it only does so because of limited access to resources and other opportunities, which make female-headed households more vulnerable. These findings are consistent with past studies which show that female-headed households are more vulnerable compared to their male-headed counterparts, especially in SSA (Milazzo and Walle, 2015). In addition, the results are consistent with Uganda's national statistics which show that women are less educated, are largely employed in subsistence agriculture (UBoS, 2019), and are less likely to own productive resources such as land (Mwesigye *et al.*, 2020). This suggests that messages and policy recommendations that are proposed in this study can be applied in many other SSA countries which are faced with female exclusion and high inequalities in access to resources, skills, and welfare.

The study also finds that weather shocks such as drought, floods, and irregular rains are deleterious to food security. The number of months the households were food insecure increased with an increase in the length of drought, floods, and irregular rains. Also, the weather shocks were associated with fewer meals per day and other measures of food insecurity. The results also showed that food insecurity is higher for households in rural areas, compared to those in urban and peri-urban areas. In addition, the results indicated that food insecurity is highest among refugees compared to nationals. These results might be explained by the fact that poverty and deprivation are higher in rural compared to urban areas, and among refugee communities.

This study joins the body of literature that calls for the enhancement of access to resources and opportunities to women as one way to close the gender gap in welfare and food insecurity. In addition, the study shows that climate change-induced weather shocks affect food security, and hence require concerted efforts. Addressing food security challenges can reduce hunger and poverty, and can enable the country to achieve Sustainable Development Goals (SDG) 1 and 2.

These findings have important policy implications. The results suggest a need for interventions to reduce the gender gap in access to resources and income generating opportunities. Specifically, there is a need to strengthen educational opportunities, enable access to productive assets for widows and divorced women, and to provide targeted safety nets for female-headed households to address vulnerabilities. Also, the results suggest the need to address weather shocks such as drought, irregular rains, and floods. Drought and irregular rains can be addressed through enhancing irrigation systems. Indeed, the government of Uganda has prioritized this intervention in the next national development plan as one way of promoting agricultural production and productivity (NPA, 2020). Together with efforts to address drought, the government needs to support communities with clear drainage systems, and by creating ridges in gardens to reduce the impact of floods, especially in lowland areas. In addition, there is a need to relocate households who have settled in wetlands since it makes them more susceptible to floods. Lastly, there is a need for safety nets and other interventions to support rural and refugee households who remain vulnerable and are more susceptible to suffering from food insecurity. Interventions can be in the form of cash and non-cash transfers especially during the off-season when the food stocks fall.

References

Akampumuza, P. and Matsuda, H. (2017) Weather shocks and urban livelihood strategies: the gender dimension of household vulnerability in the Kumi district of Uganda. *The Journal of Development Studies* 53, 953–970.

Asfaw, S. and Maggio, G. (2018) Gender, weather shocks and welfare: evidence from Malawi. *The Journal of Development Studies* 54, 271–291.

Asfaw, S., Arslan, A., Karfakis, P., Lipper, L. and Mortari, A.P. (2016) Welfare impacts of climate shocks: evidence from Uganda. Food and Agriculture Organization of the United Nations, Rome.

Bandara, J.S. and Cai, Y. (2014) The impact of climate change on food crop productivity, food prices and food security in South Asia. *Economic Analysis and Policy* 44, 451–465.

Burke, C. and Kobusingye, D.N. (2014) Women's land rights in northern Uganda: West Nile, Acholi, Lango, Teso and Karamoja. Oxfam, Kampala.

Campbell, B.M., Vermeulen, S.J., Aggarwal, P.K., Corner-Dolloff, C., Girvetz, E. *et al.* (2016) Reducing risks to food security from climate change. *Global Food Security* 11, 34–43.

Carter, N.A., Humphries, S., Grace, D., Ouma, E.A. and Dewey, C.E. (2017) Men and women farmers' perceptions of adopting improved diets for pigs in Uganda: decision-making, income allocation, and intra-household strategies that mitigate relative disadvantage. *Agriculture and Food Security* 6:18.

Challinor, A.J., Watson, J., Lobell, D.B., Howden, S.M., Smith, D.R. and Chhetri, N. (2014) A meta-analysis of crop yield under climate change and adaptation. *Nature Climate Change* 4, 287–291.

Clay, N. and Zimmerer, K.S. (2020) Who is resilient in Africa's green revolution? Sustainable intensification and climate smart agriculture in Rwanda. *Land Use Policy* 97:104558.

Davis, B., Di Giuseppe, S. and Zezza, A. (2017) Are African households (not) leaving agriculture? Patterns of households' income sources in rural Sub-Saharan Africa. *Food Policy* 67, 153–174.

Environmental Alert (2010) Climate change in Uganda: insights for long-term adaptation and building community resilience. Environmental Alert, Kampala.

EPRC (2018) Fostering a sustainable agro-industrialisation agenda in Uganda. Economic Policy Research Centre, Kampala.

FAO (2015a) Climate change and food security: risks and responses. Food and Agriculture Organization of the United Nations, Rome.

FAO (2015b) Country programming framework 2015–2019: responding to the challenges of poverty, food insecurity and climate change. Food and Agriculture Organization of the United Nations, Kampala.

Government of Uganda (1995) Constitution of the Republic of Uganda. Uganda Printing and Publishing Corporation, Kampala.

Irish Aid (2017) Uganda Climate Action: action report for 2016. Resilience and Economic Inclusion Team, Irish Aid, Kampala.

Lwasa, S., Koojo, C., Mabiriizi, F., Mukwaya, P. and Sekimpi, D. (2009) *Climate change assessment for Kampala, Uganda: a summary*. United Nations Human Settlements Program (UN Habitat): Cities and Climate Change Initiative, Nairobi.

McSweeney, C., New, M. and Lizcano, G. (2012) UNDP climate change country profiles: Uganda. United Nations Development Program, New York. Available at: www.geog.ox.ac.uk/research/climate/projects/undp-cp/UNDP_reports/Uganda/Uganda.hires.report.pdf (accessed 1 February 2021).

Mehar, M., Mittal, S. and Prasad, N. (2016) Farmers' coping strategies for climate shock: is it differentiated by gender? *Journal of Rural Studies* 44, 123–131.

Milazzo, A. and Walle, D.v. (2015) Women left behind? Poverty and headship in Africa. *Policy Research Working Paper 7331*. World Bank, Washington, DC.

Mubiru, D.N., Radeny, M., Kyazze, F.B., Zziwa, A., Lwasa, J., Kinyangi, J. and Mungai, C. (2018) Climate trends, risks and coping strategies in smallholder farming systems in Uganda. *Climate Risk Management* 22, 4–21.

Mwesigye, F., Guloba, M. and Barungi, M. (2020) Women's land rights and agricultural productivity in Uganda. In: Konte, M. and Tirivayi, N. *Women and Sustainable Human Development: Empowering Women in Africa*. Palgrave Macmillan, London, pp. 71–88.

Niles, M.T. and Salerno, J.D. (2018) A cross-country analysis of climate shocks and smallholder food insecurity. *PLoS One* 13:e0192928. doi.org/10.1371/journal.pone.0192928.

NPA (2020) The Third National Development Plan (NDP III) 2020/21 – 2024/25. National Planning Authority, Kampala.

Porter, C. (2012) Shocks, consumption and income diversification in rural Ethiopia. *The Journal of Development Studies* 48, 1209–1222.

Porter, J.R. and Semenov, M.A. (2005) Crop responses to climatic variation. *Philosophical Transactions of the Royal Society of London. Series B, Biological Sciences* 360, 2021–2035.

Republic of Uganda (1998) The Land Act, 1998. Uganda Government Printing and Publishing Corporation, Entebbe, Uganda.

UBoS (2019) Statistical Abstract. Uganda Bureau of Statistics, Kampala.

UNMA (2018) State of Climate of Uganda in 2017. Uganda National Meteorological Authority (UNMA), Kampala.

5 Managing Livelihood in Displacement: The Politics of Land Ownership and Embodied Health and Well-being by Senior Women in Kenya

E. Onyango

Introduction

Social exclusion due to land dispossession is an insidious assault on human dignity and a cause of social injustice (Syagga, 2011). When people are deprived of their right to own and use land and its associated resources, their social, spiritual, economic, and cultural well-being are negatively affected (Gilbert, 2013). In most African communities, land provides daily nourishment, sustenance, medicine, and most importantly is a form of identity through ancestral land ownership systems (Lynch, 2006; Médard, 2010). Ownership of land is associated with several benefits, such as increased self-esteem, cultural pride, a sense of stability, and continuity of family lineage (Syagga, 2006). Due to man-made and natural disasters, millions of people are deprived of the right to access and use land, as they must migrate to new areas in search of livelihoods.

Internal displacement, or the forced movement of people within their own country, results from several factors, including natural disasters, resource conflict, forced evictions, and political and ethnic violence. According to the Internal Displacement Monitoring Centre (iDMC), by the end of 2019 there were nearly 51 million people around the world living as internally displaced persons (IDP) because of conflict or climate-related disasters, with many more at chronic risk of displacement (Cazabat, 2020). In Kenya, for example, displacement due to natural disasters and political violence has rendered millions landless. In 2020, a year when Kenya and the rest of the world have dealt with the devastating effects of COVID-19, the impacts of climate change – flash floods, droughts, and locust infestations, etc. – have continued to affect people's lives and livelihoods, especially those of women and children.

Defined as the capabilities, activities, and assets, both material and social, that are essential for survival, livelihoods depend on historical and current contextual conditions, available resources, institutional process, and organizational structures (Chambers and Conway, 1992, p. 6; cited in Scoones, 2015). Like many other African countries, institutional and organizational barriers to land ownership in Kenya date back to the colonial era (Ogot and Ochieng', 1995). This era saw the displacement of many Kenyan ethnic groups, who were increasingly marginalized when forced to move out of their ancestral lands (Harbeson, 2012; Roberts, 2012). Post-independence, the new Kenya bore the marks of its colonial past as the country faced serious land tenure issues and ethnocentric governance (Harbeson, 2012). Executive powers ballooned through constitutional amendments, as various presidents illegally allocated and amassed parcels of land to themselves and their political allies. The result of these actions maintained

DOI: 10.1079/9781789247053.0005

and reinforced processes of land dispossession (Gilbert, 2013), heightened ethnic tensions, and most importantly, led to conflict and displacement, which adversely affected the livelihoods of vulnerable groups such as women.

Women constitute the majority of the world's poor and are more likely to engage in livelihood activities that depend on prevailing climate conditions (Resurrección, 2013). Women also face numerous social, economic, and political barriers that limit their engagement in decision making and resource ownership, which has direct and indirect effects on their livelihoods (Kamungi, 2002). For example, as this chapter illustrates, displaced women must navigate through gender inequality issues as well as challenges of displacement that limit their empowerment and full engagement in household provisioning.

This chapter focuses on the interrelationships between land dispossession, climate variability, and the health and well-being of displaced senior women involved in care responsibilities for their offspring and grandchildren. It forms a part of the author's larger study which aims to understand how Kenyans socially construct their health and well-being in different places over the course of their lives. In so doing, this research highlights perceived meanings of health and well-being, and identifies indicators that matter to Kenyans. The findings presented in this chapter are based on both a review of the literature on forced evictions and land tenure issues in Kenya, as well as an assessment of the impact of displacement and climate change on senior women in Nyandarua County, Central Province. The goal is to examine the link between the health and well-being of women, especially senior women, who provide care for their progeny while in displacement and in times of climate change.

Materials and Methods

Kenya is divided into 47 counties, with 290 subcounties within the eight provincial boundaries (KNBS and ICF Macro, 2015). The larger study took place between January and April 2017 in four counties within four provinces: Kisumu County (Nyanza Province), Nairobi County (Nairobi Province), Nyandarua County (Central Province), and Makueni County (Eastern Province). This research adopted a qualitative approach that included policy document reviews, 28 focus group discussions (FGD) (n = 273), and 31 in-depth interviews (IDI) with community leaders.

Recurring bouts of post-election violence and an increasing frequency of climate-related disasters negatively affect the lives and livelihoods of senior women. Grandparents must navigate the challenges of displacement and land dispossession with uncertainties that come with providing for their families and the sustainability of their family lineages. The experiences of hardship and 'worrying too much' are thus embodied and reflected in senior women's health and well-being. Evidence from this chapter draws from a review of external literature and an amalgamation of other studies, but the findings are guided primarily by insights from the author's fieldwork experiences. To give a voice to the senior women in Central Province, the findings discussed in this chapter are also supported by anecdotal accounts from the FGD and IDI participants.

The chapter is divided into two sections: the first describes the history of the land tenure system in Kenya from the colonial to the post-independence eras, which exacerbated the land dispossession problem through forced evictions and displacement. Analyzing these historical processes is essential to addressing the causes of land injustice in Kenya, as well as to illustrating how these injustices continue to affect the health and well-being of senior women, which is the focus in the second section. Understanding these issues is important in tackling deprivations and vulnerability that women and children face, and for addressing deeper root causes of poverty.

Kenya's Land Tenure Systems and Land Injustice

A historical overview of land and land tenure in Kenya

The history of land tenure from the colonial to the post-independence eras in Kenya has been characterized by lack of democracy, but also state decay, as corrupt political elites have used their authority to illegally acquire land (Harbeson, 2012, p. 19). Colonial rule in Kenya, just like other European colonies, was typified by centralized and substantially unchecked executive authority, which oversaw land tenure relations and brought all land under the authority of the colonial governor (Mutua, 2008). These actions were implemented through the Crown Land Ordinances of 1902, 1915, and 1926 which institutionalized land tenure by setting aside trust lands for the exclusive use of Africans, while allowing the colonial governor to set aside portions of land for other uses, including the potential expansion of lands owned by European settlers (Harbeson, 2012). In 1946, the acting Labor Commissioner, Wyn Harris, estimated that as many as 250,000 African squatters were already living in the forests and on other trust and settler lands (Ogot and Ochieng', 1995).

With colonization, traditional land ownership practices in Kenya, which had mostly been communal, disintegrated and a transformation of land tenure relations began, especially within the land reserves. This process led to an evolution of the land tenure system towards one of private holdings, and accelerated the divide between those with and those without land access. As the value of land in the reserves appreciated, it became inaccessible to the poor, and individuals who aligned themselves with the colonial authority became the major beneficiaries of new private holdings. The less educated Kenyans who had limited possibility of engaging with the colonial government migrated to other areas, especially to the Rift Valley (Harbeson, 2012).

While there were imminent warnings for potential unrest due to social exclusion, especially in land ownership and governmental engagement, the colonial authority ignored them. This failure led to the Mau Mau rebellion, which was seen as a radical movement to reunify Kenyans and fight for Kenya's independence from Britain (Carl and Nottingham, 1966). The rebellion triggered a response by British troops, and after 8 years of confrontation and the detention of the Kapenguria Six, the rebellion was quelled and the colonial government advocated land tenure reforms as a lasting solution. With the nationalist leaders in prison, the colonial authority initiated the Land Consolidation and Registration (LCR) program as part of the land reform initiatives.

Later known as the 'Swynnerton Plan,' the LCR program enabled a middle class group of men to secure legal title deeds to low-cost development holdings, which enabled them to use the holdings as security for loans and to engage in meaningful agriculture (Swynnerton, 1955). Although the LCR program was crucial in stimulating Kenya's agricultural development, it marginalized an entire group of people, especially women, with limited stakes in the land holdings. The other ethnic groups in the country also became ostracized as the dominant Kikuyus capitalized on the new land policy and expanded their influence (Lynch, 2006). Being the largest group in Kenya, the Kikuyus not only acquired land within the 'white highlands' in the Central Province and Rift Valley, but also in other rural and urban areas occupied by tribes which later fought them in the violence that occurred following the 1992, 1997, and 2007 elections (Roberts, 2012).

In transitioning to an independent Kenya, land tenure policy became the centerpiece of the campaign agenda for the nationalist leaders with the slogan, 'restoring stolen land' to the landless, squatters, and unemployed Africans (Harbeson, 2012, p. 21). However, the Lancaster Constitution, developed with the colonizers' influence and adopted at independence, demanded a commitment from the African political leaders to respect individual property rights for both Africans and European settlers. As such, the limitations of the LCR program and the Swynnerton Plan trickled down with Kenya's Constitution at independence (Manji, 2014). Additionally, executive land management powers were enshrined in the Constitution, as was the right to purchase land anywhere in the country without regard to the antecedent ethnic reserve.

Kenya's leadership post-independence has not been any better in creating equal communities. The unrestrained and authoritarian post-independence rule, coupled with endemic corruption and unequal representation of Kenya's ethnic communities in government, instigated ethnic tensions and abated Kenya's economic growth during the first (Jomo Kenyatta, 1963–1978) and the second (Daniel Moi, 1978–2002) presidents' administrations (Mutua, 2008). Corrupt land tenure dealings did not remain in the executive office, but penetrated through to local level governments and into government parastatals, corporations, businesses, and professional classes (Harbeson, 2012). Like a virulent cancer, corruption and land grabbing stifled the equitable distribution of resources, with those in government and their close allies amassing wealth at the expense of the mainstream Kenyan population. The outcome weakened the government's integrity, capacity, and effectiveness in addressing cycles of post-election violence (Roberts, 2012). The 2007–2008 post-election violence was one of the worst, with over 1500 dead, 3000 innocent women and children raped, and nearly half a million people internally displaced (Roberts, 2012). Many of those displaced were women and children who had also experienced displacement after the general elections of 1992 and 1997. This pattern of recurring conflict diminished livelihoods through the destruction of property, which led to displacement and land dispossession.

Displacement and land dispossession

Today, nearly 51 million people around the world have been internally displaced as a result of disasters and conflicts, the majority of whom are in the developing countries (UNHCR, 2019). In Kenya, for example, the iDMC estimates that in 2017 alone, climate-related disasters such as drought and floods led to the displacement of 3.4 million people (Cazabat, 2020). The most affected area in Kenya is the north-eastern region, where ethnic violence over natural resource access, along with decades of economic, social, and political marginalization, has fostered both conflict and displacement. Although the northern territories have borne the greatest effects of climate change, the past decade in Kenya has witnessed an increasing frequency of droughts and flash floods in other parts of the country. These extreme weather conditions and other disasters have significantly influenced the adaptation and coping capacities of affected communities. For example, 2020, one of the worst years since independence, began with a devastating drought, and continued with locust infestations that devastated the eastern and northern territories, clearing most of the vegetation and farm crops. Between April and May 2020, Kenya also experienced unusually severe rainfall that caused flooding and led to the displacement of millions in the northern, western, and coastal counties. Amid the global COVID-19 pandemic, many families have experienced high levels of food insecurity with no pastures for their animals. Pastoralist communities, such as the Samburus, have been affected and ethnic clashes are imminent as pastoralists migrate in search of alternative pastureland and water resources.

Forced evictions due to political and ethnic violence have also displaced many Kenyans, especially in Rift Valley and Central Province. Since the rise of the multiparty system in the 1990s, incidents of post-election violence in 1992, 1997, and 2007 have disrupted lives and livelihoods, as individuals have been forced to flee and abandon their homes, assets, and land. These tensions are linked to historical land tenure issues, endemic corruption in land dealings, and autocratic and ethnocentric governance. Subsequent presidents have used their executive powers to exploit land tensions as a strategy to remain in power by designating government and other land parcels to specific ethnic communities as a way of altering voter demographics (Nasongo *et al.*, 2012).

It is also worth noting that over the years, some displaced families and individuals have returned to their farms, while others have relocated to other areas and have been integrated into other local communities. The government has also created initiatives to resettle IDPs on some government lands. However, not all IDPs have benefited from these efforts, because the government resettlement conditions required displaced persons to provide evidence of having owned land previously. As such, squatters, and people whose land title deeds were destroyed or lost in the violence, did not benefit from the program. In some places (e.g. in Olenguruone) where people had been displaced, the government has gazetted the land as forest and has closed it to habitation. Environmental conservation efforts such as governmental gazettement and afforestation initiatives have also led to displacement and instability of IDPs in Kenya.

In Nyandarua County, for example, IDPs have been assigned parcels of land on the Abadare ranges where they practice agroforestry. When the assigned land is fully forested, these individuals are assigned to other parts of the range. This practice is a source of psychological stress for most IDPs given the uncertainty that surrounds their livelihoods and the human–wildlife conflicts in the area. Many of the families engaged in afforestation programs are the poorest of the poor, with no resources to relocate or buy land in other places. They tend to live in rented houses in the neighboring market centers like in Ndaragua where some have become 'street people,' living precariously as handcart operators, hawkers, domestic workers, casual laborers, and small-scale farmers engaged in the government's reforestation efforts. The combination of land dispossession, an agriculturally based economy, and heavy reliance on climate-sensitive livelihoods in times of climate variability, leave these vulnerable populations in abject poverty with limited hope of a better future.

Livelihoods, landlessness, climate change and women's health and well-being

A recent survey by iDMC revealed that displacement reduces household income by 80%, and that displaced households make on average less than US$1.95 a day (Cazabat, 2020). The survey also showed that most of these households feel less secure and are less content with their new housing than they had been with their pre-displacement housing. The new houses are smaller, yet family sizes are often larger, as many of the displaced families now include both adult offspring and grandchildren. Moreover, because the new spaces are often rentals, residency is insecure and families can be evicted for payment default (Cazabat, 2020). The negative impact of displacement on child education is another reality that most displaced households bear, as many adult offspring are unable to transition to professional life and often survive on temporary work (iDMC and NRC, 2019). Economic activities such as selling of firewood, vegetables, second-hand clothing, and manual labor do not generate income sufficient to feed, educate, and provide other basic needs. Consequently, poverty rates are high among displaced populations, and serve as a precondition for commercial sex work, early marriages, and increased chances of being infected with HIV and other sexually transmitted infections (Kamungi, 2009; Cazabat, 2020). Those living with chronic health conditions such as HIV not only experience stress due to displacement and land dispossession, but also due to the medical needs and effects of climate variability on their nutrition and that of their families.

Women are disproportionately affected in conflict, as well as in times of displacement and climatic instability. Women make up most of the world's poor and are more dependent on sensitive livelihoods threatened by climate change, such as subsistence farming. Moreover, women face socio-economic and political barriers that limit their ability to cope with the effects of climate change and forced eviction. In patriarchal societies such as Kenya, women have unequal access to resources and decision making but they are charged with the responsibilities of securing vital household supplies such as water, food, and fuel for cooking and heating – resources that depend significantly on the state of the environment. In the context of climate change and dislocation, livelihood options become erratic, and traditional food sources unpredictable and scarce (FAO *et al.*, 2019; Mbow *et al.*, 2019).

Evidence shows that the effects of conflict – including land dispossession, loss of livelihood opportunities, and general disruption of life – affect women and children significantly more than men (Opata and Singo, 2004; Kamungi, 2009; Lindo, 2010). While men may fight in armed conflict, women are usually left behind to care for children and themselves in unfamiliar environments (Cazabat, 2020). When conflict breaks down existing social support systems, caring and providing for children is a huge burden. Recent studies in the Central Province reveal increased rates of alcoholism among the men, which could be due to the effects of colonization as well as land dispossession, as men find it hard to adjust to poverty, landlessness, and inability to provide for their families (Muturi, 2014a,b). As such, men often resort to idleness, domestic violence, drunkenness, and promiscuity, and make the women feel that they are better off without the husbands; therefore they divorce or separate (KNBS and ICF Macro, 2015).

The structural injustices in the form of land dispossession and ethnocentric politics (Farmer, 2003) that have dominated Kenya's political discourse since independence limit the freedoms and choices available to displaced women. With limited livelihood options and in times of environmental change, displaced women in Kenya live in extreme poverty and are unable to support their children's education. With increased incidents of drought and flash floods, women face losses in income and harvest, which decrease their food and financial security. Due to the effects of dislocation and harsh climatic conditions, women remain preoccupied with meeting basic needs such as food and medical care, and schooling might be considered a luxury. Although Kenya has a free primary education policy, there are basic school expenses – school uniforms, books, bags, school admission fees, and other school infrastructural development needs – that are not supplied by the government, and which the parents are expected to provide (Adesina *et al.*, 2012).

Moreover, public primary schools in Kenya are characterized by overcrowding. The World Data Atlas estimates that in 2015, the pupil–teacher ratio in primary schools in Kenya was 31:1 (World Data Atlas, 2019). As such, the middle class households prefer to send their children to private schools, which are inaccessible to vulnerable populations, such as displaced and dispossessed women. As demonstrated in a recent survey in Gilgil about 75% of displaced children during the 2007 post-election violence stayed out of school as their families had lost everything and could not afford school admission fees and uniforms (Cazabat, 2020). Stressed in their new environments and having lost their parents and relatives to the violence, some children were unable to perform well and are likely to remain in the lower wage category into the future. Displaced households are also food insecure: a study of IDPs in Turkana revealed that 94% of the households were found to be severely food insecure (Gichunge *et al.*, 2020). Food insecurity not only affects current generations but also children born to mothers deficient in nutritional requirements, which compromises the physical and mental development of fetuses, and thus reduces the ability of these children to achieve their maximum potential in life.

As a means of adaptation, displaced women in small towns like in Ndaragua in Nyandarua County rent strips of farms in the forest to grow vegetables, potatoes, and cereals for their own consumption, and to pay for housing and medical costs. Most women also engage in firewood collection within government reserves in order to generate income. Some have joined self-help groups as opportunities to build their economic base and social networks. However, several challenges limit resiliency of the dispossessed households. With elevated chances of crop failure due to unpredictable climatic conditions, there is an increase in food prices that makes food inaccessible to vulnerable populations, particularly the women and girls whose health has been found to deteriorate in times of food shortages (Owoaje *et al.*, 2016). Moreover, government regulation of the land rented in these forests and regulation over firewood collection have implications for livelihood sustainability, especially for displaced households headed by women. With limited livelihood options, displaced women are unable to raise funds to sustain membership in social groups which could enhance their ability to move beyond the challenges of land dispossession and forced evictions. Based on recent fieldwork in Kenya and a review of literature, the author finds that the socio-economic effects of land dispossession in times of climate change lead to considerable psychological distress in vulnerable populations, and the effects are embodied and reflected in the health and well-being of senior women.

Senior Women's Health and Well-being: The Embodiment of Landlessness in a Climate Change Era

Land and electoral injustices that have marred Kenya's history over the years have been embodied by the marginalized populations and manifest in their health and well-being. The senior women in Kenya who bear the burden of raising the next generation in these precarious social conditions are one such disadvantaged population. These grandmothers encounter physical and emotional distress because of challenges associated with familial sustenance and caregiving for offspring and grandchildren. In this way, senior women bear a double burden of care responsibility: one for their own offspring who are unemployed, and another for their grandchildren, especially those with HIV-infected parents, or with parents that have migrated to urban centers in search of employment. The distress older women experience as a result of these tensions and constraints is even greater in contexts of displacement. In this situation, concerns about family sustainability and continuity are magnified to an even greater degree, and the emotional, physical, social, and psychological torment endured by these women are expressed in the form of health outcomes – headaches, hypertension, insomnia, schizophrenia, and burning sensations in their heads. In exploring the historical land tenure issues and in linking these issues to the health and well-being of senior women, a contextualized picture emerges that reveals how people's health can be embedded within structural systems – including the land tenure systems and ill-conceived forest conservation strategies. These factors coupled with the growing frequency of droughts and other climate-related disasters negatively affect both lives and livelihoods, and exacerbate cycles of poverty and marginalization. These issues are discussed in the following three subsections.

The ethics and economies of basic needs in displacement

> There are no lands in this community… Some of us are displaced people from other places and we have lived here for years, our parents are not here. We are surviving in this community.
>
> (Hanna, 61-year-old woman, central Kenya)

Historical land dispossession from the colonial to post-independence eras has created physical and psychological distress. These effects are embodied and reflected in the health and well-being of Kenya's senior women. The land tenure issues date back to pre-independence when Kenyans, particularly the Kikuyus, were forced out of their ancestral land to create exclusive 'white highlands' for the white settlers (Ogot and Ochieng', 1995). After independence, the political system in the country was marred by ethnocentric governance that enabled the community in power to unethically acquire land at the expense of other tribes. Middle class men who could afford to buy land did so, which left the less advantaged to remain as squatters. These factors, coupled with governmental corruption, contributed to the contentious land ownership issues, political conflict, post-election violence, and ensuing displacement (Roberts, 2012).

Kenya's displaced and dispossessed populations face increasingly dire conditions in a changing environment characterized by increasing frequency of drought and flooding. Failed rains due to climate change have not spared these vulnerable populations as their livelihoods continue to be affected. Displaced families face a number of climate-related challenges, including increased food prices, diminished supplies of firewood, crop failures, and reduced access to subsistence resources. These challenges are made even more complex for older adults when they are faced with caring for their lineage in displaced conditions. Diminished social networks and a lack of credit trustworthiness only add to these burdens. This situation has led one displaced senior woman in central Kenya, who is engaged in small-scale farming in the forest, to state:

> … Like even now we want to plant in this land from the government, we do not have the seed. So, it is again borrowing or asking your children to support. The children are unemployed, they are waiting to be taken care of too. So, if the drought continues… there are no good things and with the issue of headache, there is just a lot of sunshine.
>
> (Marie, 57-year-old woman, central Kenya)

Another senior woman who takes care of her grandchildren and young adult offspring in displacement, who faces increasingly frequent droughts, and who is also participating in a government reforestation scheme, stated:

> … But now the past 3, 4 years consecutively… we have not had any harvest. You hear people saying that in this parcel of land normally I get 5 or 6 sacks of maize but now, there is no harvest at all. Just a few people that are lucky to have money to buy the seeds and the fertilizers are the ones with some harvest. But for those of you who do not have money, where do you get help? You know for you to get money you must be having some money. *'Pesa inamulikuwa na ile nyingine. Kama hauna pesa, wewe unamulika na nini?'* [You use money to make money. If you have no money, how can you make money?]
>
> (Veronica, 72-year-old woman, central Kenya)

Reflecting on the cyclic nature of their experiences, many grandmothers feel as if they are ensnared in poverty. The challenges of displacement and climate change have reduced their ability to provide for or educate their offspring, diminished their resource access, undermined their capacity to financially invest in their farms, and reduced their household food security. Even when their offspring and grandchildren are lucky enough to attend school, they often remain unemployed as adults, and are thus incapable of contributing.

Related to both the land issue and climate change is the Kenyan government's system of forest management where IDPs and squatters are temporarily assigned deforested government land or gazetted forest land with an aim of reforestation through agroforestry. This program was part of the 2005 Kenya Forest Act initiatives that aimed to engage communities adjacent to forests with the Kenya Forest Services in forest land management. The community groups included grazers, fuelwood collectors, beekeepers, tree seedling producers, and farmers (Basciano et al., 2017). In particular, the IDPs and squatters are involved as crop and tree cultivators through the Plantation Establishment and Livelihood Schemes (PELIS). This model of managing landlessness creates social and economic uncertainties for the families. One senior participant indicated that:

> The farms that we plough are not our farms. We are given a half-acre parcel of land within the forest where there are no trees. When it is raining, you are expected to plant trees and crops. When the trees grow, we must vacate and move to an area that is deforested. We leave the lands and the forest for the government. For all these years that I have been here for 30 years, I have raised my children with this and it's very unstable. Now most of the lands are forested and we have not been shown any alternative land to move to.
>
> (Jane, 60-year-old woman, central Kenya)

This example reveals one way that landlessness affects the livelihood options available to displaced persons. Economic predicaments of sustenance provision for offspring and grandchildren can also be a source of stress, especially when older female caregivers are unable to meet subsistence needs. For example, the pressure to invest in the education of offspring and grandchildren with the hope that they will eventually be able to provide social and financial support to the family can be agonizing, particularly when unemployment rates remain high. Kenya has one of the highest unemployment rates in the world, ranked fifth in 2017, and has a youth unemployment rate of 40%. The COVID-19 pandemic has further exacerbated the unemployment crisis in the country. A recent labor force survey by the Kenya National Bureau of Statistics (KNBS) revealed that at least 300,000 Kenyans had lost their jobs between January and March 2020 and that the unemployment rate stood at 13.7% (KNBS, 2020).

With high unemployment rates, senior women must be concerned not only with educating their offspring, but also with caring for them, despite them being at an age where it is socially and culturally expected that they should be independent and able to provide for their parents and contribute to society. Additionally, with lower educational attainment, offspring who migrate to urban areas as adults remain in the working class, and any children they have must be sent back to the grandparents, often without social or financial support. As a result, seniors feel that they are constantly raising children, a frustrating situation for them, especially when they are unable to provide for their own basic needs. One senior woman reflected on her experience and said:

> ... The children leave the community for the urban areas where they will never get any good employment, they only do casual work which will not allow them to even send any financial support to us as the parents. In the process of all these, the children conceive and give birth. When this happens, they send back the grandchildren to us. Because they are not having any good employment, they leave us with their children without any support. It is like we are raising children till our death. It is so frustrating when we cannot provide for our grandchildren.
>
> (Veronica, 72-year-old woman, central Kenya)

The HIV pandemic has exacerbated these challenges and created new ones. The grandmothers are traumatized by the loss of their adult offspring, the prospect of raising orphans, and the burden of meeting basic physical, financial, and emotional needs for themselves and their grandchildren. Another participant said that she was:

> ... Taking care of orphans. One of them is mentally handicapped and she also gave birth to a child. She cannot take care of herself and I now must take care of her child as well. I must take care of these children and I do not even see

when they will grow up to be independent. If you leave them, they will just die... Here, if you are to go for casual work you are always bothered and the money you get is not enough to take care of all these needs. Now even with the drought, everything has become expensive, there are a lot of problems.

(Sally, 55-year-old woman, central Kenya)

As they care for the orphaned grandchildren, the grandparents must also battle their own chronic ailments such as hypertension, diabetes, and other health issues. In certain cases, the grandchild[ren] and/or the grandparents are also HIV-infected, and adherence to drugs becomes a major concern that generates stress and frustration. Taking 1 day off per month to collect the antiretroviral drugs at a health facility is equivalent to 'a day without food' for the grandmothers and their families. Thus, stress is ever-present when people are dispossessed of land.

Displacement, identity with land and sustenance of family lineage

... I have been in this community for over 30 years and I have no land and no house called by my name, but I live in a rental house. When I got a house that I can call mine, is that of *juakali* [temporary]. It is not my parental home, but it belongs to the government... The government can come any time to take it up... they [our children] also do not have land and they have no income...

(Marie, 57-year-old woman, central Kenya)

Women's land rights are not only crucial to their livelihoods, they also contribute to identity, dignity, and social inclusion. In patrilineal societies such as Kenya, women's right to land ownership depends on their relationship to men – usually fathers, husbands, brothers, or sons. Displacement in gender-biased societies exacerbates women's experiences of social exclusion, especially if they lose the male members of their families to incarceration, death, separation, or divorce. Displacement from ancestral lands leads the women to worry about land as a commodity for economic production as well as a form of family identity and a means for sustaining their family lineage. They consider their families as unsettled since they lack land on which to build a home. In these situations, family stability and continuity are a serious source of stress.

Moreover, the seniors in this study observed that the current generation does not adhere to instructions left by their deceased parents to preserve ancestral land, which they consider a form of cultural identity. Evidence suggests that individuals and their communities have their own values, norms, customs, and knowledge systems that affirm their identity and diversity and that these play a key role in sustainable community development and livelihoods (Daskon and Binns, 2010). Like most African communities, Kenyans, especially those from the central region, have a strong attachment to land and ethnic identity (Lynch, 2006). Additionally, because of the history of political and ethnic violence, land is considered a guarded property that should not be sold to an 'outsider' – that is, a person of a different ethnic background as was exclaimed by the older adults from Central Province:

We have one problem as seniors of this age of ours, I am talking about our young boys and not the girls... Our young boys do not know anything about the curses. If they are left land and the father said that the land should not be sold and it should be the family's property, the sons sell this to others from outside. When we see other seniors of age being dishonored by their sons, this causes us the seniors who are still alive a lot of stress. We feel that this might also happen to us when we are departed. It is a lot of stress to us. This even makes us die even earlier than expected.

(Sally, 55-year-old woman, central Kenya)

Beyond the confines of ethnic identity rooted in land, displaced seniors increasingly experience neglect and rejection by their offspring, especially when adult offspring move to the urban centers. Some become successful and forget about their roots, mainly when they do not have an ancestral land to call home.

... They go to other towns like Nairobi, Nakuru or to other places where they are not known. It is now me as the grandmother to make sure that these children have food, clothing, education, and other basic needs. Us as grandmothers, we have problems, *eeeh*, women.

(Naomi, 64-year-old woman, central Kenya)

This outcome is inconsistent with the past when offspring could take care of their older parents. There is also a lack of respect by the current generation, as was highlighted by seniors who perceived that their role as mentors was no longer appreciated. They felt left out in the opportunity to nurture the current generation and feel there is a loss in cultural identity. 'When they reject themselves, they also reject their parents and their culture, and they just come to town [urban centers].' These issues cause worry to the seniors as they see no hopes of leaving their offspring settled in a particular place and, given the attachment to land and a traditional home, the seniors experience both physical and psychosocial health challenges.

Traditional African societies were more collective, and the welfare of widows and orphaned children was regarded as a community responsibility. Following the HIV/AIDS epidemic in the 1990s and 2000s, many children were orphaned, and their grandparents were widowed and/or lost most of their older offspring. With the increased need for support, community welfare systems have degenerated, and these vulnerable groups are no longer communally cared for. Consequently, their rights are violated by extended family members who ought to provide protection. This situation is another source of distress, particularly for women who are deprived of their inheritance from their deceased spouses. One participant stated that with their caregivers:

> ... The orphaned children are suffering in this country and especially... those without land,... have no stable place to call home where they can get education, food, etc., ... so as to be a resource to the country and to themselves in the future.
>
> (Veronica, 72-year-old woman, central Kenya)

Government initiatives in the form of social welfare systems have been deficient in addressing the livelihood challenges of these vulnerable populations. In 2013, the National Social Protection Policy in Kenya established a social welfare system to support vulnerable populations including the seniors, orphaned children, and people with disabilities. This system led to the creation of the Orphans and Vulnerable Children – Cash Transfer (OVC-CT) and the Older Persons Cash Transfer Program (OPCTP) from which senior women and orphaned children in displacement could receive some financial support. However, from the author's own experience in the field, and review of prior research, indicates that these programs have made minimal impact, especially to displaced households whose members tend to lack national identification documents and who have moved to new areas and lost their existing social connections. Most importantly, the process of issuing these cash transfers has been tainted with corruption, which limits the ability of the target population to benefit. One senior woman acknowledged:

> This issue of the orphaned children let the government give them support. And this support because our people who are '*tumbo kubwa*' [big stomached – corrupt chiefs], the leaders, they are the ones that benefit from these moneys. The money does not reach these children or the older people as it is benefiting others who are known to the leaders... There is a lot of corruption and nepotism.
>
> (Marie, 57-year-old woman, central Kenya)

The combination of these factors creates tremendous challenges for seniors, especially senior women, and government efforts are not effective in providing support. Thus, displacement and land dispossession directly diminish senior women's livelihoods and lead to psychosocial distress. The experiences of stress and nutritional deficiency remain under their skin throughout their life course and those of their next of kin. These embodied experiences are expressed in the form of health and well-being outcomes – or what this chapter refers to as 'epitomes of distress.'

Epitomes of distress

> ... We suffer so much looking for income especially for educating our children and yet we have no source of income. So, we are caught in stress because we long to take our children to school, you do not see the money, you suffer, your body is affected, your heart is disturbed, your heart aches and as such you are so stressed...
>
> (Marie, 57-year-old woman, central Kenya)

Senior caregivers are distressed by economic tensions and constraints that come with increased burdens of care during difficult economic situations, characterized by high unemployment and the HIV epidemic. Seniors care for their own offspring who remain dependents as young adults due to unemployment. Likewise, they provide for their grandchildren either after the death of their adult offspring or after the adult offspring migrate into urban centers in search of employment opportunities. In either case, the seniors are faced with moral questions (i.e. that which is considered right to do) and the economic predicament of providing for the basic needs of their families. Thus, the seniors are in 'deep thought' throughout their lives as they must consistently care for their progeny. This outcome occurs despite the fact that cultural expectations dictate that adult offspring should provide social, emotional, and financial support to their elderly parents. In the narratives revealed in this study, it is evident that seniors are chronically exposed to stress due to their care responsibilities. They embody these experiences which are in turn reflected on their bodies in the form of persistent headaches, hypertension, insomnia, and burning sensations in their heads. One woman stated:

> '... We have children who are to be in school. At night, you can't sleep, we have headache in the night, the morning, lunchtime, and evening, we have nowhere to turn to'
>
> (Naomi, 64-year-old woman, central Kenya)

This finding corroborates those by Yaris (2014) which revealed that Nicaraguan grandmothers caring for children of immigrant mothers 'think too much' because of the uncertainties surrounding immigration, family unity, and the provision of basic needs for their families. The chronic distress is embodied by the grandmothers and reflected on their bodies in the form of somatic symptoms such as lack of sleep and 'brain hurting.'

As the seniors care for their progeny, they must also endure the physical and psychological torment associated with exposure to displacement, socio-cultural variability, and climate change. This exposure is embodied and reflected on their skin, a concept this chapter refers to as 'epitomes of distress.' Different health outcomes are experienced by different seniors. For example, experiences of anxiety, stress, and heartaches are common among senior women in central Kenya. In addition, the experiences of distress in the form of burning sensations in the head reflect the effects of 'thinking too much' and of having to bear the burden of raising their offspring and grandchildren. These women consider themselves to be pressed in all directions:

> ... In school for our children and grandchildren, in hospitals and with the sun in the form of the current drought. We are stressed but have nowhere to turn to. It is too much for us, a consistent headache, and you can't sleep.
>
> (Marie, 57-year-old woman, central Kenya)

Displaced senior women in Kenya also experience feelings of hopelessness as they carry the burden of caring for their progeny. Others experience persistent headaches which they associate with embodied distress. This embodied distress is revealed in the lack of sleep (insomnia) especially at night, worrying and thinking too much about the economics of basic needs, uncertainties of the future, and family instability. Other embodied distresses expressed by participants are personality disorders such as schizophrenia and other health outcomes including hypertension. The seniors observed that they are exposed to several stressors that negatively affect their health and well-being. As one participant notes:

> '... Mental health issues such as schizophrenia are affecting us as seniors in this community. Because of the stress... we end up with this kind of conditions'
>
> (Hanna, 61-year-old woman, central Kenya)

General body aches and fatigue were also linked to senior participants' distress. These psychosocial health issues are perceived to weaken the immune system, contribute to general body weakness, and increase the risk of body failure: '... We have a lot of health challenges due to the stress; we get body aches from all over the body. You cannot even explain the disease, it's all over the body, it's like the body just fails and you die' (Marie, 57-year-old woman, central Kenya).

To this end, the lived experiences of grandmothers are embodied and reflected in the epitomes of distress. The economic tensions and experiences of unstable family life due to displacement, landlessness, climate variability, and socio-cultural erosions remain key issues and potential sources of trauma. Consistent exposure to these stressors can lead to their normalization, though over time, they are reflected in senior women's bodies and affect outcomes related to health and well-being. This exposure also influences perceptions and meanings that seniors ascribe to societal health and well-being, issues which are important aspects of a healthy community and require consideration in assessment and formulation of health and well-being policies.

Conclusion

This chapter adds to the understanding of the effects of involuntary displacement on senior women engaged in care provision for their extended families, by exploring the effects of land dispossession in the context of changing climate conditions. Specifically, this work demonstrates the link between the broader land issues, climate change, and the prevailing health and well-being of displaced senior women in Kenya. While most grandparents in Kenya bear the burden of caregiving for their offspring and grandchildren, displaced senior women experience unique challenges that come with displacement and extreme climatic conditions. This study is grounded in a historical analysis of Kenya's land tenure system and land injustices from the colonial era to post-independence. This work goes further to explore how historical land dispossession in a changing climate relates to senior women's livelihood options, thereby influencing their health and well-being.

The grandmothers are distressed by the economic tensions and constraints as they face increased care responsibilities. Difficult economic situations characterized by climate variability, unemployment, and the HIV epidemic only add to the stress. The seniors must care for their own children who remain dependents as young adults due to unemployment. Likewise, they provide for their grandchildren either after the death of their adult children or after the adult children migrate into urban centers in search of new employment opportunities. In either case, the seniors are faced with the moral and economic predicament of providing for the basic needs of their families with a lack of support. The result is that seniors consistently experience psychological trauma and are in 'deep thought' their entire life as they care for their progeny. The grandmothers embody these lived experiences which are in turn reflected on their bodies in the form of persistent headaches, hypertension, insomnia, schizophrenia, and burning sensations in the head. Analysis of processes of land dispossession and its associated impacts in times of climate variability is essential to addressing the deeply rooted land injustices and to providing the evidence that experiences of land injustices negatively affect health and well-being. Understanding these issues is important in tackling deprivations and vulnerability that women and girls face, and for addressing the deeper root causes of poverty, not only among displaced senior women in the central region of Kenya, but also those in other regions of the country and other developing countries.

References

Adesina, O.A., Adeyeye, M., Adjei, J.K., Achoka, J.S., Agbehonou, E. *et al.* (2012) *Managing Conflicts in Africa's Democratic Transitions*. Lexington Books, Lanham, MD.

Basciano, T., Erskine, M.C., Golden, S.L., Hammer, K.R., Iannuzzelli, F. *et al.* (2017) They protect the forests. Who protects them? The intersection of conservation, development, and human rights of forest defenders: lessons from Kenya, Peru and Sri Lanka. *International Human Rights Clinic report*. Johns Hopkins School of Advanced International Studies. Available at: www.jscholarship.library.jhu.edu/handle/1774.2/61399 (accessed 1 February 2021).

Carl, R. and Nottingham, J. (1966) *The Myth of Mau Mau: Nationalism in Kenya*. Praeger, New York.

Cazabat, C. (2020) Measuring the costs of internal displacement on IDPs and hosts: case studies in Eswatini, Ethiopia, Kenya, and Somalia. Internal Displacement Monitoring Centre (iDMC), Geneva, Switzerland.

Chambers, R. and Conway, G. (1992) Sustainable rural livelihood: practical concepts for the 21st century. *IDS discussion papers No 296*. Institute of Development Studies, Brighton, UK.

Daskon, C. and Binns, T. (2010) Culture, tradition and sustainable rural livelihoods: exploring the culture–development interface in Kandy, Sri Lanka. *Community Development Journal: an International Forum* 45, 494–517.

FAO, IFAD, UNICEF, WFP and WHO (2019) The state of food security and nutrition in the world 2019: safeguarding against economic slowdowns and downturns. Licence: CC BY-NC-SA 3.0 IGO. Food and Agriculture Organization of the United Nations, Rome.

Farmer, P. (2003) *Pathologies of Power: Health, Human Rights, and the New War on the Poor*. University of California Press, London.

Gichunge, C., Mutiso, D. and Brynjarsdottir, J. (2020) Predictors of social support, physical health and mental health among food insecure internally displaced persons in Turkana, Kenya. *Conflict and Health* 14:58.

Gilbert, J. (2013) Land rights as human rights: the case for a specific right to land. *SUR International Journal on Human Rights* 10:18.

Harbeson, J.W. (2012) Land and the quest for a democratic state in Kenya: bringing citizens back in. *African Studies Review* 55, 15–30.
iDMC and NRC (2019) Global report on internal displacement. Internal Displacement Monitoring Centre, Geneva, Switzerland and Norwegian Refugee Council, Oslo. Available at: www.internal-displacement.org/sites/default/files/publications/documents/2019-IDMC-GRID.pdf (accessed 1 February 2021).
Kamungi, P.M. (2002) The lives and life-choices of dispossessed women in Kenya. United Nations Development Fund for Women (UNIFEM) African Women in Crisis Programme.
Kamungi, P.M. (2009) The politics of displacement in multiparty Kenya. *Journal of Contemporary African Studies* 27, 345–364. doi.org/10.1080/02589000903166713.
KNBS (2020) National economic survey, 2020. Kenya National Bureau of Statistics, Nairobi. Available at: www.dc.sourceafrica.net/documents/119905-KNBS-Economic-Survey-2020.html (accessed 1 February 2021).
KNBS and ICF Macro (2015) Kenya demographic health survey 2013–14. Kenya National Bureau of Statistics, Nairobi and ICF Macro, Calverton, MD.
Lindo, J.M. (2010) Are children really inferior goods? Evidence from displacement-driven income shocks. *The Journal of Human Resources* 45, 301–327.
Lynch, G. (2006) Negotiating ethnicity: identity politics in contemporary Kenya. *Review of African Political Economy* 33, 49–65.
Manji, A. (2014) The politics of land reform in Kenya 2012. *African Studies Review* 57, 115–130.
Mbow, C., Rosenzweig, C., Barioni, L.G., Benton, T.G., Herrero, M. *et al.* (2019) Food security. In: Shukla, P.R., Skea, J., Calvo Buendia, E., Masson-Delmotte, V., Pörtner, H.-O. *et al.* (eds) *Climate Change and Land: an IPCC special report on climate change, desertification, land degradation, sustainable land management, food security, and greenhouse gas fluxes in terrestrial ecosystems*. Intergovernmental Panel on Climate Change, United Nations, Geneva, Switzerland.
Médard, C. (2010) Indigenous' land claims in Kenya: a case-study of Chebyuk, Mount Elgon district. Institut de Recherche pour le Développement (IRD), Paris and Kenyatta University, Nairobi.
Mutua, M. (2008) *Kenya's Quest For Democracy: Taming The Leviathan*. Lynne Rienner Publishers, London.
Muturi, N. (2014a) Engaging communities in the understanding of excessive alcohol consumption in rural Central Kenya. *Journal of International Communication* 20, 98–117.
Muturi, N. (2014b) Alcohol consumption and reproductive health risks in rural Central Kenya. *Sexual and Reproductive Healthcare: Official Journal of the Swedish Association of Midwives* 5, 41–46.
Nasongo, J.W., Wamocha, L.L. and Achoka, J.S. (2012) Could forgiveness and amnesty be a panacea for Kenya's post-election conflict era? In: Adebayo, A.G. (ed.) *Managing Conflicts in Africa's Democratic Transitions*. Lexington Books, New York, 271 pp.
Ogot, B.A. and Ochieng', W.R. (1995) *Decolonization and Independence in Kenya*. Villiers Publications, London.
Opata, G.P. and Singo, S.M. (2004) The economics of displacement: a study of the changing gender roles, relations and its impact on the livelihood and empowerment of women refugees in Kenyan camps. Center for Refugee Studies, Moi University, Kesses, Kenya.
Owoaje, E.T., Uchendu, O.C., Ajayi, T.O. and Cadmus, E.O. (2016) A review of the health problems of the internally displaced persons in Africa. *The Nigerian Postgraduate Medical Journal* 23, 161–171.
Resurrección, B.P. (2013) Persistent women and environment linkages in climate change and sustainable development agendas. *Womens Studies International Forum* 40, 33–43.
Roberts, M.J. (2012) Conflict analysis of the 2007 post-election violence in Kenya. In: Adebayo, A.G. (ed.) *Managing Conflicts in Africa's Democratic Transitions*. Lexington Books, New York, pp. 141–154.
Scoones, I. (2015) *Sustainable Livelihoods and Rural Development: Agrarian Changes and Peasant Studies*. Practical Action Publisher, Nova Scotia, Canada.
Swynnerton, R.J.M. (1955) A plan to intensify the development of African agriculture in Kenya. Colony and Protectorate of Kenya, Nairobi.
Syagga, P.M. (2006) Land ownership and use in Kenya: policy prescriptions from an inequality perspective. *Readings on Inequality in Kenya: Sectoral Dynamics and Perspectives*. Society for International Development (SID), Nairobi, pp. 289–344.
Syagga, P.M. (2011) Public land, historical land injustices and the new Constitution in Kenya. Society for International Development (SID), Real Estate and Construction Management, University of Nairobi.
UNHCR (2019) Global trend: forced displacement in 2019. UNHCR, The UN Refugee Agency, Geneva, Switzerland. Available at: www.unhcr.org/5ee200e37.pdf (accessed 1 February 2021).
World Data Atlas (2019) Education statistics. World Bank, Washington, DC. Available at: www.data.worldbank.org/data-catalog/ed-stats (accessed 1 February 2021).
Yarris, K.E. (2014) "Pensando Mucho" ("Thinking too Much"): Embodied Distress among Grandmothers in Nicaraguan Transnational Families. *Culture, Medicine and Psychiatry* 38(3), 473–498.

6 Seeing Through Water: Gender, Anxiety and Livelihoods in Large-scale Infrastructural Development in the Era of Climate Change

Y.A. Braun

Introduction

In Lesotho, southern Africa, the commodification of water and the building of large dams is at the center of an extractive, export-oriented strategy to generate national revenues by selling water to the Republic of South Africa. As critics feared, the project has also affected the lives and livelihoods of tens of thousands of Basotho people since 1987, and has created especially intense pressures on rural women. An US$8 billion water scheme, the Lesotho Highlands Water Project (LHWP) was designed as a five dam mega-project, so far with three dams completed and one currently under construction.

At the center of the LHWP is one of the cleanest, most abundant sources of freshwater in the region, integral to well-being and livelihoods in the rural, agriculturally based communities of highlands Lesotho. The construction phases of large dam and infrastructural projects such as the LHWP often extend over long periods of time, creating social, environmental, cultural, political, and economic consequences in the proximate communities and landscapes. Large dams also fundamentally alter riverine systems and water resources, and generate rapid socio-environmental change in the communities where they are situated. Changing relationships to water resources reflect an underexplored dimension of the impacts of these mega-projects on rural households, a form of dispossession that negatively affects livelihoods in gendered ways. These changes can create or exacerbate gendered forms of anxiety about the present and future amid the proposed benefits of LHWP development and the realization of the effects of climate change in the region.

The diversion and sale of Lesotho's water to South Africa was planned to address rising demand for water in the region, yet climatic variability over the course of the project has raised questions about the viability and consequences of this endeavor. Within Lesotho, climatic changes have included periods of drought, erratic rainfall and hail, and earlier frosts, negatively affecting the food and water supply and damaging fields central to the livelihoods within rural communities. Over the course of the ongoing project, the shifting climate and changing relationships to riverine sources heightened water anxieties in both countries, creating unintended consequences for gendered livelihoods in Lesotho.

A narrative approach is taken in this chapter to document and analyze the gendered social and socio-environmental consequences of globalized river basin development using water as the lens to understand the depth and breadth of these changes in people's lives. The analysis is focused on people's narratives about water, both because of its centrality to the extractive logic of this US$8 billion transnational project and its centrality to the lives and livelihoods of those

 Gender, Climate Change and Livelihoods: Vulnerabilities and Adaptations (eds J. Eastin and K. Dupuy)
DOI: 10.1079/9781789247053.0006

who live in close proximity to the dams and reservoirs. In this chapter, the author addresses the convergence of these issues based on primary multi-site ethnographic field research conducted in all three active dam areas of Lesotho in 1997 and 2000–2002, as well as ongoing documentary research. In addition to daily informal interviews, periods of fieldwork included participant observation, documentary research, and over 300 formal interviews conducted with local research teams in 45 villages affected by the LHWP. Water remains central within Lesotho's national development plans and to the stability of the region even amid changing climate conditions. More locally, as water becomes more precarious within the lives of highlands residents living near the LHWP, this chapter reveals the multilayered, complex, embodied experiences of infrastructure policy and its consequences, for the everyday lives and livelihoods of people directly affected by these projects.

Consequences of Large-scale Infrastructure Development

Infrastructure projects involving large-scale dams continue to be pursued as a means towards development and poverty reduction in the Global South (Bromber *et al.*, 2014). Yet, large dams are controversial as national development solutions, and their costs and benefits for society and the environment are extensively debated (WCD, 2000; Scudder, 2006). Critics argue that the local effects of large dam projects disproportionately affect the poor and marginalized (Roy, 2000; Braun, 2006, 2020b), which can include development-induced displacement and resettlement (Hussain, 2008), and lead to organized opposition locally and globally, such as in India (Khagram, 2004) and Chile (Borgias and Braun, 2017).

While proponents argue there have been significant practical benefits to this utilitarian, developmentalist approach to rivers, and water more generally, critics maintain that it marginalizes alternative framings of these human relationships and meanings to an imperative of development and commodification of water resources (McCully, 2001). Contestation about large dams has perhaps elevated these debates more than other forms of river basin development, and has demonstrated the uneven platforms from which social actors engage in these debates. Tur *et al.* (2018, p. 125) conducted factor analysis on the content of discourses in the debate around hydroelectric dams in Brazil, noting the centrality of shared 'techno-economic rationales' across multiple actors except one: people affected by dam construction. Their analysis points to how isolated are the discourses of people directly affected, and how their full, embodied lived experiences are not centrally addressed in these debates. As people who experience the consequences of large infrastructure projects, their experiences might best be understood as forms of knowledge that reveal multiple meanings for, or alternative relationships to, natural resources in rural communities.

The social and environmental changes associated with large-scale dam development are social and economic disruptions to local rural cultures and livelihoods, which create subsequent challenges and uncertainty that may produce stress and anxiety (Workman and Ureksoy, 2017). The erasure, minimization, or absence of their knowledge and experiences (Enloe, 2007; Braun, 2011) within the dominant discourses in these debates both reflect and constitute the privileging of technocratic economic rationales over other values and approaches. This outcome highlights the need for attention to power, and for centering the multiple narratives operating to shape meaning and lived experience within these embodied landscapes as they absorb the majority of the social and environmental changes associated with large-scale dams, and infrastructure development more generally.

Feminist Political Ecology and Intersectionality

Political ecology is a multidisciplinary approach that emphasizes how environmental resources and decision making are material and political, and are affected by different interests, investments, and power (Peet and Watts, 2004). Within and beyond political ecology, feminist political ecology (FPE) brings critical attention to the role of gender, and power more generally, in the structuring of relations to resources and in the production of knowledge (Rocheleau *et al.*, 1996).

Recent scholarship in FPE has begun to explicitly engage developments in feminist theory more broadly (Buechler and Hanson, 2015), highlighting the synergies between contemporary theorizing and analysis of intersectionality, and 'how human–environment relations intersect with various social relationships, such as class, ethnicity, and gender' (Radel, 2009, p. 335). While the intersectionality approach advances efforts to capture the complexities of lived social life of individual and group identities, institutions, structures, and systems across historical, geographical, and cultural contexts (Crenshaw, 1991), the environment is not always at the forefront in the analysis. FPE brings attention to the environment, revealing how resource struggles and their consequences are gendered, and 'how lived experiences and practices are productive of, and produced through, gendered ideologies, structural power relations, and processes of both local and global change' (Truelove, 2011, p. 145). In this way environmental resources and struggles are sites for the contestation and reproduction of social differences, such as gender, race, class, age, and dis/ability (Braun, 2015).

Drawing on Yanow (1995), who argued for 'seeing built spaces – that is, policy settings – as texts, and specifically as stories,' the author of the current study focuses on local narratives that circulate about water during a period of development-induced environmental change to render visible the policy stories that people tell within this built and natural environment (Fig. 6.1). Intersectional analysis is grounded in the stories of women and men directly affected by the LHWP, their narratives of water and socio-environmental change, exposing the contradictions of development and inequality and demonstrating how social structures, relations, processes, and policies shape social experiences and livelihoods in the context of large-scale river basin development and climate change.

Research Context

Lesotho is a small, mountainous, landlocked country surrounded by South Africa on all borders with a population of approximately two million people. Over 80% of Basotho live in rural areas, and have established a diversity of gendered livelihood strategies that rely on rain-fed agriculture, livestock production, wage labor, and informal economic activities (Epprecht, 2000). Historically, Lesotho has been understood as conforming to a classic model of a dual economy, with a primary subsistence agricultural sector in the rural regions of the country and a secondary migratory labor reserve economy dependent on South Africa's mining and industrial sectors.

The mountainous and hilly landscape in much of the country has made livestock production and rain-fed agriculture in rural Lesotho challenging, with steep slopes that must be terraced against the forces of erosion and surface runoff. The most arable lands are in the lowland areas, where rainfall is lower during the growing season than in the mountains, an area with less arable land and a shorter growing season due to the risk of frost and colder temperatures. Increased climatic variability in the southern Africa region is typically associated with drought, which has been linked to widespread food shortages in Lesotho in 1992 and 2000–2003. In this latter period, climatic conditions in Lesotho have also included earlier frosts and more erratic rainfall, including periods of intense rain and hail that damaged fields and reduced food supply. The Prime Minister of Lesotho declared a state of famine in April 2002, naming these destructive rains as a primary cause, and it was estimated that 42% of the population was food insecure with some areas experiencing famine conditions by November 2002 (Mphale *et al.*, 2003). Climatic shocks contributed to the elevated risk of food insecurity and increased reliance on money for food, and this occurred at a time when rural families were enduring low purchasing power and significant decline of the historical employment opportunities that brought remittances to rural areas.

Decades of rural–urban migration within Lesotho and migration to South Africa resulted in the emergence of oscillatory migration patterns where workers – historically men to the mines but recently women for domestic work and garment industry jobs (Bennett, 2006) – moved between urban workplaces and rural homes on a regular basis. The migration of Basotho men to South African mines peaked in the late 1970s and 1980s when up to 35% of active wage earning men from Lesotho were working in South Africa (CIA, 2010), often making women *de facto* heads of household and primary

Fig. 6.1. Central Intelligence Agency (CIA), Lesotho Reference Map (https://www.cia.gov/the-world-factbook/static/d81dcdbd3aa5f4107bd95bd7f1c55ffb/LESOTHO_reference_map.jpg). Published by the Central Intelligence Agency. Creative Commons Zero.

farmers. With the retrenchment of male miners since the early 1990s (Global Policy Network, 2006), migration for work in the South African mines has decreasingly been an option for men from poorer, rural areas such as those affected by the LHWP. Local opportunities for wage labor in these remote highland areas remain limited, and most wage labor still involves migration to South Africa or to urban areas in Lesotho. Amid these changes, the importance of women's informal small-scale income generation to Basotho households has increased in significance as retrenchment reduced remittances.

These historical and contemporary migration patterns are a direct result of high poverty and unemployment rates (UN-Lesotho, 2007), as well as structural relations of inequality regionally and globally, which have also proven to be major factors in the spread of HIV and other sexually transmitted infections (STI) during the period of the LHWP. As the governments of Lesotho and South Africa and international interests were implementing the large-scale LHWP during the 1990s and 2000s, the global HIV/AIDS health crisis was taking hold in southern Africa during the same period (Braun and

Dreiling, 2010; Kenworthy, 2017). Lesotho had an estimated 31% prevalence rate by 2002 (Mphale *et al.*, 2003), with devastating impacts on life expectancy and human development (Makoae, 2010). The HIV/AIDS epidemic in Lesotho has particularly affected the productive adult population aged between 15 and 49, similar to other southern African nations, further exacerbating the challenges of poverty and development that were largely framed as the *raison d'etre* for the proposed LHWP in Lesotho.

In the early to mid-1980s, the development priorities of international agencies and the pressures of Lesotho's politically and economically dominant neighbor, South Africa, influenced the Government of Lesotho's national development plans. A neoliberal vision of markets and of development during this time stressed export-oriented and resource extractive development models, approaches which had a long history in the region (Braun, 2020a). This neoliberal, export-oriented vision in Lesotho was articulated as marketing 'white gold,' or the extraction, nationalization, and sale of Lesotho's seemingly abundant water (Hoag, 2019) to South Africa through the transnational water delivery scheme, the LHWP.

Lesotho Highlands Water Project

The LHWP is a multi-dam infrastructural and development project being undertaken by the governments of Lesotho and South Africa and is considered the most complex water scheme in Africa. The project is funded in part by the World Bank (2010), the African Development Bank, the European Investment Bank, and several European funding agencies based on a treaty signed in 1986. The LHWP is being implemented by the parastatal Lesotho Highlands Development Authority (LHDA) in Lesotho and the Trans Caledon Tunnel Authority (TCTA) in South Africa, with the whole project overseen by the Lesotho Highlands Water Commission (formerly the Joint Permanent Technical Commission). The primary purpose of the water delivery scheme is to capture, transfer, and sell water from the rural Maluti Mountains of Lesotho to the Gauteng region of South Africa through a series of five dams linked to cross-national tunnels constructed in four phases, as a means of generating national revenues for development in Lesotho (LHDA, 1986). Three dams have been completed, Katse, 'Muela (the hydropower station), and Mohale, as part of Phase 1A and 1B, completed in 1998 and 2002, respectively (Fig. 6.2).

The LHWP's Phase 1A and 1B affected an extensive geographical area in Lesotho, with concentrated effects in communities around the dams, including the submergence of arable and grazing land, and the building of associated infrastructure such as tunnels and roads. At least 20,500 residents of over 120 villages in the project areas of the Maluti mountains (Phase 1A) were affected by the LHWP (Tshabalala and Turner, 1989; Sechaba Consultants, 2000), with the Mohale Dam resettling over 400 households and affecting at least an additional 1700 people. Households reported losing important material, social, cultural, and environmental resources such as agricultural, grazing, and garden lands; access to water sources for drinking, household, and agricultural use; forests for fuel wood; river basins for soils, rocks, and grasses used in constructing houses and cultural displays, such as painting designs on houses; and wild herbs and vegetables for food and medicinal purposes; and in some cases resulting in resettlement (Braun, 2015). These losses generally burdened already poor Basotho households and created especially intense pressures on women (Braun, 2005), consistent with the social and environmental consequences of other large dams (Scudder, 2006; Tilt *et al.*, 2009). The highlands areas chosen as dam construction sites for Phase 1 contain some of the most remote communities within Lesotho, with some of the highest rates of unemployment and destitution (Tshabalala and Turner, 1989; Sechaba Consultants, 2000). Baseline studies at the beginning of the LHWP note that 60% of households in the Katse and 'Muela areas (Phase 1A) fell below the average income for each area and were considered 'very poor' (Tshabalala and Turner, 1989, p. 9; POE, 1991, p. 25). The impacts were also felt in South Africa, as demand side questions were raised about the impacts on low-income township residents who were experiencing rising water prices, among other effects, due to the cost of the LHWP (Letsie and Bond, 2010).

Based on the 1986 treaty, subsequent phases had to be renegotiated between the two countries.

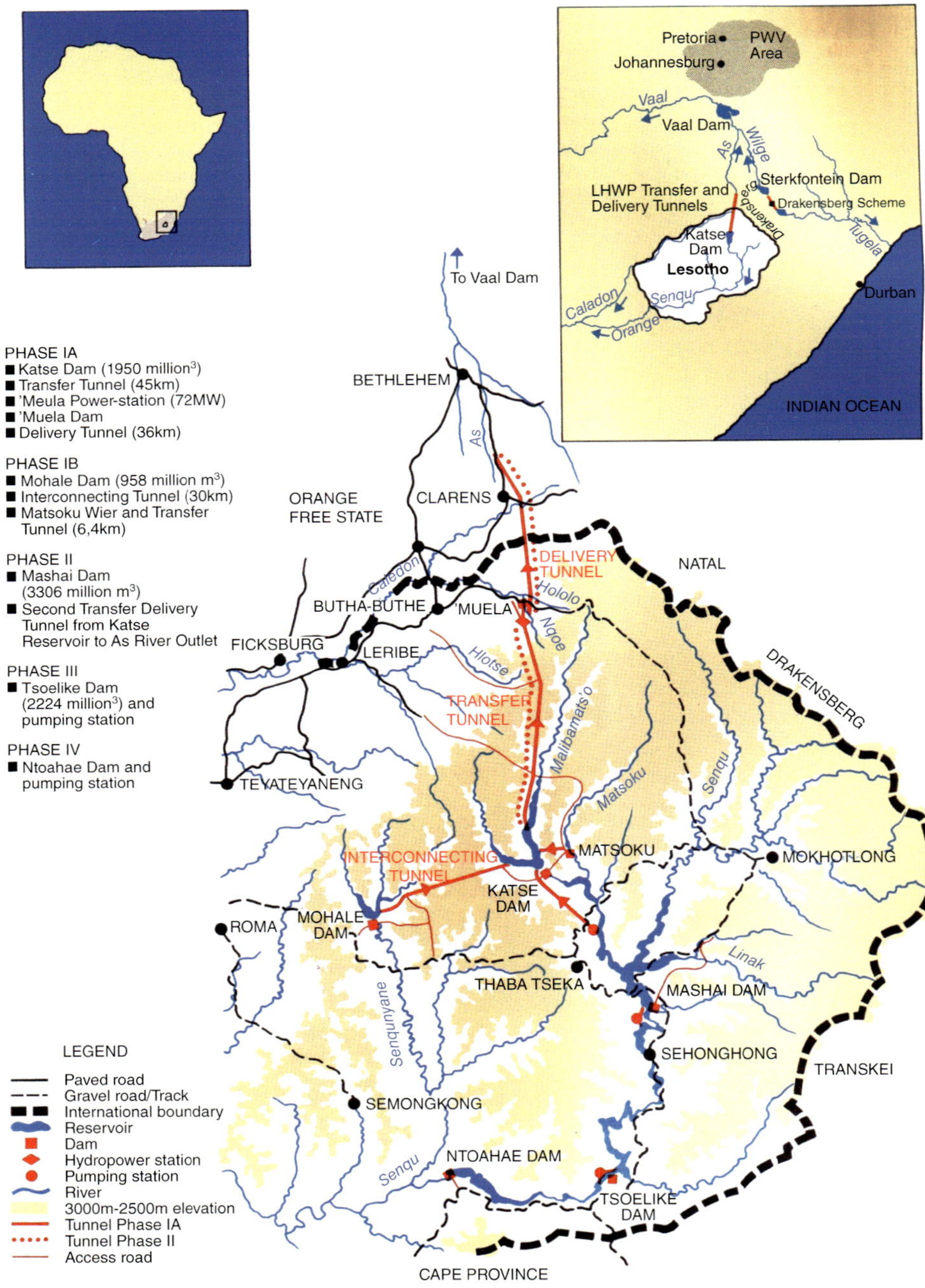

Fig. 6.2. Tambo (uploader), map of the Lesotho Highlands Water Project (https://upload.wikimedia.org/wikipedia/commons/5/51/LHWP_map_resized.jpg). Creative Commons BY-SA 3.0.

The Phase 2 agreement was formalized in 2011, and construction of the Polihali Dam is now under way with water delivery anticipated to begin in 2025. The delay between the two phases largely reflected concerns about inadequate compensation and corruption in Phase 1, and that there was less water than anticipated, leading to fewer royalties for Lesotho and continued problems of scarcity, inequities, and skyrocketing prices for water in South Africa (Bond and Ndlovu, 2010). Despite significant evidence of climatic change in the region – reduced water availability, drought, and erratic rainfall – and the record of biopsychosocial and economic consequences from Phase 1 of the LHWP, project proponents and government officials in both countries agreed to continue the LHWP, stating that the project's full potential could not be realized when all phases are completed (Braun, 2020a).

Water, climate change and livelihoods

The LHWP was planned to service water to the greater Johannesburg region, and to become the largest inter-basin transfer in southern Africa, by diverting water from the headwaters of the Senqu (Orange) river in Lesotho into the Ash tributary of the Vaal river in the Free State in South Africa. As the LHWP heightened instability of the riverine systems, changing climatic conditions in the region would raise additional questions about the calculations of both water supply and demand, and of the risks and vulnerabilities that the LHWP would induce or exacerbate (Braun, 2020b). Development and government officials in Lesotho had primarily valued water in the riverine systems through a utilitarian lens of abundance, viewing the free-flowing resource in the mountains as largely untapped potential (Hoag, 2019). The strong urban–rural divide in the country magnified this perspective, where urban elites in Maseru looked to resolve Lesotho's development dilemmas by generating national revenues through resource extractive strategies whose effects would largely be seen miles away in the rural mountain communities. Urban disdain for rural communities, seen as part of the problem of persistent poverty facing Lesotho, fueled the justification for the rapid social and environmental changes the LHWP would bring (Thamae and Pottinger, 2006).

The highlands area residents who stood to be affected were, of course, situated differently from their urban counterparts regarding their views on the risks and benefits of river basin development. The dynamic and complex realities of how rivers formed a part of the cultural, social, economic, political, and environmental fabric of their lives was largely unimagined or irrelevant within formal institutional narratives about the LHWP. Formative project-related studies even proclaimed the rivers as not being 'used' and therefore not a factor in anticipating Lesotho's future water needs (Bond, 2002, p. 143). Contrary to the narrative of abundance propelled by project proponents, other analyses suggested there may in fact be insufficient water in Lesotho to share with South Africa over time, and that water diversion through the LHWP may in fact lead to 'severe water scarcity' in Lesotho in 10–30 years (Bond, 2002, p. 143).

Local experiences of water abundance in Lesotho suggest it has been uneven at best (Hoag, 2019). For rural communities in Lesotho directly impacted by the LHWP, perspectives on water and development were always contextual and situated in place and time, inseparable from the rhythms and responsibilities of material and symbolic life in these highland areas and the larger historical, structural, and cultural dynamics in which they are embedded. Even as project proponents continually tried to depoliticize the LHWP, the massive infrastructural project reshaped the physical and biopsychosocial landscapes in project-affected areas, changing relationships to water and riverine resources for rural residents who now grappled with the meanings and consequences of new and intensified vulnerabilities and risks to their livelihoods.

Seeing Through Water: Local Narratives of Gender, Anxiety and Livelihoods

The villages in the remote and rural highlands areas directly affected by the LHWP are some of the poorest in Lesotho, which itself is one of the poorest countries in the world. These remote highlands communities, generally speaking, share extreme poverty; the absence of modern social infrastructure such as electricity, running water, and plumbing; and have limited paved roads. Most villages here are largely organized around a subsistence-oriented agricultural economy paired with migratory wage labor employment, where farming is central to livelihoods. Yet, most families report not being able to live off of farming alone (Abbot, 2002), made worse by changing climatic conditions amid LHWP-induced losses and socio-environmental changes.

Local narratives of socio-environmental change in the context of large-scale dam development in the highlands of Lesotho reveal the combined policies of action and inaction, of development planning, of intervention and neglect. The author focuses on three narratives of change that centralize water as an integral component, albeit in different and contrasting ways: water as an agent, water as a risk, and water as a commodity – regulated, controlled, and distributed within systems of meaning and hierarchies of power that attach a price to each drop, no longer free-flowing, no longer priceless.

In all the narratives, water is never simply just the background, a silent, neutral landscape in which social lives and change play out. Rather, water is part of an embodied landscape through which we see how development policies and institutions both organize gendered experiences of socio-environmental change and position local communities to absorb the consequences to their lives and livelihoods.

Water as agent: 'The river shook the earth'

In this first narrative water is understood as an agent – an actor in a relationship – where it actively shapes landscapes and forms relationships with living beings in culturally and socially responsive ways. This understanding can be revealed most sharply when these relationships are in crisis or out of balance, such as during the earthquake at Katse Dam. Several interviewees described the earthquake as when 'the river shook the earth.'

In 1996, an earthquake rocked the Katse Dam area in the middle of the night. People ran from their homes, terrified, as dishes crashed to the floor, mud walls cracked, and thatched roofs collapsed. The filling of Katse Dam, the first dam of the LHWP and the largest dam in Africa at the time, induced the seismic activity. The weight of the newly forming, massive reservoir created enormous pressure on the earth, shifting its tectonic plates and creating deep cracks that ran for miles through this highlands area.

Reservoir-induced seismicity is not uncommon when catchments begin to fill, but the LHDA did not anticipate, monitor, or communicate this outcome as a possibility to local communities. Villagers affected by the earthquake claimed that it was wholly unexpected and, as can be imagined, created tremendous panic and fear. Residents of one village most affected by the earthquake, Ha Mapeleng, were outraged and most of the community eventually demanded to be resettled further back from the reservoir. The post-earthquake demands of villagers emphasized concerns regarding safety, shelter, and communication. Compensation for earthquake damages became a necessary aspect of mitigation negotiations.

Some local villagers were less surprised by the violence from the river and earth, however. While they had not anticipated the specific form of an earthquake, some had long feared a negative response from the Water Snake believed to inhabit, or to be, the water. Long-standing concerns about the way in which the LHWP, and specifically Katse Dam, would affect the river systems were expressed in different terms: ecological, social, economic, and cultural, often combined in and across dimensions. One narrative emphasized the historical importance of the Water Snake, a powerful entity within the river who maintained positive relationships with local communities as long as he was kept happy and not exploited. As the scope of the environmental changes became clear during the creation of the reservoir, fear of the Water Snake's response to the development-induced changes of the LHWP had concerned some living proximate to the river. In the aftermath of the reservoir-induced earthquake, these residents felt certain this was a fierce blow that revealed the deep anger of the Water Snake in response to the dam and reservoir, and they feared his future actions, especially if his warnings went unheeded.

Like all beliefs, not all people shared these views or held these to be truths to the same degree. One strong believer, Thabang, from the village of Ha Mapeleng was devastated by these events and the meaning it had for him. He could not imagine the danger they all faced and he felt the violence and anxiety of being 'out of sync' with the Water Snake whom he had learned to relate to with respect and humility for over 60 years. Even as his village was eventually resettled further back from the reservoir, he refused to go, saying he must stay and 'be swallowed,' face the wrath that was his due for 'allowing this to happen'; for allowing the LHWP to happen. This resignation about the LHWP and his responsibility revealed a tragic twist, as there was no local participation in the decision making about the LHWP for the communities most directly affected by its consequences.

Thabang's wife, 'Matseliso, however, wished to move. She also knew the Water Snake was unhappy, but she felt the impossibility of changing these circumstances. For her, there was nothing they could do but to be resettled further back to hopefully escape the direct consequences of the Water Snake's anger. Their son and daughter-in-law, who previously lived with them, resettled with their four children in the newly created resettlement area of New Mapeleng. 'Matseliso wished to join them but was forbidden to do so by her husband, who was angry that his son had left. In addition, her family of origin was now cut off from her due to the massive reservoir, which she did not dare cross even if she could afford to pay for a boat ride.

The spatial divide between villages and families created by this newly formed reservoir was particularly isolating for women who typically join their husbands' families after marriage in this patrilocal community, and for the elderly who were often reluctant or fearful to cross the massive body of water. 'Matseliso begged her husband to spare their grandchildren from this isolation. She thought of leaving Thabang to live with her son and grandchildren, but, within the patriarchal, gendered laws of Lesotho she would lose the already limited rights of married women to property, and besides, she would worry about him. With her husband, she lives with the fear and anxiety that this changed landscape has upset the balance, the relationship, which people like themselves had long worked to build with the river, with the Water Snake, and what that may mean for their future. At the same time, 'Matseliso's constrained choices about how she might respond to this specific fear and these general socio-environmental changes remain circumscribed by the unchanged: the persistent patriarchal laws that undergird gendered cultural ideologies and gender relations that limit women's options and rights.

Water as risk

'Water as risk' was a related theme in local narratives about socio-environmental change. In this narrative, water is powerful, threatening, with the potential to submerge, to drown, to take away life (contemporary lives as well as ancestral spirits' graves), and to take away pain. The threat of submergence was pervasive and very real, and clearly a part of the dominant concerns of both development authorities and communities to be resettled. During the construction phases of the dams, proximate communities faced the uncertainties, anxieties, and risks associated with the impending dams and reservoirs. White posts staked across the landscape indicated future reservoir heights – or, more accurately, these stakes noted the submersion levels for communities notified of their future resettlement. White stakes circling the landscape became ominous threats, everyday anxious reminders of changes to come, resettlement to happen, losses to be borne.

After damming, the reservoirs were significantly larger than the previous river, which presented different kinds of risks and consequences. This was particularly true at Katse Dam, which now necessitated a motorized boat to cross, an option that was not always available. Those with greater resources might hire a boat or taxi to visit family in villages now made far from each other, or to go to the one clinic in the area; although youth were much more likely to imagine a boat ride with less anxiety than their elders. For most people, however, the cost of the fare and their worries about the risks were prohibitive, leaving them to loathe the reservoirs that now created an impasse between themselves and relatives. Married women in this patrilineal and patrilocal society particularly felt the effects of this further dislocation and for some, isolation. Like 'Matseliso above, they had left their family of origin upon marriage and most often had moved to a new village in the highlands, often in the same region or in the same river basin area. Now the reservoir created both a symbolic and material barrier to these familial connections, to home.

Still others described the reservoirs as a site of fear, desperation, and escape, a place some men drowned themselves when hardships resulting from the LHWP became too much. With migration for work in the South African mines decreasingly an option for men from poorer, rural areas, local opportunities for wage labor in these extremely remote highlands areas remained very limited, even in the construction era when few local men, and even fewer women, found formal work with the project despite promises of employment. The combination of migration opportunities closing, the loss of agricultural and grazing fields to the LHWP, changing climatic conditions, and the absence of job opportunities, created intense economic and social pressures on women and men who struggled to generate sustainable livelihoods. These pressures intersected with gendered ideologies about men and masculinity and cultural expectations of men being providers. The stress of extreme poverty and development-induced loss, coupled with feelings of failed expectations, was part of a tragic spiral of despair leading some men to commit suicide. Like the tragedy of over indebted farmers ingesting pesticides, neighbors described men drowning themselves in the Katse reservoir, using the reservoir that shaped their pain to take away their pain.

Water as commodity

Local people in the communities affected by dam and reservoir construction also experienced new forms of regulation that limited their access to water, as the once free-flowing resource became commodified and owned under new policy doctrines designed to generate a specific model of economic development. As each drop of water carried a price, the communities directly bore the costs of the reorganization and commodification of natural resources, including exclusionary regulations that challenged or even criminalized traditional activities, and incorporated displaced people into new consumer regimes.

Many remote highlands families who continued to live proximate to the mega-dams lost access to clean, reliable water sources as the easily accessible rivers were transformed into massive reservoirs of water to be sold. Damming the riverine system caused a series of consequences, including altering a number of the natural springs that supplied clean, fresh water to local communities. 'Mathato Lejone expressed frustration about losing the water source she relied on for years, explaining: 'This dam has brought hardship. Now our springs have vanished. These were near to us, but now we must go far to get water.' 'Mathato's experience was a common one as many people reported springs drying up around the Katse basin after the creation of the new reservoir and women talked about having to find new sources of water, sometimes walking significantly further to do so (Braun, 2005).

Households who were resettled in peri-urban or urban areas confronted struggles of a rapidly changed economic and environmental landscape that integrated them into customer-based utility systems in more cash-reliant settings. While each new house had the potential for its own outside water tap, re-settlers to these areas were surprised to find the potential ease of water access made impossible due to the high cost of water and connecting to the utility. Many households could not afford to purchase directly the water needed to meet their daily needs for drinking, cooking, bathing, and cleaning. Relative class privilege positioned some families who could afford to, to sell water by the bucket to neighbors 'off the grid' as a means to subsidize their own water bills and help their neighbors (Braun, 2015). Gender ideologies still positioned women in this context to largely be charged with the gathering of water and the household tasks associated with it. By sharing water for tasks, such as laundry, exchanging services for water, and taking turns or pooling limited resources to purchase water, women creatively managed high prices for water and water scarcity.

Water as a commodity also draws water into increasing surveillance, making it regulated space within privileged economic valuations. While development authorities planned hopefully for the remote reservoir to become a popular tourist site for waterskiing and pleasure boating, drawing reservoir water for household use and sustenance became criminal acts. This reservoir water in the highlands was now a commodity, available only to generate revenue, but not for local people to access or to even buy. Some took the risk of 'stealing' water from the reservoirs, but most did not dare – for legal or moral reasons, and as the walls of the reservoir could be quite steep and hazardous. Local women's labor was taken for granted as they must find replacement water sources, often walking further to do so, to meet the everyday needs of their families. In this commodification, the challenges associated with these environmental changes pivoted on the naturalization of women's labor as uncompensated and limitless. During this period of social and environmental change, water is never simply just the background but is always central to the dynamics of livelihoods and development within these embodied landscapes.

Conclusion

National strategies for economic development in the Global South commonly employ infrastructure projects such as mega-dams, hydroelectric power, and, as this case captures, international trade in water, or 'white gold' as it is known in Lesotho. As the remote, rural poor in Lesotho negotiate the impacts of the LHWP in their lives and livelihoods, what the author calls the 'embodied landscapes of development' (Braun, 2015), they experience the intrusion and privileging of the transnational and local interests that organize and finance projects such as the LHWP. This chapter focuses specifically on understanding the multiple narratives of water that operate in the context of this mega-project, particularly for people most directly affected by the LHWP and for whom rivers and springs were intimately tied materially and symbolically to their sustenance and livelihoods. These emerging narratives about water in the context of river basin development do not exist outside of power relations, but rather reflect and constitute dynamics of inequality, as one narrative – commodification – is privileged over other understandings of, and relationships to, water. While water is central to this story, the dominant story of 'white gold' renders water as only a commodity, when it is so much more in the lived realities of people's lives and livelihoods prior to, during, and after the LHWP. In this process, the author shows how such an approach simultaneously reduces the complex hydrological and ecological systems of this region to 'white gold,' and creates unintended gendered consequences for the human communities that rely on this resource.

For local landowners in the vicinity of the LHWP their experiences of development often entailed displacement and dispossession of their homes and lands, altered access to water and riverine resources, integral parts of residents' rural livelihoods and the fabric of their shared history. 'Seeing through water' provides lenses to reveal the operation of development initiatives in embodied landscapes, rendering visible how development-induced environmental changes – such as the transformation of the crossable river to a massive reservoir – intersect with culture, gender ideologies, and patriarchal systems to shape gendered and classed experiences of development and climate change. These seemingly individual narrations, or the policy stories people tell (Yanow, 1995), weave together a much larger tapestry that reveals the contradictions of national development strategies in the lives of those most directly affected.

'Seeing through water' reveals the importance of understanding the discourses of people directly affected by large-scale projects, discourses which are too often isolated (Tur *et al.*, 2018) or minimized. More attention to their perspectives and their livelihoods will offer an opportunity to more fully understand and take account of the changes, costs, and burdens of international river basin development that significantly reorganize resources and relationships in the places in which they live.

References

Abbot, J. (2002) Lesotho's food crisis: balancing humanitarian and development responses to food shocks. CARE International, South Africa – Lesotho. Available at: www.sarpn.org/documents/d0000153/index.php#footnote (accessed 1 February 2021).

Bennett, M. (2006) Lesotho's export textiles and garment industry. In: Jauch, H. and Traub-Merz, R. *The Future of the Textile and Clothing Industry in Sub-Saharan Africa*. Friedrich-Ebert-Stiftung, Bonn, Germany, pp. 165–177.

Bond, P. (2002) Lesotho's water, Johannesburg's thirst: communities, consumers, and mega-dams. In: Bond, P. (ed.) *Unsustainable South Africa: Environment, Development and Social Protest*. University of KwaZulu-Natal Press, Scottsville, South Africa.

Bond, P. and Ndlovu, M.M. (2010) Development dilemmas of mega-project electricity and water consumption. In: Freund, B. and Witt, H. (eds) *Development Dilemmas in South Africa*. University of KwaZulu-Natal Press, Scottsville, South Africa.

Borgias, S.L. and Braun, Y.A. (2017) From dams to democracy: framing processes and political opportunities in Chile's 'Patagonia without dams' movement. *Interface: a Journal for and About Social Movements* 9, 300–328.

Braun, Y.A. (2005) Selling the river: gendered experiences of resource extraction and development in Lesotho. *Research in Rural Sociology and Development* 10, 373–396.

Braun, Y.A. (2006) Large dams as development: restructuring access to natural resources in Lesotho. In: Jorgenson, A.K. and Kick, E.L. (eds) *Globalization and the Environment*. Brill Press, Leiden, The Netherlands, pp. 151–171.

Braun, Y.A. (2011) Left high and dry: an intersectional analysis of gender, dams, and development in Lesotho. *International Feminist Journal of Politics* 13, 141–162.

Braun, Y.A. (2015) Interrogating large-scale development and inequality: bridging feminist political ecology, intersectionality and environmental justice frameworks. In: Buechler, S. and Hanson, A.S. (eds) *A Political Ecology of Women, Water and Global Environmental Change*. Routledge, New York, pp. 19–37.

Braun, Y.A. (2020a) Lesotho's white gold: the political ecology of temporality and the economy of anticipation in resource extraction and large dam infrastructural projects. *Journal of Political Ecology* 27, 853–876.

Braun, Y.A. (2020b) Environmental change, risk, and vulnerability: poverty, food insecurity, and HIV/AIDS amid infrastructural development and climate change. *Cambridge Journal of Regions, Economy and Society* 13, 267–291.

Braun, Y.A. and Dreiling, M.C. (2010) From developmentalism to the HIV/AIDS crisis: the amplification of women's rights in Lesotho. *International Feminist Journal of Politics* 12, 464–483.

Bromber, K., Féaux de la Croix, J. and Lange, K. (2014) The temporal politics of big dams in Africa, the Middle East, and Asia: by way of an introduction. *Water History* 6, 289–296.

Buechler, S. and Hanson, A.S. (eds) (2015) *A Political Ecology of Women, Water and Global Environmental Change*. Routledge, New York.

CIA (2010) *CIA World Factbook: Lesotho*. Central Intelligence Agency, Washington, DC.

Crenshaw, K. (1991) Mapping the margins: intersectionality, identity politics, and violence against women of color. *Stanford Law Review* 43, 1241–1299.

Enloe, C. (2007) *Globalization and Militarism: Feminists Make the Link*. Rowman and Littlefield Publishers, Boulder, CO.

Epprecht, M. (2000) *'This Matter of Women is Getting Very Bad': Gender, Development and Politics in Colonial Lesotho*. University of KwaZulu-Natal Press, Pietermaritzburg, South Africa.

Global Policy Network (2006) Country Report – Lesotho. Available at: www.gpn.org/ (accessed 1 February 2021).

Hoag, C. (2019) 'Water is a gift that destroys': making a national natural resource in Lesotho. *Economic Anthropology* 6:sea2.12149.

Hussain, M. (2008) *Interrogating Development: State, Displacement, and Popular Resistance in North East India*. Sage, New Delhi.

Kenworthy, N. (2017) *Mistreated: the Political Consequences of the Fight Against AIDS in Lesotho*. Vanderbilt University Press, Nashville, TN.

Khagram, S. (2004) *Dams and Development: Transnational Struggles for Water and Power*. Cornell University Press, Ithaca, NY.

Letsie, D. and Bond, P. (2010) Debating supply and demand characteristics of bulk infrastructure: Lesotho–Johannesburg water transfer. Background Research Series, Municipal Services Project, Kingston, ON, Canada. Available at: www.municipalservicesproject.org/sites/default/files/LETSIE%20and%20BOND%20Debating%20Supply%20and%20Demand%20Characteristics%20of%20Bulk%20Infrasturcture.pdf (accessed 1 February 2021).

LHDA (1986) Lesotho Highlands Development Authority order of 1986. Lesotho Government Gazette Extraordinary, Maseru, Lesotho.

Makoae, M. (2010) *HIV/AIDS Caregiving without ARVs in Lesotho: Food, Medicine, Body and Emotions*. LAP Lambert Academic Publishing.

McCully, P. (2001) *Silenced Rivers: the Ecology and Politics of Large Dams*. Zed Books, London.

Mphale, M.M., Rwambali, E.G. and Sechaba Consultants (2003) Lesotho Food Security Issues Paper. *Working and Discussion Papers, July*. Forum for Food Security in Southern Africa, Overseas Development Institute, London.

Panel of Environmental Experts (POE) (1991) Report to LHDA and the World Bank. Maseru, Lesotho Highlands Development Authority, Lesotho.

Peet, R. and Watts, M. (eds) (2004) *Liberation Ecologies*. Routledge, London.

Radel, C. (2009) Nature / environment: natures, gendered. In: Kitchin, R. and Thrift, N. (eds) *International Encyclopedia of Human Geography*. Elsevier, Amsterdam, The Netherlands.

Rocheleau, D.E., Thomas-Slayter, B. and Wangari, E. (eds) (1996) *Feminist Political Ecology: Global Issues and Local Experiences*. Routledge, London and New York.

Roy, A. (2000) *The Algebra of Infinite Justice*. Penguin Books, New Delhi.

Scudder, T. (2006) *The Future of Large Dams: Dealing with Social, Environmental, Institutional and Political Costs*. Earthscan as elsewhere, London.

Sechaba Consultants (2000) Poverty and livelihoods in Lesotho, 2000: more than a mapping exercise. Sechaba Consultants, Maseru, Lesotho.

Thamae, M.L. and Pottinger, L. (eds) (2006) On the wrong side of development: lessons learned from the Lesotho Highlands Water Project. Transformation Resource Centre, Maseru, Lesotho.

Tilt, B., Braun, Y.A. and He, D. (2009) Social impacts of large dam projects: a comparison of international case studies and implications for best practice. *Journal of Environmental Management* 90, S249–S257.

Truelove, Y. (2011) (Re-)conceptualizing water inequality in Delhi, India through a feminist political ecology framework. *Geoforum* 42, 143–152.

Tshabalala, M. and Turner, S.D. (1989) 1988 Socio-economic census of the Lesotho Highlands Water Project, phase 1A areas, Volume I, Main Report. Lesotho Highlands Development Authority, Maseru, Lesotho.

Tur, A.A., García-Andreu, H., Ortiz, G. and Domínguez-Gomez, J.A. (2018) Discourse analysis of the debate on hydroelectric dam building in Brazil. *Water Alternatives* 11, 125–141.

UN-Lesotho (2007) United Nations development assistance framework 2008–2012. United Nations, Maseru, Lesotho.

WCD (2000) Dams and development: a new framework for decision making. Report of the World Commission on Dams. Earthscan as elsewhere, London.

Workman, C. and Ureksoy, H. (2017) Water insecurity in a syndemic context: understanding the psycho-emotional stress of water insecurity in Lesotho, Africa. *Social Science and Medicine* 179, 52–60.

World Bank (2010) Project performance assessment report: Lesotho Highlands Water Project, phase 1B (Loan 4339-LSO) Community Development Support Project (Credit 3308-LSO). World Bank, Washington, DC.

Yanow, D. (1995) Built space as story: the policy stories that buildings tell. *Policy Studies Journal: the Journal of the Policy Studies Organization* 23, 407–422.

7 Caring for Corn and Beans: Reassessing Subsistence Agriculture and Climate Change

E. Garner

Introduction

Policy and development research often posit women's lack of material resources as a reason for their vulnerability to climate change, suggesting that increased participation in income generating activities is the solution to increasing resiliency. In parallel, rural development initiatives seek to increase women's income through crop diversification and access to productive assets to improve crop yields. While both climate change and women in agriculture literature reference domestic responsibilities as barriers to increasing resilience, they maintain that subsistence farming and care work are separate divisions of women's time. Using a care work framework, the author argues that subsistence agriculture *is* care work and that not recognizing it as such minimizes opportunities for addressing gender disparities within care work and climate change policies and programs.

Effective climate change policies must consider geographic, socio-economic, and demographic characteristics (Morton, 2007), as well as have in-depth understandings of cultural differences (Adger *et al.*, 2013). As such, mitigation strategies must incorporate gender dimensions for successful and sufficient adaptation (Gabrielsson, 2015). Men's and women's respective social roles and responsibilities influence the direct effects of climate change, as well as those of climate change measures that do not adequately account for gender (Roehr, 2007). Enarson and Meyreles (2004, p. 62) reviewed gender and disaster studies literature, finding that these focused on women or the power differentials between men and women, and reflected a social vulnerability approach. A material approach is pervasive throughout climate change literature, framing women's vulnerability to climate change in terms of access to assets, despite the additional contributions of cultural norms, roles, and responsibilities (Alston, 2013).

Agriculture and climate change discussions place smallholder and subsistence farmers among the most vulnerable populations to climate change (Morton, 2007). For female farmers, the barriers to equal participation in agriculture link to unequal access to land and capital, credit, information, and agricultural education (FAO, 2011), as well as to the ability to hire labor, purchase inputs, and access marketing channels (Razavi, 2009). Time is also framed as a material asset as women may face additional burdens to participate in high-value markets due to their time-intensive dual roles in productive and reproductive labor (Jiggins *et al.*, 1998). These same barriers to increased agricultural production are identified as those that increase rural women's vulnerability to climate change (Kristjanson *et al.*, 2017), as these resources are needed to adapt production practices and build resiliency, mainly through income and nutrition.

DOI: 10.1079/9781789247053.0007

It is essential to find an appropriate framework for understanding climate change that accounts for these material barriers to increasing smallholder women's resilience without ignoring or sidelining the social construction of gender. The lens used so far, according to feminist analysis, has not yet adequately accounted for the gender context in its exploration of neither climate change (Enarson and Meyreles, 2004; Kidder *et al.*, 2014) nor agriculture (Sachs and Patel-Campillo, 2014). Further critiques find that positioning 'vulnerability' and 'resilience' as problematic dichotomies prevents a relational conceptualization to account for positionality and power dynamics (Tschakert and Tuana, 2013). Post-structural approaches call for a deeper analysis of masculinity, querying of the relationships between humans and the environment, and a greater perspective on how humans are part of nature, rather than opposite nature (Alaimo, 2009; Gaard, 2015).

A care work reframing of subsistence agriculture can address some of the feminist concerns regarding climate change policy. This reframing fits into a broader push for policy to formally recognize the gender disparities in care work, as well as the value of care work as a public good (Daly, 2002). Ferrant *et al.* (2014, p. 1) argue that policies should address the gender inequality in unpaid work as this '...is the missing link that influences gender gaps in labor outcomes.' Despite being an integral part of the economy, 'unpaid work in social production, subsistence production, family businesses, and the community is often ignored' by public institutions (Antonopoulos, 2008, p. 1). However, this could be changing especially among international agencies. For example, the United Nation's Sustainable Development Goal on Gender Equality specifically calls for formal recognition and valuing of unpaid care and domestic work through social protection policies, as well as the promotion to share responsibility within the household (SDG 5, target 5.4; UN, undated). Since care as a concept looks at the relationship between the material and ideological, broadening these initiatives to include subsistence agriculture could push policies to recognize (rural) gendered experiences of labor, the environment, and the home more fully.

To move beyond a material understanding of women's relationship to climate change, the author explores a care work framework that explicitly includes women's participation in subsistence agriculture, while expanding climate change frameworks to consider care work. The case of Honduras is used in the current study to reflect on the need for and opportunities of this framework. Through this case study, the author reconceptualizes subsistence agriculture as a vital contribution to household and community agriculture. This is done by broadening the definition of care work, and by positioning agriculture within a corporeal–material–socio-cultural framework that reconnects the various 'food domains' (Allen and Sachs, 2007). The subsequent sections, 'Theory of care work' and 'Care work, subsistence agriculture and climate change,' outline a theoretical context to care work, before analyzing subsistence agriculture and climate change through two particular points. The final section, 'Conclusion,' summarizes the analysis.

Gendered Analysis of Climate Change, Agriculture and Care Work: The Honduras Case Study

The Central American region is one of the most vulnerable regions in the world to climate change (Kawas *et al.*, 2011). Honduras in particular is one of the most vulnerable countries due to its geographic, socio-economic, and ecological conditions, making climate change one of the greatest threats to the country (CGIAR, 2015). Climate projections in Honduras are particularly concerning for agriculture. They indicate higher temperatures and lower rainfall in the future, especially in the western highlands region that is already experiencing droughts and aridity (Kawas *et al.*, 2011). Simultaneously, storms and floods have had the highest human and economic impact in Honduras (World Bank, 2009).

Previous examples of extreme weather events in Honduras include hurricane Mitch in 1998, which had countrywide impacts (DFID, 2004; World Bank, 2009; Kawas *et al.*, 2011). The impacts of the hurricane fell disproportionately on women for many reasons, including the lack of gendered considerations in mitigation planning and the responsibility to provide care for the household and others at the expense of assets (Delaney and Shrader, 2000). Further, women experienced disproportionate burdens during reconstruction due to 'their triple roles in maintaining the household, engaging in community organizing, and productive work in the informal economy' (Nelson *et al.*, 2002, p. 55; see also Casolo, 2009).

Honduras initiated several policy responses after this disaster (Kawas *et al.*, 2011), and is more recently re-evaluating climate change strategies specific for agriculture (CGIAR, 2015). Within the action frameworks, land management and water use are a top priority, as are conditional cash transfer and civic engagement programs (World Bank, 2009). Focusing on land and water are important as farmers in Honduras are particularly reliant on rainfall; only 5.7% of the total cropland is under irrigation compared to the regional average of 11.49% (World Bank, 2009). Land and water are also of major concern given the vulnerability of the staple crops – corns and beans – to fluctuations in temperature and rainfall (Kawas *et al.*, 2011; Diaz-Ambrona *et al.*, 2013).

Further, socio-economic approaches are essential to addressing the specific associated vulnerabilities, as Honduras is one of the poorest countries in the region (World Bank, 2009). Rural families are disproportionately in poverty and the majority derive their livelihoods from agriculture (Kawas *et al.*, 2011). Therefore, ecological as well as economic factors must be addressed equally to reduce climate change vulnerabilities (DFID, 2004).

While Honduras is recognized for making efforts to incorporate agriculture into the discussion of climate change (CGIAR, 2015), the integration of a gender–agriculture–climate nexus is lacking. Honduras's food security policies, the majority of which focus on nutrition, do integrate gender (six out of the seven reviewed policies) (Gumucio and Rueda, 2015, p. 50). However, in reviewing policies for the Latin American region, Gumucio and Rueda (2015) notice that policies regarding development planning, forest ecosystems, and risk management do not address gender to the same extent as food security, and climate change policies in general do not include gender to a significant extent, if at all. The emphasis on gender in relation to food security and nutrition reasserts women's positions as caretakers responsible for food preparation, rather than production.

According to the Honduras Secretary of Agriculture and Husbandry (*Secretaria de Agricultura y Ganaderia*), rural women in agriculture are mainly growing corn and beans, the primary staple crops, as well as gardening and processing household agriculture goods. Women also work in developing small- and medium-scale household and artisanal agricultural industries, such as small livestock, tortillas, vinegar processing, vegetable processing, soaps, sausages, dairy, and woodcrafts. Though these 'secondary' activities are not usually counted in the agricultural economy, women are contributing significantly to household savings. These activities generate and maintain a reserve fund, enabling the family production unit to weather negative shocks in agriculture. Further, these activities provide working capital for household agriculture to ensure the production of basic grains, to buy supplies, to rent a plow, payment of wages to laborers, etc. (SAG, undated).

There are three main reasons for this gender distinction in agriculture in Honduras: (i) gender differences in time distribution due to women's reproductive responsibility and household care; (ii) gender differences in access to production inputs and resources, principally, credit, land, education, and social networks; and (iii) gender differences resulting from institutional and market failures and discrimination (Staritz and Reis, 2013). Access to land, credit, and markets, as well as technical assistance and group membership are major barriers to increasing women's participation and decision making (Deere and Leon, 2001; Casolo, 2009). There are also significant differences in access to technology, extension, and in women's participation in the supermarket value chain as compared to the local value chain (Torres Barragan, 2012). The division of labor within the household, as well as participation in decision making and control over resources, reflects the traditional roles of men and women (Torres Barragan, 2012). In particular, land tenure and socio-cultural constraints limit their participation and power in the 'higher-value' markets and confound these barriers (Casolo, 2009; Vonderlack-Navarro, 2010).

In Honduras, traditional social norms place great pressure on women to play their reproductive role and assume their responsibility for household care (Staritz and Reis, 2013). Likewise, these norms define the division of labor and the gender division of resource access and control; women's labor concentrates in fewer sectors than that of men (IFPRI, 2013). Existing analysis of women in Honduras demonstrates that concepts of gendered responsibilities and ideals of masculinity and femininity reinforce divisions in household labor. Women have reported spending 103 hours a week in care tasks (Kidder *et al.*, 2014) and almost 8 hours a week in household gardens (IFPRI, 2013). Jo Rowlands notes the following in her analysis of women's empowerment in Honduras:

> Honduran women contend with inequalities on many fronts. The sexual division of labor leaves women carrying the burden of domestic and reproductive labor... Honduran women are expected to fit into the gender role, common throughout Latin America, of wife–mother–maintainer of the home. To be a woman is to be long-suffering and resourceful; to carry the responsibility of the welfare of your children and often of your partner... Women endure with fatalism machismo and violence in various forms.
>
> (Rowlands, 1997, p. 34)

Research into the potential of land reform movements (Casolo, 2009) and microcredit loans (Vonderlack-Navarro, 2010) for women's empowerment exemplify the limitations that delineations of roles and responsibilities in Honduras create. For example, Vonderlack-Navarro (2010) found that ignoring women's household relations in economic empowerment initiatives facilitated and limited women's use of loans and repayment strategies. Women's increased income threatened men's traditional identity as breadwinner,' altering gendered power relations (Vonderlack-Navarro, 2010, p. 124). Through her analysis, Vonderlack-Navarro recommends that programs must include men's perceptions, minimize interventions that expose vulnerable women to greater financial stress, and be 'conducted in conjunction with – not in replacement of – other health, social, and educational initiatives' (Vonderlack-Navarro, 2010, pp. 132–133). Her recommendations for improving microcredit programs in Honduras would also improve gender considerations in agriculture and climate change. Casolo's (2009, pp. 394–395) review of land reform and entitlement programs in Honduras reveals similar conclusions, warning against policy and development initiatives that fail to engage gender, as well as class, history, and location, in a substantive way.

Additional insight into care work in Honduras comes from Oxfam's initiative to improve development practitioners' approach to care work (Kidder *et al.*, 2014). A series of mixed-gendered workshops and focus groups showed that 'the patterns of care here are determined by the strong influence of the church, the collaboration between families, and by beliefs that reproduce inequalities' (Kidder *et al.*, 2014, p. 504). The report notes that participants understood the link between poverty and care work; however, staff resisted efforts to redistribute care work within the household to maintain men's participation. Further, the group solutions focused on increasing technology to lighten the household workload; this resulted in a community public works project. As such, the focus on a community-level project minimized the intervention's direct impact on intra-household dynamics.

Theory of Care Work

The theory of care work is a broad field that encompasses many different perspectives on the structuring of domestic and effective labor. A narrow definition of care work emphasizes face-to-face (or personal) contact responding to a particular physical or emotional need; this definition excludes the manufacturing of a physical product or work in the service industry where emotion is circumstantial (England and Folbre, 1999, p. 40; Daly and Lewis, 2000). However, others approach care work in a broader sense. Joan Tronto (2012 in Esquivel, 2014, p. 427), in particular, expands the definition to include activities of caring not just for people but also for objects and the environment, with the goal of improving our world and our ability to live in it well.

Research regarding care work has shifted between paid and unpaid work as caregiving labor itself has shifted across these realms (Daly and Lewis, 2000). Today, the care literature has expanded to include: (i) paid care work outside the home; (ii) paid care work within the home; and (iii) the individual balance between paid work outside the home and unpaid care work within (England, 2005). In addition, care work can include unpaid care activities outside of one's household such as for friends, neighbors, next of kin, or for the community as a whole (Esquivel, 2014, p. 428), as well as taking place in the public sphere of markets and the state (Esquivel, 2013, p. 13).

In addition to the conceptual framing of household responsibilities, care work also builds from and interrogates dichotomies that correlate directly to family farm operations: productive/reproductive labor and responsibilities; public/private market engagement; and rational/emotional decision making. These dichotomies position women within private space, leaving public (economic and political) spaces as masculine. The division between reproductive and productive labor and its value then reproduce within these distinct spaces and continue to define feminized spaces even when labor is paid, formal, and in public spaces (Folbre, 2001). They further ignore the ways that unpaid work is interlinked with paid work rather than distinct, in fact subsidizing the economy as it goes unnoticed (Antonopoulos, 2008). Analysis of private–emotional and public–rational division of spheres pervades political (Phillips, 1991) and economic (Folbre, 2001) literatures, as well as environmental analysis and ecofeminist literature (Shiva, 1989). While the literature varies greatly in how it approaches these dichotomies, care work literature in general challenges the positioning of women as the natural and obvious caretakers, due to biological, emotional dispositions (Beneria, 1979). Literature regarding agricultural production seldom bridges the production/reproduction dichotomy. This continues despite the recognition by the International Year of Family Farming and the Food and Agriculture Organisation (Garner and de la O Campos, 2014) regarding the need to understand family farm operations as an integrated unit dependent on reproductive and productive labor.

After a review of the various forms and understandings of care, Esquivel (2014) outlines the transformative approach to care aimed at 'restructuring' the underlying framework (citing Fraser, 1995). To do so, care is a concept that intertwines the economic benefits of care with the social and gender relations of social interaction, and the policies that shape the provision of care (Esquivel, 2014, p. 433). A transformative approach requires the 'Triple R Framework': recognize, reduce, and redistribute care (Elson, 2008 in Esquivel, 2013). According to Esquivel (2014), recognition involves both the accounting of time spent in care labor as well as the acknowledgment of the contributions of care work as a public good. Reduction of care work involves increasing efficiencies of time use through improved industries and technologies, while redistribution of care work can take place within the household, calling for men's greater involvement in caring activities (Esquivel, 2014).

Subsistence agriculture as care work

Building from this framing, a clear argument can be made that subsistence agriculture is care work. A look at family farms demonstrates that women's family obligations and productive activities are deeply interconnected. Even women's participation in off-farm employment can be motivated as support for the family farm. Further, the propensity of women farmers to claim the identity of farmer's wife (Brasier *et al.*, 2014) positions women within the household operation as unpaid support and denotes an aspect of care or affection. Further, housework in the Global South can be a completely different type of activity, where collecting water and food production depend on cultural traditions and household maintenance (Lee-Smith and Trujillo, 1992, p. 79).

In introducing a corporeal–material–socio-cultural framework, Allen and Sachs (2007) have already done the hard work of reconnecting the various 'food domains' of food work in order to support the emergence of a new field of 'feminist food studies' (citing Avakian and Haber, 2005). The domains that Allen and Sachs (2007, pp. 1–2) identify are 'women's food work in the labor market (material), their responsibility for food-related work in the home (socio-cultural), and their relationship with eating (corporeal).' By reconnecting the different aspects of women's food work – both productive and reproductive, in private and public spaces – Allen and Sachs (2007) bridge the gap between women's work laboring in fields for food production and their work preparing food in the house for consumption. As such, the nexus of food domains can be used to locate subsistence agriculture within care work, contextualizing women's relationship to climate change as a result.

In regard to the *material* domain, Allen and Sachs (2007) focus on women's paid labor in the production, processing, distribution, and retailing of the food industry. Lockie and Kitto (2000) identify the gap in connecting consumption and production as a system; value chain literature also contributes to this approach in seeking to connect actors and actions along supply networks (Barrientos and Ware Barrientos, 2002). While this work connects transnational systems of food production and situates women and gender within them, subsistence agriculture provides a space to move beyond formal labor in food and look at the connections on a smaller scale. Here, consumption and production is traced within a household and within communities along low-value market chains. Further, women's bodies in production become the space for this transference, as the *corporeal* is the traditional site for analysis. This also frames the socio-cultural motivations and regulations of participating in subsistence agriculture as a household responsibility and duty.

Within the *socio-cultural* realm in particular, women's decisions are family-oriented, choosing foods that will please others (especially husbands), that reflect well on their ability as mothers and caregivers, and that reproduce cultural norms (Allen and Sachs, 2007). While there is debate surrounding how and why women choose particular crops (Carr, 2008), the existing analysis on women's decision making around food preparation and diet/nutrition could provide insight. This discussion can prove to be especially useful for projects and policies aimed at diversifying women's agricultural production beyond traditional crops in order to improve empowerment through increased income.

As Allen and Sachs (2007) identify, the three domains of material, corporeal, and socio-cultural cannot be left in isolation. Through these connections, women's role in food preparation (long recognized as care labor) is not separate from understanding women's role in food production. Subsistence agriculture provides a substantial example of where the three intersect and provide the basis for understanding subsistence agriculture as care work.

Theories of care work introduce opportunities to reassess the value of women's participation in agriculture and feminized spaces within agriculture, which provide a deeper context for understanding the potential impacts of climate change. While the initial push and reassertion of women in care, and care as a feminized labor space, was from notions of 'natural' dispositions, the care work framework challenges these dichotomies (i.e. love *and* money coexist), and in doing so reassesses the value of care work with particular focus on its contributions as a public good. Starting from the point that subsistence agriculture is care work due to its position within a process of food provision, we can use this framework to understand the value of subsistence agriculture as a feminized space, reinterpreting gendered assessments of risk and vulnerability to climate change.

Care Work, Subsistence Agriculture and Climate Change

The care work framing is beneficial in understanding subsistence agriculture as well as women's position in response to climate change. Using Honduras as a case study, this framework provides a new analysis for understanding: (i) women's participation in subsistence agriculture as the result of socio-economic and gendered systems; and (ii) reassessing the value of subsistence agriculture and its benefits as a public good.

Moving beyond 'natural' disposition and material analysis

First, the care framework helps to understand how discourses on women's natural disposition to care for others have reproduced feminized spaces of labor – for example as nannies, nurses, or in other care industries – but that this disposition is also socially reproduced and not necessarily innate. Literature on care work references the initial push in Western contexts for women to be excluded from wage labor (Beneria, 1979) and involved in care work as based on biological or historical contexts (Folbre, 2001). However, there are 'discursive and material' consequences and explanations for this continued gendered delineation (England and Lawson, 2005, p. 78). Folbre (2001) offers two arguments against women's sole involvement in care work, or its natural fit. First, the transition in some economies to focus on brainpower over brawn has included an increased participation of women in paid labor activities, exemplifying that there are other processes informing (and creating alternatives to) women's participation in care (Folbre, 2001). Second, care work is not improved through subordination or lack of choice in participation, nor does choice happen outside 'moral values and social pressures' (Folbre, 2001, p. 7).

Using this framework to analyze subsistence agriculture enables movement beyond the natural disposition discourse of women farmers (Shiva, 1989), as well as material differences in assets in agricultural production (FAO, 2011). In doing so, this framing includes a socio-cultural context for the gendered divisions observed in Honduras. The working assumption in policy and research is that women in the western highlands region of Honduras are participating in subsistence crops (corn and beans) due to the material reasons of lack of assets or established market participation (Staritz and Reis, 2013). Including the consideration of gendered roles and responsibilities in the household, cultural values around subsistence crops, and the social relevance of these crops and their production, deepens the understanding of women's and men's engagement in agriculture. Policy and research can improve their approaches by not assuming that these divisions are natural or solely material, but rather are socially constructed and produced through social institutions and mechanisms beyond the market.

In addition, there is reason to believe that men and women contribute to the production of crops across the household; however, men are more likely to be responsible for high-value, market-oriented crop production than women (Carr, 2008). The care work approach brings into question the process that may have positioned women as more likely to participate in subsistence agriculture, as well as the role of choice to diversify to other crops if those material barriers were eliminated. While this does not suggest that there is a natural connection between women and subsistence agriculture, feminine identity development may be more connected to these crops and the environment than is currently acknowledged. Policy and program design must consider these questions given the apparent stark divisions between masculine and feminine spaces in Honduras, as well as other countries and regions.

This consideration is particularly important to address in climate change policy. First, we can discuss women's position as 'caretakers' for the environment, but within a social context that is not void of the moral or social obligations noted in the care work framework. Second, previous critiques of hurricane Mitch preparation, for example, included over-reliance on women to provide care for their families and others during recovery. Further, by not including subsistence agriculture in recovery calculations for drastic climate events, climate change policy risks understating the time burden and responsibilities that household care involves. Third, subsistence crops, including corn and beans in western Honduras, can be more susceptible to climate changes (Kawas *et al.*, 2011). By not accounting for gendered divisions in agriculture, policy can underestimate the situated resilience that women and men experience.

Within this framing, we can also start to reconceptualize vulnerability. First, a woman's participation in agriculture is no longer a demonstration of her natural inclination to be closer to the environment or solely because of her access to material resources. Rather, it is a result of the socio-cultural context that divides masculine and feminine responsibilities. This framing also opens discussions on the development of vulnerabilities shaped by masculinity and the concern that concrete heteronormative, gendered family ideals and identities minimize resiliency strategies and risk management. This framing further identifies that there is more at risk than material impacts of climate change, as the socio-cultural context has created these distinct identities. As demonstrated by the research on microcredit loans (Vonderlack-Navarro, 2010), strongly held gender identities can moderate impacts of interventions. As such, climate change policy and programs in Honduras and beyond must reframe gendered vulnerabilities to climate change.

Value of subsistence agriculture

Much of the literature on care work focuses on the differentiated value and attention given to work that happens in private spaces or that involves care. Feminist expansion of the term 'work' began to diversify sites of analysis beyond the formal work experience and into the home and other unregulated sites (England and Lawson, 2005, p. 78). In particular, the 'devaluation framework' documents the penalties of participating in feminized labor ('cost of care'), for the participation of both women and men in predominantly female occupations (England, 2005, p. 382). This process occurs through cultural ideas and underestimations of women (England, 2005). The cultural ideal is also applied to men, who are still perceived as breadwinners and deserving of higher pay (England, 2005). Thus, when women make up the majority of an occupation a cultural association is made and the relative wages of the entire occupation are lowered, regardless of education and other requirements (England, 2005).

Second, the framing of care as a public good recognizes that while there are direct recipients of its benefits, others receive indirect benefits through social return (England, 2005). Education is the classic example of how care work benefits society more greatly than direct market impact can measure. In addition, feminist analysis expands the concept of public good beyond the Marxist labor theory that value only goes to the employer (England, 2005). Care work creates emotional connections that reinforce reciprocity and trust, possibly eliciting a solidarity that allows groups to outcompete others (Folbre, 2001, p. 16). Since the value of public goods diffuses beyond the immediate beneficiaries, it makes the market value of care more difficult to assess (England, 2005). These two points lead to additional theoretical understandings of how women's labor and work that is considered feminine are devalued, often citing that intrinsic reward is prioritized over monetary compensation for this labor, despite the value that it provides both for the recipient as well as society (England, 2005).

Assessing the value of subsistence agriculture, both on market and in perception, is an important application of the care framework. The dual-market system positions export-oriented crop production as high-value, making participation in this market a high priority for development policy and programs despite the associated risk (Vanhaute, 2011). The western highlands of Honduras demonstrate that there remain significant barriers to smallholder production in these systems (Jansen *et al.*, 2006; Blandon *et al.*, 2009, p. 974), which disproportionately affect women due to cultural and intra-household gender inequalities (Speizer *et al.*, 2005; Casolo, 2009; Vonderlack-Navarro, 2010). This socio-economic context reinforces women's primary participation in 'low-value' markets, increases the risk of 'high-value' market participation (Vanhaute, 2011), and leaves women more vulnerable to shocks (de la O Campos and Garner, 2014).

Within this context, women participate in and create local food systems by selling or giving excess corn and bean production to friends, family, and neighbors. Previous research has analyzed the impact of local food systems in the USA, mainly in terms of community development and neoliberal resistance (Hinrichs and Allen, 2008), as well as women's agricultural contributions to their communities (Trauger *et al.*, 2009). Women's motivations in agriculture can be categorized as civically minded, both in terms of caring for the environment as well as through non-financial motivations (Jarosz, 2011). Women's agricultural impact and engagement of 'alternative' markets can serve to both question 'modern' and 'conventional' market systems, as well as the dichotomy between public and private returns (Trauger *et al.*, 2009).

The care work framework provides insights into this classification of low-value agriculture. First, women's primary participation in the low-value food networks could influence a devaluation of these market systems due to the application of cultural ideals. In Honduras, for example, the traditional identity of breadwinner is still very masculine (Vonderlack-Navarro, 2010), as are access and ownership to essential inputs, such as land (Casolo, 2009) and technologies (Torres Barragan, 2012). The application of this cultural ideal would prioritize masculine markets, which are also associated with greater profit (i.e. their productivity and impact can be measured), as men's earnings are for the household and women's activities are secondary. This view materializes the 'cost of care' for women and men participating in low-value agriculture. Building from this, the second point is that subsistence agriculture is either uncompensated through markets, or compensated by friends, family, and/or neighbors in informal markets and not necessarily in monetary units. As such, this production provides a service, both to the household and to the community. The assumption that intrinsic reward is compensation enough (England, 2005) also feeds into the devaluation of subsistence crops.

In the absence of a fuller understanding of the value that local food systems create, the care framework allows us to reframe vulnerability and agency in this context. First, the predicted volatility of international food markets (de la O Campos and Garner, 2014) increases the associated risk of participating in them (Vanhaute, 2011) as climates become less predictable. As such, subsistence agriculture can provide a risk management strategy through diversification of livelihoods and crops. It most likely already does, and deserves recognition for the role that it plays (Lee-Smith and Trujillo, 1992). In addition, we can predict that the relationships and social networks created and reinforced through the interactions of subsistence agriculture may also contribute support and information networks; both of which are critical in times of disaster recovery. This reframing of subsistence agriculture can also promote a shifting conception of women's vulnerability in rural areas, identifying agency within the context of subsistence agriculture, specifically, and care work more generally.

Conclusion

There is a need for a new approach to climate change. As demonstrated by the case in Honduras, it is essential to consider the nexus of gender and agriculture in climate change policy and action. However, these policies must move beyond the material understandings of gendered divisions in agriculture and in vulnerability to climate change. Instead, they must consider the socio-cultural context that shapes and produces gendered roles and responsibilities.

One step forward is to reframe subsistence agriculture as care work. Doing so can expand existing materialist explanations of the gender-differentiated impacts of climate change on agriculture as it accounts for socio-cultural context. In doing so, care work identifies the limitations of the assets-based solutions promoted in literature, as well as assertions of women's natural responsibility to care for, and connect to, the environment alongside of other care duties. Further, a care framework helps to re-evaluate the conceptions of vulnerability and work to identify agency by understanding the process of delineated spaces and reassessing the value of 'low-value' agriculture.

However, full analysis also needs to consider the limitations of applying the care framework to subsistence agriculture. As care is already a space that is undervalued and under-recognized, expanding the concept of care work *could* further marginalize women's participation in agriculture, consistent with the theory itself. Rather than using a care framework to promote greater attention to subsistence agriculture, framing this agricultural participation as such could result in its collapse with other domestic responsibilities. This inclusion into the domestic space may further limit policy and program attention to supporting women in agriculture, given the hesitation of policy to address intra-household contexts (Esquivel, 2014), despite international calls to recognize and value unpaid care and domestic work through formal institutions (e.g. SDG target 5.4; UN, undated).

Additionally, it is not clear if the care framing can substantially address the structural issues that feminist critiques of development, agriculture, and climate change present. This framing does offer opportunities to position the household within a global context, to better situate gender identity formation, and question dichotomies. However, despite certain exceptions, the existing literature minimally challenges heteronormative household composition, addresses intersectionality, offers solutions for the redistribution of care work, or addresses government inclusion at different levels. All of these are recommendations for promoting a feminist vision of food systems (Sachs and Patel-Campillo, 2014). Further, the attention to value does not automatically move the conversation beyond capital assets, and policy design may be inclined to reinforce and compensate subsistence agriculture like care work. As such, moving forward, we need better accounting of and inclusion of care work and agriculture, as well as a move beyond materialist connections of women and the environment.

The major strength of the care work approach is being able to reassess and reframe the value of subsistence agriculture, and women's care work, in order to improve policy design. The ultimate goal is to better address gender divisions of labor in policy solutions as well as to better incorporate care work into policies that address rural and agricultural contexts. The need for a new feminist framework is clear. Analyzing the opportunities of care work is one step towards achieving that.

References

Adger, W., Barnett, J., Brown, K., Marshall, N. and O'Brien, K. (2013) Culture dimensions of climate change impacts and adaptation. *Nature Climate Change* 3, 112–117.

Alaimo, S. (2009) Insurgent vulnerability: masculinist consumerism, feminist activism, and the science of global climate change. *Kinder, Kon, og Forskning, Denmark* (*Women, Gender, and Research*) 3, 22–35.

Allen, P. and Sachs, C. (2007) Women and food chains: the gendered politics of food. *International Journal of Sociology of Agriculture and Food* 15, 1–2.

Alston, M. (2013) Introducing gender and climate change: research, policy and action. In: Alston, M. and Whittenbury, K. (eds) *Research, Action and Policy: Addressing the Gendered Impacts of Climate Change*. Springer, New York, pp. 3–16.

Antonopoulos, R. (2008) The unpaid care work – paid work connection. *Working Paper No. 541*. The Levy Economics Institute, New York.

Avakian, A.V. and Haber, B. (2005) Feminist food studies: a brief history. In: Avakian, A.V. and Haber, B. (eds) *From Betty Crocker to Feminist Food Studies: Critical Perspectives on Women and Food*. University of Massachusetts Press, Amherst, MA, pp. 1–28.

Barrientos, A. and Ware Barrientos, S. (2002) Extending social protection to informal workers in the horticulture global value chain. *Social Protection Discussion Paper Series No. 0216*. World Bank, Washington, DC.

Beneria, L. (1979) Reproduction, production and the sexual division of labor. *Cambridge Journal of Economics* 3, 203–225.

Blandon, J., Henson, S. and Cranfield, J. (2009) Small-scale farmer participation in new agri-food supply chains: case of the supermarket supply chain for fruit and vegetables in Honduras. *Journal of International Development* 21, 971–984.

Brasier, K., Sachs, C.E., Kiernan, N.E., Trauger, A. and Barbercheck, M.E. (2014) Capturing the multiple and shifting identities of farm women in the north eastern United States. *Rural Sociology* 79, 283–309.

Carr, E.R. (2008) Men's crops and women's crops: the importance of gender to the understanding of agricultural and development outcomes in Ghana's central region. *World Development* 36, 900–915.

Casolo, J. (2009) Gender levees: rethinking women's land rights in north eastern Honduras. *Journal of Agrarian Change* 9, 392–420.

CGIAR (2015) Estrategia de cambio climatico de Honduras se prepara para el future con planes regionals de adaptacion. CGIAR Research Program on Climate Change, Agriculture and Food Security (CCAFS), 17 June 2015.

Daly, M. (2002) Care as a good for social policy. *Journal of Social Policy* 31, 251–270.

Daly, M. and Lewis, J. (2000) The concept of social care and the analysis of contemporary welfare states. *The British Journal of Sociology* 51, 281–298.

de la O Campos, A.P. and Garner, E. (2014) Women's resilience to food price volatility: a policy response. Food and Agriculture Organization, Rome.

Deere, C.D. and Leon, M. (2001) Who owns the land? Gender and land-titling programmes in Latin America. *Journal of Agrarian Change* 1, 440–467.

Delaney, P.L. and Shrader, E. (2000) Gender and post-disaster reconstruction: the case of hurricane Mitch in Honduras and Nicaragua. *Draft report for LCSPG/LAC Gender Team*. World Bank, Washington, DC.

DFID (2004) Adaptation to climate change: making development disaster-proof. Key Sheet 06. Department for International Development, London.

Diaz-Ambrona, C.G.H., Gigena, R. and Mendoza, C.O. (2013) Climate change impacts on maize and dry bean yields of smallholder farmers in Honduras. *Revista Iberoamericana de Estudios de Desarrollo* (*Iberoamerican Journal of Development Studies*) 2, 4–22.

Elson, D. (2008) The three Rs of unpaid work: recognition, reduction and redistribution. Paper presented at the Expert Group Meeting on Unpaid Work, Economic Development and Human Well-Being. United Nations Development Programme, New York.

Enarson, E. and Meyreles, L. (2004) International perspectives on gender and disaster: differences and possibilities. *The International Journal of Sociology and Social Policy* 24, 49–93.

England, K. and Lawson, V. (2005) Feminist analyses of work: rethinking the boundaries, gendering and spatiality of work. In: Nelson, L. and Seager, J. (eds) *A Companion to Feminist Geography*. Blackwell, Malden, MA, pp. 77–92.

England, P. (2005) Emerging theories of care work. *Annual Review of Sociology* 31, 381–399.

England, P. and Folbre, N. (1999) The cost of caring. *The Annals of the American Academy of Political and Social Science* 561, 39–51.

Esquivel, V. (2013) Care in households and communities. *Background Paper on Conceptual Issues*. Oxfam Research Reports. Oxfam Publication, Oxford, UK.

Esquivel, V. (2014) What is a transformative approach to care, and why do we need it? *Gender and Development* 22, 423–439.

FAO (2011) Women in agriculture: closing the gender gap for development. *State of Food and Agriculture*. Food and Agriculture Organization, Rome.

Ferrant, G., Pesando, L.M. and Nowacka, K. (2014) Unpaid care work: the missing link in the analysis of gender gaps in labor outcomes. *Issues Paper*. OECD Development Centre, Organisation for Economic Co-operation and Development, Paris.

Folbre, N. (2001) *The Invisible Heart: Economics and Family Values*. The New Press, New York.

Fraser, N. (1995) From redistribution to recognition? Dilemmas of justice in a 'postsocialist' age. *New Left Review* I, p. 212, July–August.

Gaard, G. (2015) Ecofeminism and climate change. *Women's Studies International* 49, 20–33.

Gabrielsson, S. (2015) Gender Matters: adaptive capacities to climate variability and change in the Lake Victoria Basin. In: Inderberg, T.H., Eriksen, S., O'Brien, K. and Sygna, L. (eds) *Climate Change Adaptation and Development: Transforming Paradigms*. Routledge, Abingdon, UK, pp. 83–97.

Garner, E. and de la O Campos, A.P. (2014) Identifying the 'family farm': a discussion of the concepts and definitions. *ESA Working Paper No. 14–10*. Food and Agriculture Organization, Rome.

Gumucio, T. and Tafur Rueda, M. (2015) Influencing gender-inclusive climate change policies in Latin America. *Journal of Gender, Agriculture and Food Security* 1, 42–61.

Hinrichs, C.C. and Allen, P. (2008) Selective patronage and social justice: local food consumer campaigns in historical context. *Journal of Agricultural and Environmental Ethics* 21, 329–352.

IFPRI (2013) Evaluation of Feed the Future Intervention ACCESO-Honduras: preliminary baseline results. International Food Policy Research Institute, Washington, DC.

Jansen, H.G.P., Rodriguez, A., Damon, A., Pender, J., Chenier, J. and Schipper, R. (2006) Determinants of income-earning strategies and adoption of conservation practices in hillside communities in rural Honduras. *Agricultural Systems* 88, 92–110.

Jarosz, L. (2011) Nourishing women: toward a feminist political ecology of community supported agriculture in the United States. *Gender, Place and Culture* 18, 307–326.

Jiggins, J., Samanta, R.K. and Olawoye, J.E. (1998) Chapter 9 – Improving women farmers' access to extension services. In: Swanson, B.E. (ed.) *Improving Agricultural Extension*. Food and Agriculture Organization, Rome.

Kawas, N.O., Ferman, E. and Wiese, K.K. (2011) Evaluacion de la vulnerabilidad a nivel municipal en Honduras. *Working Document, IISD and Tegucigalpa*. United Nations Development Programme, Ottawa, Canada.

Kidder, T., Mapandi, Z. and Ortega, H. (2014) Not 'women's burden': how washing clothes and grinding corn became issues of social justice and development. *Gender and Development* 22, 495–513.

Kristjanson, P., Bryan, E., Bernier, Q., Twyman, J., Meinzen-Dick, R. *et al.* (2017) Addressing gender in agricultural research for development in the face of a changing climate: where are we and where should we be going? *International Journal of Agricultural Sustainability* 15, 482–500.

Lee-Smith, D. and Trujillo, C.H. (1992) The struggle to legitimize subsistence: women and sustainable development. *Environment and Urbanization* 4, 77–84.

Lockie, S. and Kitto, S. (2000) Beyond the farm gate: production–consumption networks and agri-food research. *Sociologia Ruralis* 40, 3–19.

Morton, J.F. (2007) The impact of climate change on smallholder and subsistence agriculture. *Proceedings of the National Academy of Sciences of the United States of America* 104, 19680–19685.

Nelson, V., Meadows, K., Cannon, T., Morton, J. and Martin, A. (2002) Uncertain predictions, invisible impacts, and the need to mainstream gender in climate change adaptations. *Gender and Development* 10, 51–59.

Phillips, A. (1991) *Engendering Democracy*. Pennsylvania State University Press, University Park, PA.
Razavi, S. (ed.) (2009) *The Gendered Impacts of Liberalization: Towards 'Embedded Liberalism'?* Taylor & Francis, New York, pp. 1–36.
Roehr, U. (2007) Gender, climate change and adaptation: introduction to the gender dimensions. *Background Paper for the Both Ends Briefing*. United Nations Environment Programme, Nairobi, Kenya.
Rowlands, J. (1997) *Questioning Empowerment: Working with Women in Honduras*. Oxfam Publication, Oxford, UK.
Sachs, C. and Patel-Campillo, A. (2014) Feminist food justice: crafting a new vision. *Feminist Studies* 40, 396–410.
SAG (undated) Politica para la equidad de genero en el agro hondureno: 1999–2015. Secretaria de Agricultura y Ganaderia, Honduras.
Shiva, V. (1989) *Staying Alive: Women, Ecology and Development*. Zed Books, London.
Speizer, I., Whittle, L. and Carter, M. (2005) Gender relations and reproductive decision making in Honduras. *International Family Planning Perspectives* 31, 131–139.
Staritz, C. and Reis, J.G. (2013) Global value chains, economic upgrading, and gender: case studies of the horticulture, tourism, and call center industries. World Bank, Washington, DC.
Torres Barragan, R. (2012) Análisis con enfoque de género de la cadena de valor de la papa, en los municipios de Intibucá y La Esperanza, Intibucá, Honduras. Graduate project presented in partial fulfilment of Grado Academico de Licenciatura, Department of Environment and Development. Universidad de Zamorano, Honduras.
Trauger, A., Sachs, C., Barbercheck, M., Brasier, K. and Kiernan, N.E. (2009) 'Our market is our community': women farmers and civic agriculture in Pennsylvania, USA. *Agriculture, Food and Human Values* 26, 43–55.
Tronto, J.C. (2012) Democratic care politics in an age of limits. In: Razavi, S. and Staab, S. (eds) *Global Variations in the Political and Social Economy of Care*. Routledge, New York, pp. 29–40.
Tschakert, P. and Tuana, N. (2013) Situated resilience: reframing vulnerability and security in the context of climate change. In: Dugard, J., St Clair, A.L. and Gloppen, S. (eds) *Climate Talk: Rights, Poverty and Justice*. Juta and Company, Cape Town, South Africa, pp. 75–96.
UN (undated) Sustainable Development Goal 5: Gender Equality. United Nations, New York. Available at: www.un.org/sustainabledevelopment/gender-equality/ (accessed 8 February 2021).
Vanhaute, E. (2011) From famine to food crisis: what history can teach us about local and global subsistence crisis. *The Journal of Peasant Studies* 38, 47–65.
Vonderlack-Navarro, R. (2010) Targeting women versus addressing gender in microcredit: lessons from Honduras. *Affilia* 25, 123–134.
World Bank (2009) Honduras: country note on climate change aspects in agriculture. World Bank, Washington, DC.

8 Climate Change, Livelihoods and Domestic Violence in Indonesia

J. Eastin

Introduction

It is conventional wisdom that climate change poses greater risks to the marginalized and impoverished. Access to subsistence resources declines when climate instability disrupts local livelihoods, and when minimal household savings and a lack of livelihood alternatives undermine adaptation and crisis coping. Women comprise a relatively greater share of the world's poor and, as other studies in this volume demonstrate, face a gender-specific and inequitable constellation of climate change-related burdens and vulnerabilities. These can range from challenges associated with personal and familial care, to sustenance provision, to threats to personal safety and security, among others.

Domestic violence, already among the most critical threats to women worldwide, stands to worsen as climate change progresses. Rising climatic instability disrupts local livelihoods, exacerbates economic hardship, and undermines subsistence, all of which can aggravate psychological pressures that manifest in aggression towards intimate partners and other household members (Gonzales de Olarte and Llosa, 1999). In the context of discriminatory gender norms and laws that reinforce male domination, women become especially vulnerable targets (Phillips and Jenkins, 2016; Enarson *et al.*, 2018). Because poverty is a key factor in explaining patterns of domestic violence, and because climate change disproportionally affects the poor and those who rely on climatically sensitive industries, the threat should be acute in economically underdeveloped and agriculturally dependent regions (Hertel and Rosch, 2010).

This analysis estimates the impact of climate-induced livelihood shocks on the prevalence of domestic violence in 12 provinces in Indonesia, a nation reliant on agriculture (especially rice production) for food and income security. Examining interannual variability in the Standardized Precipitation and Evapotranspiration Index (SPEI), a drought severity estimator that records the effects of precipitation and temperature on water demand, enables specification of these livelihood shocks. Measurements are taken during December, the core month of the Indonesian rice-planting season, to better approximate the livelihood impact of climate variability (Caruso *et al.*, 2016). Domestic violence data is drawn from the National Violence Monitoring System Indonesia (NVMS, undated), an innovative project led by the Coordinating Ministry for Human Development and Cultural Affairs that gathers information on violent incidents from local newspapers across the archipelago. Coupling these data with detailed information on the timing and severity of climatic disturbances enables estimation of the effects of livelihood shocks on incidents of domestic violence.

 Gender, Climate Change and Livelihoods: Vulnerabilities and Adaptations (eds J. Eastin and K. Dupuy)
DOI: 10.1079/9781789247053.0008

Findings indicate that positive and negative anomalies from long-term December SPEI averages lead to increases in reported incidents of domestic violence in the following year. In other words, when rice-planting seasons are both drier-than-average *and* wetter-than-average, women are more likely to be victimized. A reduction in the average December SPEI by one standard deviation (std. dev.) in $year_t$ leads to a 9.8% increase in domestic violence reports in $year_{t+1}$. Similarly, a 1 std. dev. increase in the December SPEI elevates reports of violence by 7.3% the following year. Wealth also appears to be a mitigating factor, as the positive impact of SPEI anomalies on incidents of domestic violence rises in proportion to the share of people experiencing poverty in a given province.

The following section reflects on the theoretical and empirical linkages between household economics, livelihoods, and domestic violence. Attention is then directed towards evaluating the impact of climate change and natural disasters on livelihoods – especially agrarian livelihoods – and the implications of these phenomena for the incidence of domestic violence in vulnerable regions. The following sections detail data, methodological strategy, and findings, and the final section concludes with a discussion of key policy implications and directions for future research.

Economics, Livelihoods and Domestic Violence

A household's economic status has long received attention as a predictor of domestic and intimate partner violence. Poor women are more likely to experience abuse and less capable of escaping from it. Numerous factors contribute to and sustain this relationship, including: a lack of resources available to flee abusive partners; an inability to participate in the formal labor force because of the physical, emotional, and psychological impacts of abuse; burdens of care for children and the elderly; the potential loss of material possessions; and the threat of homelessness, among others. One study finds that 63% of female welfare recipients in a large US state have suffered domestic violence (Tolman and Raphael, 2000), while survey statistics indicate that cutting a woman's annual income by two-thirds can make her five times more likely to be victimized by an abusive partner (BOJ, 1994).

Economists have linked the problem of domestic violence to disparities in household bargaining relations between women and men (Gelles, 1976; Manser and Brown, 1980; Tauchen *et al.*, 1991; Srinivasan and Bedi, 2007; Anderson and Eswaran, 2009; Eswaran and Malhotra, 2011). A woman's economic contribution to the household and her control over household resources determines her power; the greater the level of control, the greater her ability to exit and the greater the costs (to the husband) of her doing so. When these contributions decline, women are more vulnerable to victimization and possess fewer alternatives for escape. Empirical research offers support for these contentions: Farmer and Tiefenthaler (1997) find violence in marriage to be proportional to women's access to income and financial support outside the home; the greater the access, the lower the likelihood of violence. Similarly, Aizer (2010) finds that increases in female labor force opportunities reduce female hospitalizations for assault. Anderberg *et al.* (2016) find the converse to also be true; when women lose their jobs and income, they are more likely to be abused.

While feminist scholars have criticized these models for gender stereotyping and for inadequate attention to the impact of social norms, customs, and culture in household decision making (Katz, 1997), prior scholarship indicates that livelihoods, especially livelihood shocks, do play a role in domestic violence. Studies in the natural disaster literature find that rapid losses of income, assets, and means of subsistence – the core components of livelihood – correspond with an increased risk of abuse in affected households (Sety *et al.*, 2014). This threat is pronounced in homes experiencing poverty, as a lack of savings, access to relief resources, and minimal livelihood alternatives increase the shock's duration and magnitude (True, 2012). For example, Neumayer and Plümper (2007) evaluate the gendered impact of natural disasters on human life expectancy across a sample of 141 countries from 1981–2002. They find these events to have greater effects on women's mortality than men's, because greater familial burdens and local gender norms can make women more likely to die during a disaster, and because a relative lack of social and economic resources makes them more vulnerable to violence in the aftermath. Case-based analyses generate similar conclusions: following severe flooding in New Zealand in 2004, local social service agencies reported a trebling in their domestic violence caseloads (Houghton, 2009). The primary reason cited in interviews with victims was 'financial stress' induced by lost earnings, employment, childcare, and possessions (Houghton, 2009). Similarly, in the southernmost portion of Mississippi, among the poorest regions in the USA, reports of physical assault against female intimate partners almost doubled (from 4.2% to 8.3%) in the aftermath of hurricane Katrina while reports of violence against men were unchanged (Schumacher *et al.*, 2010).

This study is part of a growing literature addressing the relationship between climate change and domestic and intimate partner violence (Abiona and Koppensteiner, 2018; Epstein *et al.*, 2020; Cools *et al.*, 2020). Similar to prior works, the present study employs climate data to approximate the impact of livelihood shocks, a method which provides insight into the direct impact of climate variability, and enables us to avoid endogeneity concerns associated with other data sources (Epstein *et al.*, 2020). This study also builds on prior research in important ways: it is the first to apply these analytical techniques to a population outside of sub-Saharan Africa. While analyses of this region are important and contribute much to our understanding of climate change and domestic violence, it is necessary to evaluate causal processes in other cultural and ecological contexts to assess their generalizability. The present study also differs in that it finds that both negative and positive climate anomalies (drier- *and* wetter-than-normal conditions) increase the incidence of domestic violence. This finding might be, at least in part, an artifact of case selection, as the production of rice, the principal subsistence crop in Indonesia, is sensitive to both droughts and floods (Lansigan *et al.*, 2000; Rosenzweig *et al.*, 2002; Kotera *et al.*, 2014). Finally, this study is the first to incorporate data on evapotranspiration – the process by which water is conveyed to the atmosphere via soil evaporation and plant transpiration – into the analysis. Including this phenomenon can enhance the accuracy of climate change approximation because climate change is characterized by rising temperature variability, which increases the evapotranspiration process and magnifies drought severity and duration, especially in the tropics (Bathiany *et al.*, 2018).

Hypothesis

This study argues that climate shocks in Indonesia elevate the incidence of domestic violence via their impact on agriculture and agrarian livelihoods. Those relying on agriculture as a primary income source in Indonesia – approximately 41% of the population – suffer when climatic stress diminishes earnings through job loss and reduced crop yields (BPS, 2020). The impact can reduce food security, especially for subsistence farmers, but also for the broader population when scarcity elevates local food prices. Food already consumes 70% of household budgets for half the population, with rice comprising the largest share – over 25% of total household expenditures for the poorest quintile (McCulloch, 2008; BPS, 2020). Thus, even minor reductions in yields or inflation in local rice markets can have dire effects. These impacts are anticipated to exacerbate social and psychological pressures – stress, anxiety, depression, trauma, substance abuse – commonly associated with domestic and intimate partner violence, which in turn should increase its incidence within affected regions (Foa *et al.*, 2000).

This study also anticipates climate-induced livelihood shocks to increase the prevalence of domestic violence by reducing the financial capabilities of victims to flee. Data from the Indonesian Peasant Union indicate that women comprise approximately 70–80% of the agricultural workforce in Indonesia (quoted in FAO, 2019).[1] Thus, when climate change harms agricultural livelihoods, women can experience disproportionate losses. These losses should not only reduce women's household economic contributions, and hence their bargaining power, but also their ability to draw on the economic resources necessary to escape abusive situations. Aggravating factors include a relative (to men) lack of control over productive household assets and technologies, limited access to external credit markets, difficulties associated with dependent care, and cultural institutions that promote female submission, discourage the reporting of violence to authorities, and restrain female participation in the formal labor force (Hayati *et al.*, 2011).

Taken together, these insights inform the following hypothesis:

> H1: Climate-induced livelihood shocks increase the incidence of domestic violence.

Data and Methods

In order to test the impact of livelihood shocks on the incidence of domestic violence, this study draws on cross-sectional time-series data from 12 provinces in Indonesia from 2005–2015. The selected time period and provinces correspond to the availability of data on domestic violence.

Dependent variable

The dependent variable in this analysis, *Domestic Violence*, is drawn from the NVMS (undated), a project led by the Coordinating Ministry for Human Development and Culture and the Korea Economic and Peace-building Transitions Trust Fund. NVMS extracts incident data on domestic and other forms of violence from printed local newspapers across the archipelago and collates them into monthly and annual reports.[2]

One drawback of relying on newspaper data, especially when compared to surveys, is that actual incidents of domestic violence are likely to be significantly undercounted. Only the most egregious violations should receive even local media attention, and as numerous studies have demonstrated, victims of violence, especially in highly patriarchal societies, are unlikely to report their experiences for fear of retribution, social stigmatization, and economic costs, among other factors (Bennett *et al.*, 2011; Nilan *et al.*, 2014). However, while the data set contains only a fraction of the actual violent incidents, data collection procedures across provinces are consistent, which reduces the likelihood of unit-specific bias. Moreover, the time-series nature of data collection can be advantageous for modeling causal relationships with a time-sensitive component to the causal process.

NVMS data on domestic violence record incidents of physical and sexual violence committed by one family member against another living under the same roof. While victims and perpetrators in the data set are not delineated by gender, it is safe to assume that in the vast majority of cases, perpetrators are men and the victims are women or children. A survey taken in 2016 with 9000 respondents from 24 provinces finds levels of violence against women in Indonesia at or above global averages. These data suggest that one-third (33.4%) of women aged 15–64 have been victimized by physical or sexual violence in their lifetime, with 94% experiencing these outcomes in the past 12 months (UNFPA, 2017). Indeed, *Komnas Perempuan*, the National Commission on Violence Against Women, reports domestic violence to be the most common form of violence against women in the country (Jones, 2018). While no similar statistics exist for men, the culture in Indonesia – the largest Muslim nation in the world – remains highly patriarchal, indicating that female on male abuse is unlikely to be common, and perhaps even more unlikely to be reported (Aisyah and Parker, 2014; Riyani, 2016; Putra *et al.*, 2019).

Table 8.1 includes the names of all provinces in this study; the cumulative incidence of domestic violence in each province from 2005–2015; the average annual provincial population; the average annual number of violent incidents normalized by population; and the average annual percentage of the population living in poverty. As Table 8.1 indicates, exposure to domestic violence ranges from a low of 0.82 to a high of 5.16 incidents per 100k people. By this metric, domestic violence is most prevalent in Papua (5.16) and Papua Barat (3.47), which are also the poorest two provinces in the sample.

Independent variable

The primary independent variable, *SPEI Anomaly*, measures deviations from long-term climate trends in order to approximate the impact of climate change. Data for this variable are drawn from the SPEI, a multi-scalar drought index designed to enable drought monitoring and analysis across a wide range of climates. The SPEI is statistically robust, and is computed with data

Table 8.1. Domestic violence statistics in Indonesia (2005–2015).

Province	Incidents (total)	Population (ann. avg.)	Incidents/100k people (ann. avg.)	% Poverty (ann. avg.)
Aceh	418	4,609,091	0.82	20.42
Jakarta region	1136	9,200,000	1.12	3.87
Kalimantan Barat	1064	4,809,091	2.01	9.33
Kalimantan Tengah	420	2,245,455	1.70	6.98
Lampung	716	7,400,000	0.88	17.55
Maluku	316	1,472,727	1.95	24.29
Maluku Utara	142	984,521	1.31	9.10
Nusa Tenggara Barat	512	4,390,909	1.06	20.31
Nusa Tenggara Timur	1124	4,836,364	2.11	22.65
Papua	1547	2,727,273	5.16	33.73
Papua Barat	332	870,077	3.47	31.55
Sulawesi Tengah	418	2,527,273	1.50	17.09

on precipitation and potential evapotranspiration (PET). The SPEI data in this analysis are taken from the Global SPEI database (v2.5), which provides monthly time-series data on drought conditions across the globe at 0.5 degrees of spatial resolution (Beguería *et al.*, undated_b). The data on precipitation and PET that are used to calculate the SPEI come from the Climatic Research Unit of the University of East Anglia. All SPEI measurements in this study are taken during December, the core month of the Indonesian rice-planting season.

The inclusion of PET into the calculation of SPEI provides the index with a key advantage over other drought indices such as Standardized Precipitation Index, or over the use of precipitation data alone, because it enables one to account for effects of temperature variability and temperature extremes on the outcome (Tirivarombo *et al.*, 2018). This accounting is particularly important for assessing the impacts of droughts on livelihoods in the context of rising climate instability, as rising temperatures can increase stress on crops and vegetation via evapotranspiration.[3]

SPEI Anomaly is calculated at the Indonesian provincial level to maintain consistency with other variables in the analysis. Grid cell estimates from the Global SPEI database were aggregated to the provincial level using ArcGIS. If the majority of a given grid cell is located within a province, the estimate is included in that province's mean calculation. The raw SPEI estimates were transformed into statistical deviations from long-term December means. The formula for calculating this measure is:

$$(x_{it} - x_i)/\sigma_i \quad (8.1)$$

where x_i is the mean December SPEI estimate over the years 1990–2015 in province *i*, x_{it} is the SPEI estimate for province *i* at time *t*, and σ_i is the std. dev. of province *i*. This process results in a standardized *SPEI Anomaly* estimate with a mean of 0, a std. dev. of 0.98, and a range from −3.05–2.78. The duration of Global SPEI data used to calculate this variable enables a more robust measure of average climate conditions than simply calculating a panel mean. Figure 8.1 offers a visualization of SPEI variability among the provinces under study in this analysis. The 12 provinces included in the data set are each highlighted according to their average annual deviations from long-term SPEI means between 2005–2015. The darker provinces include those with average scores above the SPEI mean (wetter), while the lighter provinces have average scores below the mean (drier).

The expectation is that both positive and negative deviations from long-term SPEI means in the month of December will have adverse implications for food security and agrarian livelihoods. While the effects of below-average precipitation are somewhat obvious, too much rainfall can be equally destructive. In Indonesia, the majority of rice planting takes place from October–December, just prior to the peak monsoon months of December/January. When newly planted seedlings are exposed to an excess of moisture – higher precipitation and lower evapotranspiration – during this time period, the result can undermine plant rooting and

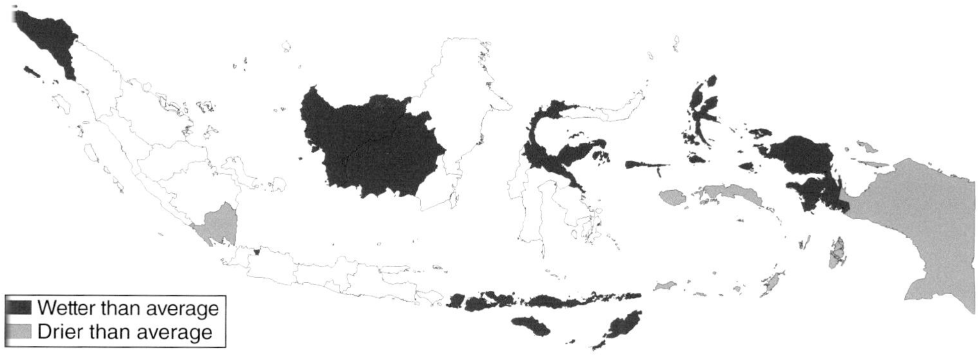

Fig. 8.1. Standardized Precipitation and Evapotranspiration Index (SPEI) variability for Indonesia. Author's own diagram.

diminish tillering, which reduces overall yields, and can result in a failed harvest (De Datta, 1981). If this excess moisture occurs in the form of flooding, then the likelihood of a harvest failure rises even further. This scenario is becoming increasingly common in Indonesia as climate variability alters traditional monsoon patterns (e.g. see Naylor *et al.*, 2007; Jakarta Post, 2014). Beyond direct reductions in income and food availability for rice growers, a reduced harvest can also lead to higher prices in local markets, which reduces food access for the broader population. This effect is magnified when delays in the rice harvest result in a pushback or cancellation of a second, smaller dry season crop, which is normally planted in April (Caruso *et al.*, 2016).

Control variables

The control variables included in the model reflect those commonly associated with domestic violence: *Poverty* is calculated as the percentage of a provincial population that lives in impoverished conditions; *Unemployment* records provincial unemployment rates; and *Education* is the mean number of years of schooling for an average person in a given province. The author anticipates that *Poverty* and *Unemployment* will positively correlate with the outcome, and that *Education* will exhibit an inverse relationship. The natural logarithm of provincial *Population* is also included as a scalar for population size. Finally, the author includes a variable designed to capture pre-existing levels of gender equality in a given province/year, as provinces with greater gender empowerment might experience lower levels of domestic violence, all things equal. The *Gender Empowerment Measure* (*GEM*) is designed to measure a province's level of gender empowerment. It is calculated using metrics of female decision making power in society, female labor force participation, and the proportion of legislative/parliamentary seats held by women. Data for all control variables were taken from *Badan Pusat Statistik* (BPS), the Indonesian governmental agency responsible for collecting statistical information. Table 8.2 provides descriptive statistics for all variables in this analysis.

Methodology, Results and Discussion

This study employs a series of negative binomial regressions to estimate the impact of livelihood shocks in year_t on domestic violence in year_{t+1}. Negative binomial models are appropriate for evaluating over dispersed count data; they are similar to Poisson regressions in that they employ a linear combination of independent variables to predict the log of the outcome, but they contain an extra parameter (alpha) that corrects for the over dispersion (Cameron and Trivedi, 2013). All models use province-clustered standard errors to account for the possibility that observations within each cluster are correlated over time. In robustness testing, the author experimented with a variety of alternative specifications to evaluate model sensitivity, including running Poisson models with random effects,

Table 8.2. Information and descriptive statistics on each variable included in the model.

Variable	Description	Source	Time	Obs	Mean	SD	Min	Max
SPEI Anomaly	Standardized Deviation of the mean December SPEI[a] measures across the period 1990–2015	Global SPEI database	1990–2015	312	0	0.98	−3.05	2.78
Domestic Violence	Number of domestic violence incidents/year	NVMS[b]	2005–2015	132	61.7	51.1	1	233
Population (ln)	Natural logarithm of provincial population	*Badan Pusat Statistik* (BPS)	2005–2015	132	14.92	0.75	13.27	16.12
Unemployment	Provincial unemployment rate	BPS	2005–2015	131	6.71	2.91	2.41	15.25
Education	Mean years of schooling	BPS	2005–2015	132	7.68	1.24	5.59	10.90
Poverty	Provincial poverty rate	BPS	2007–2015	108	18.07	9.56	3.48	40.78
GEM	Gender Empowerment Measure	BPS	2010–2015	72	63.80	7.70	47.97	79.93

[a]SPEI = Standardized Precipitation and Evapotranspiration Index
[b]NVMS = National Violence Monitoring System Indonesia

and unit and time fixed effects, ordinary least squares, and negative binomial models with alternative measures for each of the control variables. The author also ran models on using a population-normalized version of the dependent variable. None of these techniques alters the substance of the findings. Additionally, Akaike Information Criterion (AIC) and Bayesian Information Criterion (BIC) scores indicate a suitable model selection among these various alternatives.

Table 8.3 provides the coefficients and Z-scores for the variables included in each of the model estimations. The *SPEI Anomaly* variable and its quadratic term are included to evaluate the impact of drier-than-average and wetter-than-average years on incidents of domestic violence. Model 1 reports results from regressions run on the full panel, with only the *SPEI Anomaly* and *Population* variables included in the equation. Model 2 adds controls for *Unemployment*, *Poverty*, and *Education*. Model 3 adds an additional control for *GEM*. Reduced data availability for control variables in Models 2 and 3 diminishes the sample sizes.

The results in Table 8.3 provide support for the hypothesis that climate-induced livelihood shocks elevate the incidence of domestic violence in Indonesia. The coefficients on the quadratic form of *SPEI Anomaly* in Models 1–3 are all positive and highly statistically significant ($p \leq 0.05$), suggesting that deviations from long-term climatic means in either direction – drier and wetter – positively influence the outcome. These results are robust to a variety of model specifications and controls. While drought-induced livelihood shocks have previously been associated with increases in domestic violence, this finding suggests that an abundance of precipitation during the rice-planting season in Indonesia can generate similar outcomes.

Figure 8.2 presents a visualization of the curvilinear relationship between *SPEI Anomaly* and *Domestic Violence*. The vertical axis displays the predicted number of violent incidents in an 'average' province/year, while the horizontal displays standardized deviations from long-term mean SPEI score. As this figure indicates, a decrease in SPEI by 1 std. dev. increases the predicted number of violent incidents by 9.8% (from 41 to 45), while an increase in SPEI of the same amount increases the predicted outcome by 7.3%. A 2 std. dev. decrease (increase) in SPEI has a much larger effect, increasing the predicted number of incidents by 36.6% (31.7%).

This study also finds a robust relationship between poverty and domestic violence, as the coefficient on *Poverty* is positive and statistically

Table 8.3. Negative binomial regressions.

	Model 1	Model 2	Model 3
SPEI[a] *Anomaly*	−0.021	0.079	−0.008
	(0.71)	(1.97)**	(0.15)
SPEI Anomaly^2	0.086	0.074	0.070
	(4.15)***	(3.35)***	(1.96)**
Population (ln)	0.557	0.693	0.588
	(3.80)***	(5.53)***	(4.50)***
Poverty		0.033	0.029
		(2.14)**	(2.16)**
Unemployment		0.022	−0.025
		(0.32)	(0.36)
Education		−0.069	−0.034
		(0.39)	(0.22)
GEM[b]			−0.002
			(0.15)
Constant	−4.374	−6.749	−5.116
	(1.95)*	(2.89)***	(2.57)**
N	131	108	72
Provinces	12	12	12
$P > \chi^2$	0	0	0

[a]SPEI = Standardized Precipitation and Evapotranspiration Index
[b]GEM = Gender Empowerment Measure
Negative binomial regressions with province-clustered robust standard errors. Dependent variable is *Domestic Violence*. Displayed are variable coefficients with Z-scores in parentheses. *p<0.1; **p<0.05; ***p<0.01

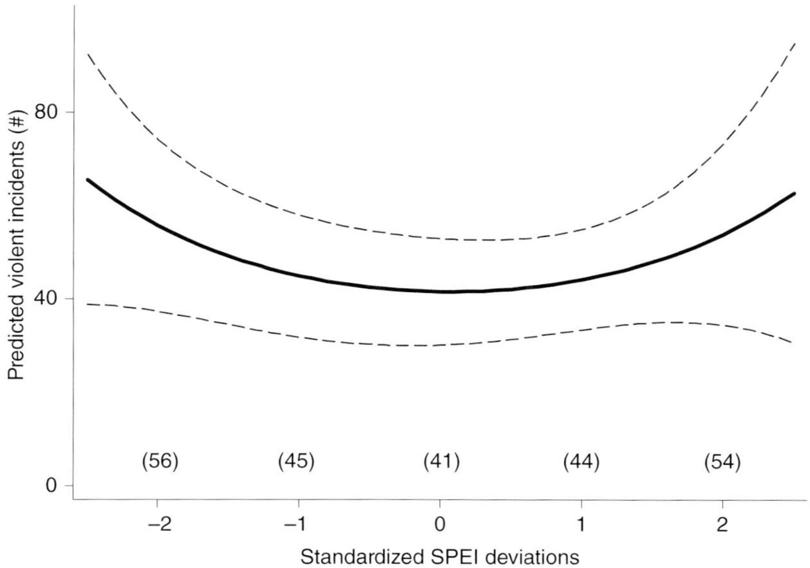

Fig. 8.2. Marginal effects plot to show the relationship between *SPEI Anomaly* [Standardized Precipitation and Evapotranspiration Index] and *Domestic Violence*. Dotted lines represent 95% confidence intervals. Author's own diagram.

significant. Specifically, an increase by 1 std. dev. in the percentage of provincial population living in poverty (from 18–28%) increases the predicted incidence of domestic violence by a substantial degree (39%). Moreover, while poverty appears to have a direct impact on the outcome, the logic of H1 also suggests that poverty can magnify the impact of climate change on livelihoods and consequently, on domestic violence. Economically marginalized populations are more vulnerable to climate change than those with greater resource access: they are more likely to live in environmentally exposed locations, participate in livelihoods that are more sensitive to adverse environmental changes, and possess fewer resources to adapt. To determine if climate change interacts with poverty, the author constructed models that include an interaction term, *SPEI Anomaly × Poverty*. Table 8.4 presents the coefficients and Z-scores from a sparse model that includes only population and poverty (Model 4), and one that incorporates the remaining control variables (Model 5).

Table 8.4. Negative binomial regressions.

	Model 4	Model 5
SPEI[a] *Anomaly*	0.257	0.129
	(3.16)***	(1.64)
SPEI Anomaly × Poverty	−0.009	−0.008
	(3.11)***	(2.63)***
Population (ln)	0.724	0.595
	(5.48)***	(4.27)***
Poverty	0.039	0.028
	(3.58)***	(2.03)**
Unemployment		−0.014
		(0.20)
Education		−0.077
		(0.50)
GEM[b]		−0.001
		(0.05)
Constant	−7.649	−4.974
	(3.75)***	(2.34)**
N	108	72
Provinces	12	12
$P > \chi^2$	0	0

[a]SPEI = Standardized Precipitation and Evapotranspiration Index
[b]GEM = Gender Empowerment Measure
Negative binomial regressions with province-clustered robust standard errors. Dependent variable is *Domestic Violence*. Displayed are variable coefficients with *Z* scores in parentheses. *p<0.1; **p<0.05; ***p<0.01

The results in Table 8.4 indicate that poverty does indeed interact with climate change to the detriment of victims of domestic violence. The higher the level of provincial poverty, the more likely that climate variability during the rice-planting season will increase the incidence of domestic violence the following year. Figure 8.3 displays a marginal effects plot of this interaction, with the plot lines representing absolute values of *SPEI Anomaly*, and the small circles representing incidents of *Domestic Violence* in the data set. Here we can see that greater deviations from long-term climatic means – and the ensuing effects on livelihoods and food security – increase the threat of victimization to a greater extent in poorer areas than in wealthier ones. The convergence of these lines at around 35% provincial poverty most likely reflects the influence of outlying observations rather than a substantive causal process.

Surprisingly, none of the additional control variables – *Education*, *Unemployment*, and *GEM* – exert a statistically significant impact on *Domestic Violence* in any of the models presented. One reason for this finding is that the poverty variable might be capturing some of these variables' effects. Indeed, when *Poverty* is removed from the regression equation, the *Education* and *Unemployment* variables are often statistically significant, and exhibit a negative and positive relationship, respectively, with domestic violence, which is in line with the author's expectations. Removing *Poverty* does not alter the sign or statistical significance of the *GEM* variable in any of the models presented.

Conclusion

This study uses data from the Global SPEI database and the NVMS to model the relationship between climate change and domestic violence in Indonesia. It finds that positive and negative deviations from long-term climate averages, when occurring in December – the core month of the Indonesian rice-planting season – increase the incidence of domestic violence in the following year. This relationship likely reflects the negative impact of climate shocks on agricultural sectors and livelihoods, an outcome which aggravates the emotional and psychological preconditions

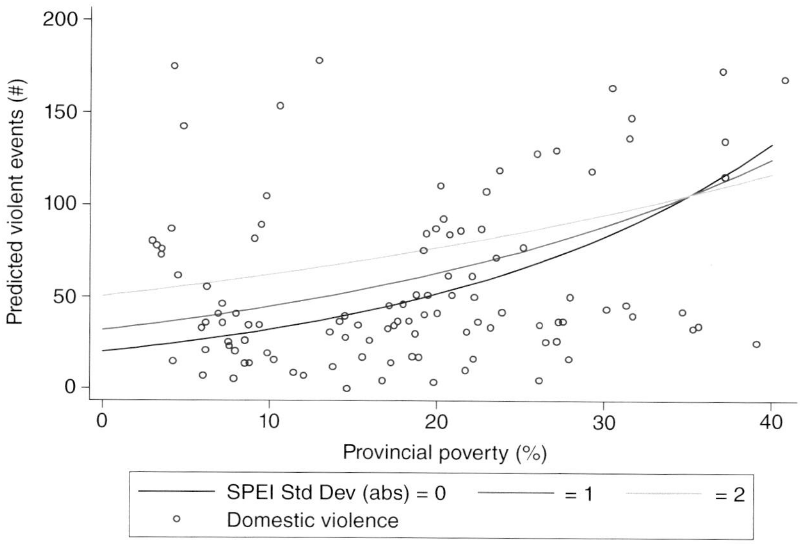

Fig. 8.3. Marginal effects plot of the interaction between *SPEI Anomaly* [Standardized Precipitation and Evapotranspiration Index] and *Domestic Violence*, conditional on the level of poverty in an Indonesian Province. Author's own diagram.

for domestic violence and abuse, disproportionately diminishes women's bargaining power in the household, and reduces women's ability to escape abusive situations. These effects are especially prominent in areas with higher levels of poverty, further illustrating the economic dimension of the causal process.

The findings presented here suggest a need for additional resources to be devoted to providing women with safe spaces to flee abusive partners, to encouraging women to report instances of victimization, and to mandating higher levels of policing of domestic violence and the prosecution of offenders. While many jurisdictions would benefit from such efforts, they are especially crucial in less economically developed regions where predominant livelihoods are vulnerable to climate shocks.

In the future, research should focus more closely on evaluating the contextual conditions that exacerbate or reduce the impact of climate shocks on domestic violence. The present analysis finds poverty to be a strong mediating factor, although there are other cultural, economic, political, legal, and social characteristics of society that could plausibly play a role. Similarly, scholars should also evaluate these dynamics in other states and regions. While it is plausible to expect the processes discussed here to be operable to some degree in other regions, empirical tests could provide confirmatory evidence that these conclusions are indeed generalizable. Finally, a crucial area of future research concerns appropriate policy interventions to combat the scourge of domestic violence. While providing women with resources to escape abuse, education, and more robust policing are certainly important steps to halt the spread of domestic violence, policy makers should also consider providing additional assistance with livelihood diversification training in order to help mitigate the economic stresses that lead to such violence, as well as to facilitate economic development and increase resilience to climate change.

Notes

[1] While formal labor statistics place this figure at around 26%, much of women's work in Indonesian agricultural sectors remains under-reported because their contributions are not always classified as formal employment.

[2] For detailed description on incident identification and other data collection procedures, see NVMS Methodology (undated).

[3] For detailed description of the advantages of SPEI in relation to alternative measures, as well as an overview of its computational method, see Beguería *et al.* (undated_a) 'About the SPEI.'

References

Abiona, O. and Koppensteiner, M.F. (2018) The impact of household shocks on domestic violence: evidence from Tanzania. *Discussion Papers in Economics 16/14*. Division of Economics, School of Business, University of Leicester, UK.

Aisyah, S. and Parker, L. (2014) Problematic conjugations: women's agency, marriage and domestic violence in Indonesia. *Asian Studies Review* 38, 205–223.

Aizer, A. (2010) The gender wage gap and domestic violence. *The American Economic Review* 100, 1847–1859.

Anderberg, D., Rainer, H., Wadsworth, J. and Wilson, T. (2016) Unemployment and domestic violence: theory and evidence. *Economic Journal (London)* 126, 1947–1979.

Anderson, S. and Eswaran, M. (2009) What determines female autonomy? Evidence from Bangladesh. *Journal of Development Economics* 90, 179–191.

Bathiany, S., Dakos, V., Scheffer, M. and Lenton, T.M. (2018) Climate models predict increasing temperature variability in poor countries. *Science Advances* 4: eaar5809.

Beguería, S., Latorre, B., Reig, F. and Vicente-Serrano, S.M. (undated_a) About the SPEI. Available at: www.spei.csic.es/home.html (accessed 1 February 2021).

Beguería, S., Latorre, B., Reig, F. and Vicente-Serrano, S.M. (undated_b) Global SPEI Database. Available at: www.spei.csic.es/database.html (accessed 1 February 2021).

Bennett, L.R., Andajani-Sutjahjo, S. and Idrus, N.I. (2011) Domestic violence in Nusa Tenggara Barat, Indonesia: married women's definitions and experiences of violence in the home. *The Asia Pacific Journal of Anthropology* 12, 146–163.

BOJ (1994) Domestic violence: violence between intimates, selected findings from violence against women. *National Crime Victimization Survey Report, NCJ-149259*. Bureau of Justice Statistics. Available at: www.bjs.gov/content/pub/pdf/vbi.pdf (accessed 1 February 2021).

BPS (2020) Statistics Indonesia. Badan Pusat Statistik. Available at: www.bps.go.id/ (accessed 1 February 2021).

Cameron, A.C. and Trivedi, P.K. (2013) Regression analysis of count data, 2nd edn. *Econometric Society Monograph No. 53*. Cambridge University Press, Cambridge, UK.

Caruso, R., Petrarca, I. and Ricciuti, R. (2016) Climate change, rice crops, and violence. *Journal of Peace Research* 53, 66–83.

Cools, S., Flatø, M. and Kotsadam, A. (2020) Rainfall shocks and intimate partner violence in sub-Saharan Africa. *Journal of Peace Research* 57, 377–390.

De Datta, S.K. (1981) *Principles and Practices of Rice Production*. Wiley, New York.

Enarson, E., Fothergill, A. and Peek, L. (2018) Gender and disaster: foundations and new directions for research and practice. In: Rodríguez, H., Donner, W. and Trainor, J.E. (eds) *Handbook of Disaster Research*. Springer, New York, pp. 205–223.

Epstein, A., Bendavid, E., Nash, D., Charlebois, E.D. and Weiser, S.D. (2020) Drought and intimate partner violence towards women in 19 countries in sub-Saharan Africa during 2011–2018: a population-based study. *PLoS Medicine* 17:e1003064.

Eswaran, M. and Malhotra, N. (2011) Domestic violence and women's autonomy in developing countries: theory and evidence. *Revue Canadienne d'Economique* (*The Canadian Journal of Economics*) 44, 1222–1263.

Farmer, A. and Tiefenthaler, J. (1997) An economic analysis of domestic violence. *Review of Social Economy* 55, 337–358.

FAO (2019) Country gender assessment of agriculture and the rural sector in Indonesia. Food and Agriculture Organization, Rome. Available at: www.fao.org/3/ca6110en/ca6110en.pdf (accessed 1 February 2021).

Foa, E.B., Cascardi, M., Zoellner, L.A. and Feeny, N.C. (2000) Psychological and environmental factors associated with partner violence. *Trauma, Violence and Abuse* 1, 67–91.

Gelles, R. (1976) Abused wives: why do they stay? *Journal of Marriage and Family* 38, 659–668.

Gonzales de Olarte, E. and Llosa, P.G. (1999) Does poverty cause domestic violence? Some answers from Lima. In: Morrison, A.R. and Loreto Biehl, M. (eds) *Too Close to Home: Domestic Violence in the Americas*. Inter-American Development Bank, Washington, DC. Distributed by the Johns Hopkins University Press, Baltimore, MD, pp. 35–49.

Hayati, E.N., Högberg, U., Hakimi, M., Ellsberg, M.C. and Emmelin, M. (2011) Behind the silence of harmony: risk factors for physical and sexual violence among women in rural Indonesia. *BMC Women's Health* 11:52.

Hertel, T.W. and Rosch, S.D. (2010) Climate change, agriculture, and poverty. *Applied Economic Perspectives and Policy* 32, 355–385.

Houghton, R. (2009) Chapter 8 – 'Everything became a struggle, absolute struggle': post-flood increases in domestic violence. In: Enarson, E. and Dhar Chakrabarti, P.G. (eds) *Women, Gender and Disaster: Global Issues and Initiatives*. Sage, New Delhi, pp. 22–111.

Jakarta Post (2014) Floods ruin thousands of hectares of crops. *Jakarta Post, Wednesday 31 December 2014, online*. Available at: www.thejakartapost.com/news/2014/12/31/floods-ruin-thousands-hectares-crops.html (accessed 1 February 2021).

Jones, B. (2018) Why stigma against victims of domestic violence persists in Indonesia. *The Conversation*, 22 September 2018. Available at: www.theconversation.com/why-stigma-against-victims-of-domestic-violence-persists-in-indonesia-103177 (accessed 1 February 2021).

Katz, E. (1997) The intra-household economics of voice and exit. *Feminist Economics* 3, 25–46.

Kotera, A., Nguyen, K.D., Sakamoto, T., Iizumi, T. and Yokozawa, M. (2014) A modeling approach for assessing rice cropping cycle affected by flooding, salinity intrusion, and monsoon rains in the Mekong Delta, Vietnam. *Paddy and Water Environment* 12, 343–354.

Lansigan, F.P., de los Santos, W.L. and Coladilla, J.O. (2000) Agronomic impacts of climate variability on rice production in the Philippines. *Agriculture, Ecosystems and Environment* 82, 129–137.

Manser, M. and Brown, M. (1980) Marriage and household decision making: a bargaining analysis. *International Economic Review* 21, 31–44.

McCulloch, N. (2008) Rice prices and poverty in Indonesia. *Bulletin of Indonesian Economic Studies* 44, 45–64.

Naylor, R.L., Battisti, D.S., Vimont, D.J., Falcon, W.P. and Burke, M.B. (2007) Assessing risks of climate variability and climate change for Indonesian rice agriculture. *Proceedings of the National Academy of Sciences of the United States of America* 104, 7752–7757.

Neumayer, E. and Plümper, T. (2007) The gendered nature of natural disasters: the impact of catastrophic events on the gender gap in life expectancy, 1981–2002. *Annals of the Association of American Geographers* 97, 551–566.

Nilan, P., Demartoto, A., Broom, A. and Germov, J. (2014) Indonesian men's perceptions of violence against women. *Violence Against Women* 20, 869–888.

NVMS (undated) National Violence Monitoring System, Indonesia. Available at [intermittent access]: www.snpk.kemenkopmk.go.id/Home/Index?lang=enandranddo=320c451a-ee0d-4e69-81e8-37a7d00721a8anduserid=41076266 (accessed 1 February 2021).

NVMS Methodology (undated) – National Violence Monitoring System, Indonesia. Available at [intermittent access]: www.snpk.kemenkopmk.go.id/About/Methodology?lang=enandranddo=61d1ea19-b8e3-4c30-a9b6-3da7cf5d5f31anduserid=30437701 (accessed 1 February 2021).

Phillips, B. and Jenkins, P. (2016) Gender-based violence and disasters: South Asia in comparative perspective. In: Racioppi, L. and Rajagopalan, S. (eds) *Women and Disasters in South Asia: Survival, Security and Development*. Routledge, London.

Putra, I.G.N.E., Pradnyani, P.E. and Parwangsa, N.W.P.L. (2019) Vulnerability to domestic physical violence among married women in Indonesia. *Journal of Health Research* 33, 90–105.

Riyani, I. (2016) The silent desire: Islam, women's sexuality and the politics of patriarchy in Indonesia. Unpublished PhD Dissertation submitted to The University of Western Australia, Perth, Australia. Available at: www.pdfs.semanticscholar.org/503d/9173c5b7383b0b5afa50f760963fdf60ce56.pdf (accessed 1 February 2021).

Rosenzweig, C., Tubiello, F.N., Goldberg, R., Mills, E. and Bloomfield, J. (2002) Increased crop damage in the US from excess precipitation under climate change. *Global Environmental Change* 12, 197–202.

Schumacher, J.A., Coffey, S.F., Norris, F.H., Tracy, M., Clements, K. and Galea, S. (2010) Intimate partner violence and Hurricane Katrina: predictors and associated mental health outcomes. *Violence and Victims* 25, 588–603.

Sety, M., James, K. and Breckenridge, J. (2014) Understanding the risk of domestic violence during and post natural disasters: literature review. In: Roeder, L. (ed.) *Issues of Gender and Sexual Orientation in Humanitarian Emergencies*. Springer, New York, pp. 99–111.

Srinivasan, S. and Bedi, A.S. (2007) Domestic violence and dowry: evidence from a south Indian village. *World Development* 35, 857–880.

Tauchen, H.V., Witte, A.D. and Long, S.K. (1991) Domestic violence: a non-random affair. *International Economic Review* 32, 491–511.

Tirivarombo, S., Osupile, D. and Eliasson, P. (2018) Drought monitoring and analysis: Standardised Precipitation Evapotranspiration Index (SPEI) and Standardised Precipitation Index (SPI). *Physics and Chemistry of the Earth, Parts A/B/C* 106, 1–10.

Tolman, R.M. and Raphael, J. (2000) A review of research on welfare and domestic violence. *The Journal of Social Issues* 56, 655–682.

True, J. (2012) *The Political Economy of Violence Against Women*. Oxford University Press, Oxford, UK.

UNFPA (2017) New survey shows violence against women widespread in Indonesia. United Nations Population Fund, 10 May 2017. Available at: www.unfpa.org/fr/node/16015 (accessed 1 February 2021).

9 Gender and Climate-Smart Agriculture in Africa

A.A. Babugura

Introduction

Climate-smart agriculture (CSA) is offered as a solution to reducing greenhouse gas (GHG) emissions and addresses the vulnerability of agriculture to the impacts of climate change. It is an approach that promises to increase productivity and improve food security in a sustainable manner, while also building resilience to climate change, contributing to mitigation, and enhancing rural livelihoods. It encompasses a wide range of technologies and practices, such as: improved water management techniques; agroforestry for carbon sequestration; conservation tillage; integrated crop–livestock management; weather forecasting; and early warning system. Given Africa's widely reported fragile agricultural systems – subject to a combination of climatic and non-climatic drivers and stressors – African governments have embraced CSA as a solution to transforming Africa's agriculture (FAC, 2011; Williams *et al.*, 2015).

Africa's agricultural systems are deemed to be the most vulnerable to the impacts of climate change on the basis that they remain rain-fed and underdeveloped (Moyo and Pereira, 2019). In addition, the rural economy is strongly rooted in agriculture with the majority of farmers being small scale with inadequate financial resources or access to infrastructure (Moyo and Pereira, 2019). Approximately 85% of Africa's agricultural output is produced by small-scale farmers, a large percentage of which are women (AGRA, 2018). Women consist of nearly 50% of the labor force in Africa's agriculture (FAO, 2015).

Gender shapes the lives of both female and male African farmers. The term 'gender' is widely defined as the socially constructed differences between women and men, girls and boys, and the unequal power relationships that result (UN Commission on the Status of Women, 2008). Gender in Africa's agricultural sector is well documented, and the sector is engulfed by major gender disparities. Studies (e.g. FAO, 2011; Quisumbing *et al.*, 2014; UN Women, 2016; UN Women, 2018) have demonstrated that compared to male farmers, female farmers have less access to productive resources, financial capital, advisory services, information, technologies, and markets. Female farmers are also often excluded from participating in decision making processes, undermining their ability to contribute to decisions. Despite the constraints they face, African female farmers continue to play a vital role in improving Africa's agricultural production and food security at national, community, and household levels. They undertake the bulk of the work to produce, process, and market food. Women are at the front line of agriculture. As food producers and stewards of natural resources in their households and communities, women are also well positioned to be effective agents of change for both mitigation and adaptation, if given the opportunity to do so (UN Commission on the Status of Women, 2008).

 Gender, Climate Change and Livelihoods: Vulnerabilities and Adaptations (eds J. Eastin and K. Dupuy)
DOI: 10.1079/9781789247053.0009

Under a changing climate, constraints faced by female farmers are expected to further hinder their ability to adapt, as well as heighten, the existing gender inequalities in agriculture. These inequalities have major implications for successful adoption and implementation of CSA in African countries. For African countries to reap the intended benefits of CSA, several challenges at the heart of agricultural development debates need to be overcome. Among the critical issues, is closing the existing gender gaps in agriculture. If gender is carefully re-evaluated in Africa's CSA interventions, there is potential to address existing gender inequalities and provide a range of benefits for both female and male African farmers in adapting to climate change. Doing so will also contribute to improved agricultural production and food security in African countries, while working towards the second Sustainable Development Goal: 'End hunger, achieve food security and improved nutrition, and promote sustainable agriculture.' Gender-responsive CSA also promises to yield important economic, social, and environmental benefits for African countries. Effective gender mainstreaming in CSA is therefore critical to ensuring that existing gender inequalities are not reinforced. Without effective gender mainstreaming, CSA will become a distant dream and another empty slogan.

'Gender mainstreaming' is defined as:

> A process of assessing the implications for women and men of any planned action, including legislation, policies or programmes, in all areas and at all levels. It is a strategy for making women's as well as men's concerns and experiences an integral dimension of the design, implementation, monitoring and evaluation of policies and programmes in all political, economic and societal spheres so that women and men benefit equally and inequality is not perpetuated. The ultimate goal is to achieve gender equality.
>
> (UNDP, 2006)

Considering the above context, the interrelated issues of gender and CSA within the African context are explored in this chapter. The importance and goal of mainstreaming gender into CSA is emphasized. The chapter draws on knowledge from CSA good practice and innovative approaches to highlight some successes and lessons learned from African countries. Having set the context, the following section presents an overview of gender within the African context. This section is followed by a brief overview of the CSA concept. Merging the two concepts in the section that follows, attention is given to gender considerations and CSA within the African context. Thereafter, opportunities for gender-sensitive actions in CSA within the African context are discussed. Gender-equitable CSA best practices in Africa are then presented, followed by lessons learned from best practices. The chapter concludes by emphasizing the importance of gender in CSA within the African context.

Gender in the African Context

Understanding gender is critical to appreciating its importance in climate change-related interventions such as CSA, as well as in other spheres of development. There is a tendency to interpret gender as a term that translates directly into 'women' or 'girls.' It is important to note that gender is not about women and girls; it is not interchangeable with 'women' and not an alternative to the word 'sex.' Gender is about social norms and power structures that affect the lives and opportunities available to different groups of men and women, boys, and girls (Kangas *et al.*, 2014). It is about socially ascribed roles, responsibilities, access, and control over resources and relations between women and men, girls and boys.

In all its diverse displays, gender has been one of the central organizing principles of African societies. Most traditional African societies are patriarchal. This is a social system of male dominance and female subordination (Bassey and Bubu, 2019). Men hold power and dominate in roles of family and political leadership, moral authority, social privilege, and control of property. Women are governed by the dictates of culturally defined customs and principles. These dictates infiltrate all levels of society and social institutions, including civil and customary legal systems. This situation makes legal systems less able to realize the rights afforded to all citizens (Mama, 2001; Bassey and Bubu, 2019). Customary laws, for example, afford women minor status, while customary practices often violate the physical and emotional integrity of women (Mama, 2001). In cases where states are divided along the tropes of civil and customary legal systems, the execution of both are deeply gendered (Mama, 2001).

Culturally defined customs and principles that govern female–male relations, present challenges to gender equality in traditional African societies. For many, it is unthinkable to oppose culture and cultural structures. Gender roles and obligations are well defined with clear boundaries and are taught to children at an early age. Girls and boys are socialized into the gender roles expected of them, their responsibilities, and positions in the home and in society. Females are second-class citizens. They are excluded in crucial decision making. Their responsibilities are centered on the home; they are expected to be submissive home keepers. Women produce agricultural crops, process and prepare food, work for wages in agriculture, collect firewood and water, care for family members, and maintain the home. While their activities are critical for the well-being of the household, they are not in economically active employment (Bassey and Bubu, 2019). Males, on the other hand, are heads of the household: they foster discipline in the home; they are official custodians of land and lineage; and they make broad management decisions of land allocation, labor organization, cropping/animal rearing patterns, and income expenditures. The call for gender equality therefore stems from acknowledging the disadvantaged positions in which women and girls find themselves due to pre-existing inequalities. To effectively eradicate gender inequality in African societies, culturally defined customs and principles that govern female–male relations must be considered.

Given that gender is socially constructed, the way gender roles are defined in Africa can be challenged. There is a need to constantly question, review, and demystify gender stereotypes that are culturally constructed to discourage women and girls from effectively participating in the development of countries on the African continent. Gender equality is a fundamental human right enshrined in the Universal Declaration of Human Rights (1948). It is a right that should be enjoyed by both females and males. However, in traditional African societies where norms reflect deeper social structures sustained and strengthened by social institutions, gender equality is complex and sensitive. In some cases, men will see it as encouraging disobedience, eroding traditions, stripping away status, and creating confusion and friction across generations and between women and men. These are not the intentions for gender equality; the call for gender equality is about the interests, needs, and priorities of both women and men, girls and boys, being equally valued and considered, recognizing the diversity within society.

As a powerful driver for growth and social justice (UN Women, 2014), gender equality is one of the fundamental requirements for combating climate change (UNFPA and WEDO, 2009; AWGGCC, 2017); hence, a prerequisite for successful adoption of CSA. In African countries where gender is a complex and sensitive issue in traditional societies, women and girls continue to face profound gender inequalities which do great harm to individuals and nations (UNDP, 2016).

The Concept of Climate-Smart Agriculture

CSA is defined as agriculture that sustainably increases productivity and resilience (adaptation), reduces/removes GHGs (mitigation), and enhances achievement of national food security and development goals (broader development) (FAO, 2010). It is envisioned to achieve a 'triple win': food security, mitigation against avoidable climate change, and adaptation to inevitable climate shifts. CSA integrates the three dimensions of sustainable development (economic, social, and environmental), and consists of three main pillars:

- sustainably increasing agricultural productivity and incomes;
- adapting and building resilience to climate change; and
- reducing and/or removing GHG emissions, where possible.

Neufeldt *et al.* (2011) point out that CSA can also have different meanings depending on the scale at which it is being applied. At the local scale, for example, it could mean providing opportunities for higher production through improved management techniques such as more targeted use of fertilizers. At national scale it could signify providing a framework that incentivizes sustainable management practices, while at global scale it could imply setting rules for the global trade of biofuels. For smallholder farmers in developing countries, the concept infers opportunities for greater food security and increased income combined with greater resilience, which to them may be greater incentive for adopting CSA than its mitigation opportunities.

The technical foundation on which CSA principles are based is not new. Traditionally, African farmers have used practical techniques, such as intercropping, mulching, crop rotation, integrated crop–livestock management, conservation agriculture (CA), agroforestry, improved grazing, organic farming, ecoagriculture, and improved water management (Boto *et al.*, 2012). While these techniques are not new, they are documented as suitable for adapting under CSA (see Table 9.1 for comparison). CSA can however be distinguished from the above related concepts in that: (i) it encompasses multiple goals; and (ii) it delivers benefits, potentially, for all three strands of sustainable development (economic, social, and environmental).

Since its launch in 2009 by the Food and Agriculture Organization (FAO) of the United Nations, the concept of CSA has gained attention at international and national levels, growing into a central model for agricultural development. It has captured the interest of various heads of states, regional organizations, national and international non-governmental organizations (NGO), and private institutions, as well as civil society (FAO, 2012). During the UN Framework Convention on Climate Change (UNFCCC) COP17 climate negotiations in Durban, South Africa, CSA was widely promoted as the future of African agriculture and as part of the solution to climate change in African countries.

Africa's agricultural sector is already constrained by challenges such as degraded ecosystems, declining soil fertility, land tenure insecurity for smallholder farmers, poor access to markets, inadequate funding, and inadequate infrastructure development (Williams *et al.*, 2015). Climate change is expected to exacerbate these challenges, severely compromising food security, economic growth, and livelihoods for millions of women, men, girls, and boys (Lobell *et al.*, 2011; Barkhordarian *et al.*, 2013; IPCC, 2014). African leaders have therefore embraced CSA as a solution to transforming Africa's agriculture.

According to Rosenstock *et al.* (2019), about US$1 billion has already been committed to climate-smart programming in Africa. Most African governments have formed CSA task forces such as the Africa Climate-Smart Agriculture Alliance, the East African Regional Climate-Smart Agriculture Alliance, and the West African Regional Alliance (Rosenstock *et al.*, 2019). In its 23rd African Union Summit (Malabo, June 2014), the African Heads of State and Government committed to addressing agriculture and climate change-related issues. They endorsed the NEPAD program on agriculture and climate change, which included components on gender empowerment, intensified support to smallholder farmers, and the establishment of an African CSA Coordination Platform. This was adopted as the Africa Climate-Smart Agriculture Vision 25X25; it aims to support at least 25 million farm households in practicing CSA by 2025 (NEPAD, 2019). At the 3rd Africa Climate-Smart Agriculture Alliance Forum, held in March 2019 in Dakar, Senegal, delegates agreed to accelerate the adoption of CSA. In 2015, the Common Market for Eastern and Southern Africa (COMESA) launched the COMESA Climate-Smart Agriculture Partnership aimed at working with governments to launch national CSA Programs. The concept of CSA is undoubtedly appealing, particularly if intended benefits are realized.

Gender Consideration in Climate-Smart Agriculture in the African Context

Reasons for considering gender in CSA initiatives in Africa are multiple and compelling. All farmers – female and male – require fair access to resources and opportunities to adapt and build resilience to climate change. However, in most African countries, female farmers, especially those in rural communities, face various constraints that heighten their vulnerabilities to the impacts of climate change. As noted above, gender is one of the central organizing principles of African societies. Structures of gender are deeply embedded in Africa's agricultural sector. They are an interlocking mixture of cultural norms, beliefs, and practices. Roles, rights, and

Table 9.1. Existing concepts and principles on which climate-smart agriculture (CSA) is based.

Concept	Definition	Principles
Ecoagriculture	Sustainable agriculture and associated natural resource management systems that embrace and simultaneously enhance productivity, rural livelihoods, ecosystem services, and biodiversity. It comprises a wide range of systems and practices that integrate productivity goals with the provision of ecosystem services (IISD, 2004).	• Enhance rural livelihoods • Conserve or enhance biodiversity and ecosystem services • Develop more sustainable and productive agricultural systems (Scherr and McNeely, 2009)
Agroforestry	Land use systems and technologies where woody perennials (trees, shrubs, palms, bamboos, etc.) are used on the same land management unit as agricultural crops and/or animals, either in some form of spatial arrangement or temporal sequence. In agroforestry systems, there are both ecological and economical interactions between the different components (Esterhuyse, 1994).	• Eradicate hunger through basic systems of pro-poor food production • Reduce rural poverty through market-driven, locally led tree cultivation systems that generate income and build assets • Promote pro-poor food production in disadvantaged areas based on agroforestry methods of soil fertility replenishment and land regeneration • Conserve biodiversity technologies, innovative institutions, and better policies • Protect watershed services • Enable rural poor to adapt to climate change and to benefit from emerging carbon markets, through tree cultivation
Conservation agriculture	Conservation agriculture (CA) is an approach that encourages the proactive and sustainable use of agricultural natural resources. CA aims to ensure the effective and sustainable use and management of natural resources through minimum disturbance of the soil. CA integrates the management of soil, water, and biological resources to maintain and enhance land productivity and reduce the level of risk.	• Disturb soil as little as possible • Keep soil covered • Mix and rotate crops (International Institute of Rural Reconstruction and the African Conservation Tillage Network, 2005)
Organic farming	A holistic production management system that avoids use of synthetic fertilizers, pesticides, and genetically modified organisms, minimizes pollution of air, soil, and water, and optimizes the health and productivity of interdependent communities of plants, animals, and people (Nemes, 2009).	• Increase long-term soil fertility • Control pests and diseases without harming environment • Ensure that water stays clean and safe • Use resources which farmer already has, so farmer needs less money to buy farm inputs • Produce nutritious food, feed for animals, and high quality crops to sell at good price

Continued

Table 9.1. Continued.

Concept	Definition	Principles
Sustainable agriculture	'The management and conservation of the natural resource base and the orientation of technological and institutional change in such a manner as to ensure the attainment and continued satisfaction of human needs for the present and future generations' (Ongley, 1996).	• Conserve resources • Environmentally non-degrading • Technically appropriate • Economically and socially acceptable • Conservation tillage, biodiversification (e.g. intercropping and crop rotation), improved crop varieties, use of animal manure, complementary use of organic fertilizer, and soil and water conservation structures

entitlements in production, division of labor, expectations, obligations, and responsibilities of females and males are governed by gender. Gender determines people's participation in agricultural production. It determines their constraints in time, mobility, and resources in performing their roles. It determines issues of land rights, access to technologies, capital, labor, support, services, and credit. Gender creates obstacles for Africa's female farmers and affects their ability to effectively participate in the agriculture sector (Doss, 2011; FAO, 2011).

In Africa, women in agriculture are said to receive only 7% of agricultural extension services and less than 10% of credit offered to small-scale farmers. Based on evidence from Burkina Faso, Kenya, Nigeria, and Zambia, women are less likely to use purchased inputs such as fertilizers and improved seeds, and are less likely to make use of mechanical tools and equipment. Within livestock farming systems, gender gaps in access to the service and input delivery systems in livestock production have been noted (Njuki and Sanginga, 2013; Mulema *et al.*, 2017). At the same time, women in Africa play a key role in improving agricultural productivity and food security in farming communities. They work as farmers on their own account, as unpaid workers on family farms in crop and livestock production, and as paid or unpaid laborers on other farms and agricultural enterprises (Doss, 2011; Quisumbing *et al.*, 2014; Jost *et al.*, 2016).

Women's roles in food production and natural resource management have helped them to become more successful in facilitating climate mitigation and adaptation strategies (UN Commission on the Status of Women, 2008). FAO (2011) highlights that closing the gender gap in agriculture would raise total agricultural output in developing countries by 2.5–4% and reduce the number of hungry people by 12–17% globally. These figures are the equivalent of 100 to 150 million people. Neglecting gender in agriculture on the other hand carries substantial costs. It is for these reasons that studies often focus on women and emphasize the importance of bringing their challenges to the forefront.

Paying attention to gender is therefore essential for meeting the objectives of CSA within the African context. It can promote greater gender equality, improve food security, promote economic and social development while enhancing climate change adaptation and mitigation. It can open opportunities for cost-effective and transformative approaches to CSA development in Africa. This requires a careful re-evaluation of gender in the current agricultural practices and any proposed CSA initiatives. Differences in women's and men's realities relating to participation in agriculture, in terms of their labor contribution, responsibilities, priorities, and access to resources and services in different African countries must be continuously re-evaluated. This is bearing in mind that the realities of women and men in agriculture vary within different countries and change depending on economic and social forces that transform gender.

A study conducted in two rural communities, Nkhamenya and Kabudula, in Malawi revealed that labor productivity is closely linked to the division of tasks and duties in households (Murray *et al.*, 2016). This study established that caring for children and food crops are typically women's tasks. Production of the tobacco crop is undertaken by both women and men. Selling of cash crops is however the men's task, even though the processing is done by all household members (women, men, girls, and boys). Girls are required to assist with household tasks and food production, while boys assist in construction. Women and girls were found to have greater workloads than men and boys. For example, it was found that women spend 8–10 h a day on agricultural activities alone, with an additional 5–6 h of household tasks and other non-agricultural tasks to complete each day (Murray *et al.*, 2016). Women and girls are therefore overburdened with workload and have no time to rest or participate in other productive activities.

Depending on the circumstances, women are sometimes requested to assist men with tasks considered as mainly for men, which in the rural communities of Nkhamenya and Kabudula (Malawi) include the transporting of agricultural produce, and postharvest processing of tobacco. Men on the other hand do not partake in tasks that are mainly considered as female tasks, such as planting staple food crops like maize, beans, soya, and groundnuts. This disparity highlights that changes in gender roles are also unequal, resulting in increased labor burden for women (Murray *et al.*, 2016).

In a study commissioned by UN Women to assess gender gaps in agricultural productivity across four African countries, Ethiopia, Malawi, Uganda, and Tanzania, it was found that female farmers have lower rates of agricultural productivity than male farmers. This was attributed to several drivers, one of which being lack of access to male family labor. Unequal access to male family labor was reported to be one of the most important factors in Ethiopia, Malawi, and Tanzania. Income constraints were said to limit the capacity of female smallholders to hire male wage labor. In Tanzania it was established that lack of male family labor was responsible for the entire gap in agricultural productivity. In Ethiopia and Malawi, lack of male family labor was responsible for about 45% of the agricultural productivity gap (UN Women *et al.*, 2015). In Malawi, Uganda, and Tanzania, lack of cash income was found to translate into women's lower use of modern farm technologies, resulting in low agricultural productivity. Low access to cash income was attributed to the heavy demand of women's time performing unpaid farm labor for their husbands (UN Women *et al.*, 2015).

In Kenya, the introduction of the new land law and policy reforms in 2011, which include the land reform support – part of the Matrimonial Property Bill – and the declaration of equal rights in the land policy, are helping to create an enabling environment for transforming gender. In some parts of the country some women are slowly starting to gain rights in community land tenure systems. In other regions (e.g. Kitengela, near Nairobi) women have representation and roles in community land use planning. The majority of women in most parts of Kenya, however, continue to face constraints due to discriminatory cultural norms and beliefs. For example, despite changes to the land policy, women's names are still not included on land titles. Lack of access to and ownership of land constrains their ability to participate in sustainable agricultural development initiatives (Rubin, 2012).

Female farmers in Kenya continue to face constraints in accessing productive assets. They experience time poverty due to roles such as fetching water and wood fuel over long distances. This puts pressure on time for other productive activities. Female farmers in most parts of the country continue to have limited decision making power which in turn reduces their ability to invest in climate change adaptation and mitigation activities such as tree planting and carbon sequestration. They face constraints in public engagement; hence, they do not participate in local official meetings. If they are present, they are not encouraged to speak in the meetings. They are also not able to participate in local governance to influence decisions that might address climate change (Rubin, 2012).

In a study conducted by the Food, Agriculture and Natural Resources Policy Analysis Network (FANRPAN) in partnership with the National Agriculture Marketing Council (NAMC), poor access to CSA technologies and inputs in South Africa was found to hinder the progress and productivity of women involved in agriculture. Over 50% of the 384 smallholder women farmers who were interviewed did not have access to CSA technologies, improved seeds, and other inputs (Mzansi Agriculture Talk News, 2020). It is hoped that the findings of the study will inform development interventions, improving the conditions of women farmers to maximize on the intended benefits of CSA.

There is a lot of work to be done in African countries to effectively adopt gender-transformative CSA. Gender-transformative CSA should seek to transform gender roles and promote more gender-equitable relationships between women and men. Success requires challenging the underlying causes of gender inequality that are rooted in broad political, economic, and socio-cultural structures (Morgan, 2014). It involves advancing gender equality through activities targeting changes in three domains, which need to be considered simultaneously, to ensure more robust, equitable, and sustainable impacts (CARE, 2016). These domains are highlighted as follows:

- *Building agency* refers to the ability for individuals to make their own choices and to act upon them. This domain addresses power relations that exist between females and males within households and communities to provide equal opportunities that improve their well-being. Agency includes self-esteem and aspirations of both females and males, as well as knowledge, skills, and capabilities they need to flourish. This is key to empowerment. Given that men usually have greater authority over decisions related to agricultural production, as well as more control over resources and services, the choices to adopt CSA practices are undertaken in gender-unequal contexts. Building agency requires changing this narrative.
- *Changing gender relations* concerns altering the power relations through which women and men live their lives – the ability of women and men to create, participate in, and benefit from networks. This domain relates to promoting collaborative and accountable relationships between women and men. It is about creating relationships with institutions and networks at higher levels to support and address specific needs and priorities of women and men. It is about how women and men organize themselves to access and control productive assets.
- *Transforming structures* addresses the political, cultural, economic, and social structures within which women and men live. This domain supports the transformation of discriminatory social norms, customs, values, and exclusionary practices (all within the non-formal sphere), and laws, policies, procedures, and services (in the formal sphere).

All three domains highlighted above are closely linked. Formal sphere and non-formal sphere structures can isolate women, prevent them from developing critical relations, and make their isolation permissible as culturally appropriate. These structures further restrict women's agency; weak agency makes it difficult for women to develop relations. Improving women's agency may have some influence on structures and relations but may only go so far, given that cultural norms and laws can erode access to resources and limit benefits (Farnworth *et al.*, 2013; CARE, 2016). As new CSA technologies and practices are introduced in African countries, it is important to ensure action across all three domains – agency, relations, and structures – are considered to avoid exacerbating the existing unequal power relations, which in turn would hinder adoption to CSA. Given that women make up nearly half of Africa's agricultural labor force, they have a major role to play in the success of CSA, and their participation is required to implement CSA.

Opportunities for Gender-Sensitive Actions in Climate-Smart Agriculture

There are several opportunities for mainstreaming gender concerns into CSA to ensure gender-responsive interventions with tangible results in Africa's agricultural sector. This demands that the specific needs, priorities, realities, interests, and capabilities of women and men are recognized and effectively addressed in the design and implementation of CSA, to warrant equal benefit for women and men. Opportunities to effectively address gender concerns in CSA lie in how African countries harmonize and translate their commitment to gender equality.

Africa's commitment to the principle of gender equality is enshrined in Article 4 (l) of the Constitutive Act of the African Union. The continent's commitment is further reflected through the African Union Strategy on Climate Change, the Protocol to the African Charter on Human and Peoples' Rights on the Rights of Women in Africa, and the Solemn Declaration on Gender Equality in Africa. African states are also signatories to numerous national, regional, and international legal instruments aimed at eliminating discrimination based on gender. These, for example, include the Dakar Platform for Action (1994); the Beijing Platform for Action (1995); the Convention on the Elimination of All Forms of Discrimination Against Women (1979); the African Plan of Action to Accelerate the Implementation of the Dakar and Beijing Platforms for Action for the Advancement of Women (1999); the UN Resolution 1325 (2000) on Women; African Development Bank's Gender Strategy; and the Gender Action Plan (2017).

In 2013, the continent established the African Working Group on Gender and Climate Change to coordinate and provide leadership for Africa's engagement in regional and global gender and climate change processes. The Women and Gender Program on Climate Change established in 2014, is aimed at engaging women in climate change-related actions (Babugura, 2019). Africa also launched the Inclusive Climate Change Adaptation for a Sustainable Africa (ICCASA) project aimed at understanding and integrating gender in planning for and implementing climate adaptation plans across Africa (Nyasimi *et al.*, 2018).

These commitments should not be merely symbolic; they need to be accompanied by concrete actions with tangible results. Gender equality will need to be mainstreamed in all Africa's CSA procedures, policies, action plans, financing, institutional structures, programs, projects, and monitoring and evaluation of all CSA activities. Gender mainstreaming requires political will and profound and systemic change involving transformation of gender relations and societal structures, as well as addressing structural obstacles such as patriarchal social norms and discriminatory laws and customs that disempower women and exclude them from equally participating in and contributing to CSA.

Gender-Responsive Climate-Smart Agriculture Best Practices in Africa

While much more work remains to be done to advance gender equality in Africa's agriculture sector, several successful gender-responsive CSA interventions that transform gender inequalities have been developed and implemented (FAO and CARE, 2019). Presented in this section are three examples of gender-responsive CSA best practices in Africa. They demonstrate how gender equality can be enhanced in CSA interventions. Lessons from successful practices can be used to inform the development and implementation of gender-transformative CSA interventions in different contexts.

Gender, Climate Change and Agriculture Support Program

In recognition of the gender gaps in Africa's agriculture sector, the key role women play in agriculture, and the brunt they bear due to climate change consequences, the African Union Development Agency (AUDA) NEPAD established the Gender, Climate Change and Agriculture Support Program (GCCASP), supported by the Swiss development organization, Helvetas. The program is aimed at supporting women smallholder farmers and other vulnerable groups to better cope with the effects of climate change, as well as to enhance their adaptive capacity and resilience. Through policy change and access to CSA technologies, it seeks to empower women smallholder farmers to participate in CSA and deal with the effects of climate change efficiently and effectively. Identified areas of intervention include closing institutional gaps, capacity building of women smallholder farmers, strengthening women's platforms, and investing in the upscaling of successful and innovative practices. The benefits are said to include improving female farmers' access to land, inputs, credit, and technologies, and enhancing women's participation in decision making at the local, national, and regional economic levels in both the social and political spheres. The program is piloted in five African countries, namely Rwanda, Niger, Malawi, Cameroon, and Ethiopia.

Under the GCCASP, Ethiopia implemented a project entitled 'Empowering Women for Improved Food and Nutrition Security' (EWIFNS). The project was implemented from 1 August to 15 December 2019. It worked directly with women as agents of change and focused on empowering them in food and nutrition security and resilience to climate change-related shocks. The project introduced labor-intensive homestead orchard development (HOD) practices. To avoid creating a double workload for women, collaboration between women and men at the household level was encouraged. Helvetas facilitated the voluntary participation of husbands of female beneficiaries. This process created the needed supportive environment to empower the women. Women were equipped with leadership and negotiation skills and supported to make decisions within their households. Women participating in the project were able to influence their husbands and other family members to collaborate with them to get work done. They had sufficient knowledge and confidence to generate their own income. They were able to make decisions on household consumption, spending, and saving at the household level (Golda and Dereje, 2020).

Consequently, 50% of the women's husbands participating in the project received conceptual and practical training on HOD. Men played a role in the successful construction of the ring basin infiltration pit and in permaculture gardening, which included intensive soil treatment practices before planting seeds and seedlings. The men assisted their wives with double digging the ring basin and together planted different seasonal vegetables and fruits. Perennial plants such as *geshos*, coffee, mangos, and lemons were planted. Families were able to feed themselves and generate income (Golda and Dereje, 2020). The project was viewed as a success, demonstrating the effectiveness and benefits of women and men working together in CSA-related projects. This was reflected in the increased household productivity, and women's increased control over income and decision making. Wives and husbands were making joint decisions on household expenditures. Household work was being shared. Central to the success of the project was the involvement of male members of the household. The collaboration between women and men benefited the entire household, while also promoting the adoption of improved CA practices.

Project successes regarding gender mainstreaming

- The project enhanced women's agency by developing their leadership skills; increasing their access to services, assets, and inputs; creating a more enabling environment for gender equality; and improving the extension and market systems.
- Women's agency was also enhanced by empowering them to have greater control over agriculture production, income, and decision making at the household and community level.
- The project was able to build agency and transform gender relations by engaging wives and husbands in collaborative agricultural activities. This changed the male spouses' perceptions of women's capabilities and roles within agriculture. The project was also able to redistribute the gender roles within the households by encouraging men to be more involved in household chores. This reduced the women's workload, allowing them more time to rest, and increased income opportunities.
- The project facilitated in creating a supportive environment to empower the women. This opened room for transforming structures, allowing for a shift in gender social norms at the household and community level. Men accepted women's empowerment where it led to household development – improved agricultural production, improved nutrition due to increased production of nutritious foods, and increased income generated from surplus production. Through the project, women and men became more aware of gender inequalities and acted together to address them, particularly in changing the roles and responsibilities within households and the community. Men understood and appreciated women's contributions in the household. Men also recognized the positive effects of their own contributions in the household, resulting in increased cooperation between women and men in the household.

Similar successes were reported for the 'Pathways Towards Women's Empowerment' project implemented in Malawi, Ghana, Mali, and Tanzania (see FAO and CARE, 2019).

Conservation Agriculture Scaling-Up project

The Conservation Agriculture Scaling-Up (CASU) project has been implemented by FAO in 48 districts of Zambia's nine provinces, across four agroecological regions. The project aimed at targeting over 300,000 smallholder farmers with at least 40% being women. A total of 268,137 farmers participated in the project. Females made up 49.1% and 50.9% were male. The local small-scale farmers were supported to adopt CA practices to enhance agricultural productivity and output, while strengthening the agricultural systems' resilience. The result also enhanced the capacity of local small-scale farmers to adapt to climate change. One of the foci within the project, was addressing barriers to rural women's socio-economic empowerment, such as their lack of access to labor-saving technologies and productive resources. A gender analysis was conducted to inform gender-responsive practices promoted by the project (FAO, 2018).

Project successes regarding gender mainstreaming

- The needs of women and men in the project were informed by a gender analysis. The project was therefore able to address identified gender-based constraints, build agency, change relations, and transform structures.
- Building agency, changing relations, and transforming structures within the project enabled female and male farmers to:
  - ✓ Have relatively equal access to and control over decent employment, income, land, and other productive resources.
  - ✓ Participate equally in decision making and accept leadership roles within the project. For example, both female and male farmers had the opportunity to be included in the project as lead farmers and follower farmers. Women lead farmers were accepted and respected by male follower farmers. They were also supported by their families. It was reported that women lead farmers were active and confident in meetings. Generally female farmers demonstrated increased confidence and skills. It was acknowledged within communities that men and women can do all tasks interchangeably. Women were involved in planning and implementation of farming.
  - ✓ Have equal access to inputs and services for agricultural development. For example, they had access to extension services and input packages containing facilitating inputs such as legume seed. Improvement in their yields was noted.

While the project recorded some successes in terms of gender mainstreaming, there were reported cases where men refused to allow their wives to farm land separately and insisted on making all production and purchasing decisions for the household. These were men who did not think it was culturally acceptable for women to take on those roles. This is a reflection that in some areas it can be difficult for women to break through social norms; hence, more work is required. Despite such difficulties, overall, meaningful involvement of women in the project was relatively successful. Gender transformation in several districts with positive results was noted.

PROMARA project

The PROMARA project in Kenya (2010–2012) was implemented to protect the country's largest surviving forest, the Mau Forest Complex. The project was designed to help with the restoration of forests and water catchment in the upper Mara basin of the Mau Forest Complex. A people-centered and gender-sensitive approach was applied to strengthen land and resource tenure and community-based resource management. The project sought to address several gender issues that hinder Kenyan women's ability to participate in sustainable agricultural development initiatives (Rubin, 2012). Through the PROMARA project, opportunities were sought to:

- increase the participation and representation of women in decision making within communities;

- address security of land tenure for women; and
- address discriminatory social norms that restrict women's opportunities in numerous ways (Rubin, 2012).

Through capacity and skills development, the project aimed to change the attitudes and behavior of women and youth groups towards the exploitation of forest resources. In turn, it also improved their capacity to adapt to climate change. Critical to achieving this goal, was reforming gender-biased institutional arrangements, to enhance women's access to information, mobilizing resources, strengthening capacity, and building linkages between policy makers and other institutions (Farnworth *et al.*, 2013). The project established woodlots and tree nurseries. Fruit trees were planted in homesteads and schools with the support of women and self-help groups. These communities now benefit economically from the project activities. Beekeeping and small farm enterprises were established and, in some cases, strengthened. A cooperative to develop technical expertise and market outlets was founded, which has provided training to community members in riverbank protection, soil conservation, and reforestation (Farnworth *et al.*, 2013).

Project successes regarding gender

- Gender was successfully mainstreamed throughout the project cycle. A gender analysis was conducted to inform the overall implementation of the project and the gender strategy. The gender strategy informed the gender-responsive activities within the project under the specific socio-cultural context. Gender indicators were established to assess outputs and impact. This process allowed for understanding the range of gender issues that must be considered in the project design, implementation, and monitoring and evaluation.
- The building of women's agency was achieved. Women were strongly empowered through targeted special initiatives. They were equipped with knowledge, skills, and capabilities to improve their ability to adapt to climate change. Women were trained and supported to create tree nurseries and to sell seedlings as an income-earning activity. They were trained on beekeeping and improved cookstoves. A new cooperative with over 300 members was formed to help market farm produce and to advocate for and conduct a range of conservation efforts in the Mau forest.
- A conducive environment for wider reforms was created allowing for transformation of structures and changing relations. The reforms included strengthening the statutory and customary land rights of women and youth. It was reported that the project team worked with community elders and local leaders to secure customary land rights based on constitutional provisions, legislation, and policies. Men were encouraged to involve women in decision making and to share household work. The success of the project was attributed to the reform of gender-biased institutional arrangements.

Lessons from Best Practices

The above best practices demonstrate the importance of mainstreaming gender in CSA activities. They reveal multifaceted benefits observed across the projects.

- Creating an enabling environment in which gender concerns can be addressed, fosters collaboration between women and men to achieve widespread desired outcomes.
- Selecting appropriate implementation strategies and methods will ensure that both women and men benefit from CSA intervention.
- Investing in women elevates their status as champion farmers and agents of change.
- Transforming gender roles and unequal power relations, greatly increases women's access to agricultural productive resources and services, while promoting CSA practices and building capacities to respond to the changing climate.
- It is vital to transform gender roles and unequal power relations which stem from cultural norms, institutionalized through social, cultural, political, and economic legislative structures. This ensures that the discrimination of women in CSA is not

legitimized. Women are enabled to overcome their limited access to productive resources, labor-saving technologies, and poor participation in decision making.

- The acquisition of knowledge and skills enhances women's capabilities to thrive in the practice of CSA, allows them to become better negotiators and competitors in the market, builds their confidence, and encourages their participation in decision making processes.
- Engaging men in CSA activities targeting women is crucial to the success of women's empowerment efforts and promotes more equal relationships and decision making at the household and community level. Men can better appreciate the benefits of women's empowerment which extended beyond agricultural and financial gains. For example, equitable intra-household decision making, stronger partnerships, and greater harmony in the household.

Most importantly, the highlighted cases are testimony that with the will to transform discriminatory social norms, customs, values, laws, procedures, and unequal power relations that shape gender inequalities in Africa's agricultural systems, the vision of CSA is possible. This requires investing time, effort, and resources to understand context-specific gender dynamics, to enable the design and successful implementation of effective gender-responsive CSA activities in Africa. Women's and men's differing needs, preferences, and constraints must be adequately identified and addressed from the development of CSA policies to the design and implementation of interventions.

Transforming set cultural norms and traditions within the African context is not easy, but with commitment and willingness from all levels of society (e.g. household, community, political, civil society, organizational, institutional, local government) and appropriate approaches, gender transformation is possible. Lessons from existing CSA interventions illustrate that this is conceivable. Currently, gender transformation in Africa's agriculture is happening at a slow pace and on a small scale. It needs to be escalated to a larger scale and with concrete long-term benefits.

Conclusion

The importance of gender consideration in CSA within the African context cannot be over emphasized. There is no doubt that gender shapes women's and men's abilities to engage and adopt CSA practices. The fundamental role women play in Africa's agriculture and the many challenges they face compared to men are acknowledged and documented. As CSA interventions are developed to address the effects of climate change, there is an opportunity to address gender inequalities in Africa's agricultural system. It is important to note that women and men need different strategies and technologies to enable their adoption of CSA. This is an opportunity to tap into women's potential as effective innovators with the ability to identify and design new technologies and improve on existing ones to address their needs. Without the active and meaningful participation of women, Africa will not only go hungry but the opportunity to adapt, build resilience to climate change, and reduce GHG emissions will be missed. To ensure effective gender-equitable strategies that will yield concrete results in Africa, CSA should therefore be informed by gender.

References

AGRA (2018) Africa agriculture status report: catalyzing government capacity to drive agricultural transformation (Issue 6). Alliance for a Green Revolution in Africa, Nairobi. Available at: www.agra.org/wp-content/uploads/2018/09/AASR-2018.pdf (accessed 22 July 2020).

AWGGCC (2017) Gender analysis of the Paris Agreement and implications for Africa. African Working Group on Gender and Climate Change. Available at: www.idl-bnc.drc.dspacedirect.org/bitstream/handle/10625/56476/IDL-56476.pdf?sequence=2&isAllowed=y (accessed 14 May 2020).

Babugura, A. (2019) Gender equality in combatting climate change: the African context. South African Institute of International Affairs, Johannesburg, South Africa. Available at: www.africaportal.org/publications/gender-equality-combatting-climate-change-african-context/ (accessed 1 February 2021).

Barkhordarian, A., von Storch, H. and Bhend, J. (2013) The expectation of future precipitation change over the Mediterranean region is different from what we observe. *Climate Dynamics* 40, 225–244.

Bassey, S.A. and Bubu, N.G. (2019) Gender inequality in Africa: a re-examination of cultural values. *Cogito (Bucharest)* 11, 21–36. Available at: www.search.proquest.com/openview/3b37ac66dafed57d2353d063b122ab9b/1?pq-origsite=gscholar&cbl=1316371 (accessed 3 February 2021).

Boto, I., Biasca, R. and Brasesco, F. (2012) Climate change, agriculture and food security: proven approaches and new investments. *Rural Development Briefing No. 29*. Brussels Development Briefings, Brussels.

CARE (2016) Gender equality and women's voice: guidance note. CARE International, Geneva, Switzerland.

Doss, C. (2011) The gender asset and wealth gaps: evidence from Ecuador, Ghana, and Karnataka. Indian Institute of Management, Bangalore, India.

Esterhuyse, C.J. (1994) Agroforestry. In: van der Sijde H.A.N.B. (ed.) *Forestry Handbook*. South African Institute of Forestry, Pretoria.

FAC (2011) Participatory Impact Pathways Analysis (PIPA). *Document for submission to FAC's Mid-term Review Team*. Future Agricultures Consortium, Brighton, UK.

FAO (2010) Gender dimensions of agricultural and rural employment: differentiated pathways out of poverty. Food and Agriculture Organization of the United Nations, Rome.

FAO (2011) State of food and agriculture: women in agriculture, closing the gender gap for development. Food and Agriculture Organization of the United Nations, Rome. Available at: www.fao.org/docrep/013/i2050e/i2050e00.htm (accessed 25 May 2020).

FAO (2012) Developing a climate-smart agriculture strategy at the country level: lessons from recent experience. *Background Paper for the Second Global Conference on Agriculture, Food Security and Climate Change. Hanoi, Vietnam, 3–7 September 2012*. Food and Agriculture Organization of the United Nations, Rome.

FAO (2015) Learning tools on nationally appropriate mitigation actions in agriculture, forestry and land use sector. Food and Agriculture Organization of the United Nations, Rome. Available at: www.fao.org/3/a-i4642e.pdf (accessed 30 July 2020).

FAO (2018) Promoting gender-responsive adaptation in the agriculture sectors: entry points within National Adaptation Plans. *FAO Briefing Note*. Food and Agriculture Organization of the United Nations, Rome.

FAO and CARE (2019) Good practices for integrating gender equality and women's empowerment in climate-smart agriculture programmes. CARE International, Geneva, Switzerland and Food and Agriculture Organization of the United Nations, Rome. Available at: www.insights.careinternational.org.uk/media/k2/attachments/Good_Practices_Integrating_Gender_Equality_Women_Empowerment_Climate-Smart_Agriculture_Programmes.pdf (accessed 29 July 2020).

Farnworth, C., Sundell, M.F., Nzioki, A., Shivutse, V. and Davis, M. (2013) Transforming gender relations in agriculture in Sub-Saharan Africa. Swedish International Agricultural Network Initiative (SIANI), Stockholm. Available at: www.siani.se/wp-content/uploads/2017/10/front_cover_and_foreword.pdf (accessed 29 October 2020).

Golda, A. and Dereje, M. (2020) Engaging men in HOD practice as an approach to women empowerment. African Union Development Agency (AUDA) – New Partnership for Africa's Development (NEPAD), Midrand, South Africa. www.nepad.org/news/engaging-men-hod-practice-approach-women-empowerment (accessed 19 July 2020).

IISD (2004) *International Ecoagriculture Conference and Practitioners' Fair: Summary Report, 27 September – 1 October 2004*. International Institute for Sustainable Development, Winnipeg, MB, Canada. Available at: www.enb.iisd.org/events/international-ecoagriculture-conference-and-practitioners-fair/summary-report-27-september-1 (accessed 1 February 2021).

IPCC (2014) *AR5 Climate Change 2014: Impacts, Adaptation and Vulnerability*. Technical Summary. Intergovernmental Panel on Climate Change, Cambridge University Press, Cambridge, UK and New York. Available at: www.ipcc.ch/site/assets/uploads/2018/02/WGIIAR5-TS_FINAL.pdf (accessed 30 June 2020).

Jost, C., Kyazze, F., Naab, J., Neelormi, S., Kinyangi, J. *et al.* (2016) Understanding gender dimensions of agriculture and climate change in smallholder farming communities. *Climate and Development* 8, 133–144. doi.org/10.1080/17565529.2015.1050978.

Kangas, A., Haider, H., Fraser, E. and Browne, E. (2014) *Gender: Topic Guide* (revised edn). Governance and Social Development Resource Centre (GSDRC), University of Birmingham, UK.

Lobell, D.B., Bänziger, M., Magorokosho, C. and Vivek, B. (2011) Nonlinear heat effects on African maize as evidenced by historical yield trials. *Nature Climate Change* 1, 42–45. Available at: www.researchgate.net/publication/50367042_Nonlinear_heat_effects_on_African_maize_as_evidenced_by_historical_yield_trials (accessed 7 July 2020).

Mama, A. (2001) Challenging subjects: gender and power in Africa contexts. Plenary address, Nordic Africa Conference – 'Beyond Identity: Rethinking Power in Africa', Uppsala, Sweden, 4–7 October 2001. *African Sociological Review* 5, 63–73.

Morgan M. (2014) Measuring gender transformative change. *Program Brief: AAS-2014-41*. CGIAR Research Program on Aquatic Agricultural Systems, Penang, Malaysia.

Moyo, B. and Pereira, L. (2019) Transforming Africa's food systems to withstand a +2°C world. Heinrich Böll Stiftung, Cape Town, South Africa. Available at: www.za.boell.org/en/2019/12/16/transforming-africas-food-systems-withstand-2-degc-world (accessed 3 February 2021).

Mulema, A.A., Farnworth, C.F. and Colverson, K.E. (2017) Gender-based constraints and opportunities to women's participation in the small ruminant value chain in Ethiopia: a community capitals analysis. *Community Development* 48, 351–369.

Murray, U., Gebremedhin, Z., Brychkova, G. and Spillane, C. (2016) Smallholder farmers and climate smart agriculture: technology and labor-productivity constraints amongst women smallholders in Malawi. *Gender, Technology and Development* 20, 117–148. doi.org/10.1177/0971852416640639.

Mzansi Agriculture Talk News (2020) UN Women calls for government and private sector to aid women in agriculture and agri-business. Available at: www.mzansiagritalk.com/archives/3830 (accessed 13 July 2020).

Nemes, N. (2009) Comparative analysis of organic and non-organic farming systems: a critical assessment of farm profitability. Food and Agriculture Organization of the United Nations (FAO), Rome. Available at: www.fao.org/family-farming/detail/en/c/282591/ (accessed 1 February 2021).

NEPAD (2019) *Concept Note: 23rd Africa Climate Smart Agriculture Alliance*. New Partnership for Africa's Development, Midrand, South Africa. Available at: www.nepad.org/programme/climate-smart-agriculture (accessed 6 July 2020).

Neufeldt, H., Kristjanson, P., Thorlakson, T., Gassner, A., Norton-Griffiths, M., Place, F. and Langford, K. (2011) Making climate-smart agriculture work for the poor. *ICRAF Policy Brief 12*. World Agroforestry Centre (ICRAF), Nairobi.

Njuki, J. and Sanginga, P.C. (eds) (2013) *Women, Livestock Ownership and Markets: Bridging the Gender Gap in Eastern and Southern Africa*. Routledge, Abingdon, UK and New York. Available at: www.reliefweb.int/sites/reliefweb.int/files/resources/Women%2C%20Livestock%20Ownership%20and%20Markets.pdf (accessed 20 July 2020).

Nyasimi, N., Freeman, K. and Elhassan, D. (2018) Africa's commitment to integrating gender in climate change adaptation polices and initiatives. CGIAR Research Program on Climate Change, Agriculture and Food Security (CCAFS), Copenhagen. Available at: www.ccafs.cgiar.org/blog/africa%E2%80%99s-commitment-integrating-gender-climate-change-adaptation-polices-and-initiatives#.XyrhkigzY2w (accessed 26 July 2020).

Ongley, E.D. (1996) Control of water pollution from agriculture. *FAO Irrigation and Drainage Paper No. 55*. Food and Agriculture Organization of the United Nations, Rome.

Quisumbing, A.R., Meinzen-Dick, R.S., Raney, T.L., Croppenstedt, A., Behrman, J.A. and Peterman, A. (2014) *Gender in Agriculture: Closing the Knowledge Gap*. Springer, Dordrecht, The Netherlands. doi.org/10.1007/978-94-017-8616-4.

Rosenstock, T.S., Rohrbach, D., Nowak, A. and Girvetz, E. (eds) (2019) *The Climate-Smart Agriculture Papers*. doi.org/10.1007/978-3-319-92798-5_1.

Rubin, D. (2012) USAID/East Africa: Report on the Regional Workshop on Gender, Agriculture, and Global Climate Change. Available at: www.culturalpractice.org/wp-content/uploads/5-2012-7.pdf (accessed 29 October 2020).

Scherr, S.J. and McNeely, J. (2009) Biodiversity conservation and agricultural sustainability: towards a new paradigm of 'ecoagriculture' landscapes. *Philosophical Transactions of the Royal Society London. Series B, Biological Sciences* 363, 477–494.

UN Commission on the Status of Women (2008) Gender perspectives on climate change. *Issues paper for interactive expert panel: Emerging issues, trends and new approaches to issues affecting the situation of women or equality between women and men. 52nd session of the Commission on the Status of Women, United Nations, New York, 25 February to 7 March 2008.* Available at: www.un.org/womenwatch/daw/csw/csw52/issuespapers/Gender%20and%20climate%20change%20paper%20final.pdf (accessed 19 July 2020).

UN Women (2014) World survey on the role of women in development 2014: gender equality and sustainable development. United Nations, New York. Available at: www.unwomen.org/en/digital-library/publications/2014/10/world-survey-2014 (accessed 9 July 2020).

UN Women (2016) Female farmers and inclusive growth in Vietnam. United Nations, New York. Available at: www.asiapacific.unwomen.org/en/digital-library/publications/2017/01/female-farmers-and-inclusive-growth (accessed 1 February 2021).

UN Women (2018) The cost of gender gap in agricultural productivity in Ethiopia. United Nations, New York. Available at: www.africa.unwomen.org/en/digital-library/publications/2018/04/study-of-cost-of-gender-gap-eth (accessed 1 February 2021).

UN Women, UNDP-UNEP PEI and World Bank (2015) The cost of the gender gap in agricultural productivity in Malawi, Tanzania, and Uganda. UN Women, United Nations Development Programme – United Nations Environment Programme (Poverty-Environment Initiative) and the World Bank, New York and Washington, DC.

UNDP (2006) Gender equality: evaluation of gender mainstreaming in UNDP. United Nations Development Programme, New York. Available at: www.web.undp.org/evaluation/documents/eo_gendermainstreaming.pdf (accessed 1 February 2021).

UNDP (2016) *Africa Human Development Report 2016*. United Nations Development Programme, New York. Available at: file:///C:/Users/agnesba/Downloads/AfHDR_2016_lowres_EN.pdf (accessed 5 May 2020).

UNFPA and WEDO (2009) Policy that supports gender equality. United Nations Population Fund, New York and Women's Environment and Development Organization, New York. Available at: www.unfpa.org/sites/default/files/pub-pdf/climateconnections_2_policy_0.pdf (accessed 9 April 2020).

Williams, T.O., Mul, M., Cofie, O., Kinyangi, J., Zougmore, R. *et al*. (2015) Climate smart agriculture in the African context. *Background paper for Feeding Africa: an Action Plan for African Agricultural Transformation, Dakar, Senegal, 21–23 October 2015*. Available at: www.afdb.org/fileadmin/uploads/afdb/Documents/Events/DakAgri2015/Climate_Smart_Agriculture_in_the_African_Context.pdf (accessed 10 July 2020).

10 Gender Differences in Awareness and Adoption of Climate-Smart Agriculture Practices in Bangladesh

E. Bryan, E. Kato and Q. Bernier

Introduction

Bangladesh is one of the most vulnerable countries in the world to the adverse impacts of climate change given its high exposure to such impacts, high sensitivity to these impacts, and limited adaptive and coping capacity (ND-GAIN, 2018). While climate change will affect all sectors of the country, the agriculture sector will be hit particularly hard. Warming temperatures, sea level rise leading to increased soil salinity, changing rainfall patterns, and the increased frequency and intensity of extreme weather events, such as flooding, coastal cyclones, and regional droughts, are projected to significantly constrain future agricultural production (Yu *et al.*, 2010; Ruane *et al.*, 2013; Rahman *et al.*, 2018). Analysis of historic climate trends already shows significant increases in temperature, drier conditions during the pre- and post-monsoon periods, and wetter conditions during the monsoon period, trends that are predicted to continue into the near future (Rahman *et al.*, 2018). Climate change is expected to reduce rice production in all three growing seasons and have significant negative impacts on fisheries production due to the increased frequency and intensity of coastal cyclones and flooding (Ruane *et al.*, 2013; Thomas *et al.*, 2013).

Climate-smart agriculture (CSA) is widely promoted to reduce the adverse impacts of climate change on the agriculture sector. CSA encompasses a set of practices and technologies that increase adaptive capacity and resilience to climate variability and change, while improving food and livelihood security and mitigating greenhouse gasses (FAO, 2010, 2013). The identification of an appropriate set of climate-smart practices at the family farm level depends on the climate challenges that farmers face, the type of farming system, and local agroecological conditions. While Bangladesh has 30 agroecological zones, each with different farming systems and climate challenges, most studies identify a range of climate-smart practices for two distinct regions of the country – the drought prone areas and the coastal zones. Most commonly identified CSA practices include changing crop types, altering planting dates, adoption of stress-tolerant crop varieties, homestead gardening, improved agronomic practices (such as cover cropping, crop rotation, and minimum tillage), technological innovations like laser leveling, crop diversification, and improved animal husbandry practices (Rahman *et al.*, 2009; Thomas *et al.*, 2013; Saroar and Filho, 2016; Alam *et al.*, 2017; Kabir *et al.*, 2017; Taneja *et al.*, 2019). Moreover, because agricultural production is already highly input intensive in many parts of the country, strategies that increase resource use efficiency are also considered climate-smart, such as increasing irrigation efficiency (Shah, 2009) and improving fertilizer application precision and efficiency (Rasul and Thapa, 2004; Ahmed

DOI: 10.1079/9781789247053.0010

et al., 2013). Other important practices relate to improved livestock management, fisheries and aquaculture, and agroforestry (Rahman *et al.*, 2009). As 39% of the workforce is heavily engaged in agriculture, according to World Bank (2020), more widespread adoption of such climate-smart technologies and practices at the farm level is critical to secure the livelihoods of poor producers in the face of climate change.

While many studies have identified key climate-smart practices and examined constraints to adoption of these practices at the farm level, it is also important to consider *who* in the household is adopting and benefiting from climate-smart practices. Adaptation by both women and men is necessary to increase the resilience of agricultural households and to ensure that the needs and preferences of both women and men are being met, particularly given their distinct livelihood roles. Similar to much of Asia, family farming in Bangladesh can be characterized as a joint activity, unlike the farming systems with separate plots managed by women and men, which are found in parts of Africa (Boserup, 1970; Lele, 1986). However, women and men do have distinctive roles in agriculture, with men providing most of the labor for field crop agriculture and women more intensively engaged in livestock production and cultivating home gardens (Ali, 2005). However, women also work in field crops, and men also have livestock-related tasks, so the gender roles in agriculture are not immutable. Moreover, while women's contribution to the sector tends to be underappreciated (Kabeer, 1994; Rahman, 2000), women's role in agriculture has increased significantly over the last two decades, and women currently account for 45% of the agricultural labor force (BBS, 2018). The growing literature on gender and climate change shows that women and men experience climate shocks and stressors differently, have different capacities to respond to climate change, and have different preferences for how they respond to climate change impacts (Huyer, 2016; Jost *et al.*, 2016; Bryan *et al.*, 2017; Kristjanson *et al.*, 2017). The gender gap in adoption of climate-smart practices is attributable to a number of factors, including gender differences in perceptions of and exposure to climate change, access to information and technologies for adaptation, access to resources, limited decision making authority, and social norms that limit options for women (Katungi *et al.*, 2008; Tall *et al.*, 2014; Twyman *et al.*, 2014; Bernier *et al.*, 2015; Jost *et al.*, 2016; Bryan *et al.*, 2017). In particular, women's lack of access to information on climate change and climate-smart practices stands out as an important barrier to adoption of CSA in many contexts (Katungi *et al.*, 2008; Beauman and Dillon, 2014; Tall *et al.*, 2014; Bernier *et al.*, 2015; Jost *et al.*, 2016).

Women in Bangladesh face greater vulnerabilities due to climate change and have more limited adaptive capacity (Ahmad, 2012). For example, they own and control significantly smaller stocks of productive assets than men in Bangladesh (Sraboni *et al.*, 2014; Roy *et al.*, 2015), and are significantly disadvantaged in terms of land ownership (Kieran *et al.*, 2015). Despite these challenges, studies show that women do contribute to climate change adaptation in distinct ways that often relate to their gendered roles in agriculture, including by adopting practices like homestead garden production, livestock care, grain and food preservation and storage, and planting roadside trees (Chaudhury *et al.*, 2012; Corcoran-Nantes and Roy, 2018; Khalil *et al.*, 2020). Recent research suggests that closing the gender gap in agriculture would result in positive agricultural outcomes, including increased efficiency in agricultural production (Seymour, 2017). Moreover, including women in agricultural decision making would further accelerate adoption of climate-smart practices, such as greater crop diversification (De Pinto *et al.*, 2019).

This chapter focuses on a range of practices that have been identified as climate-smart and appropriate for adoption at the family farm level in the context of Bangladesh, based on input from stakeholders and a review of the literature, as well as a review of ongoing agricultural interventions aimed at enhancing agricultural productivity and climate resilience in the country. Sex-disaggregated data from two communities in Bangladesh are used to assess the gender differences in access to different sources and types of agricultural and climate information. The gender dimensions of awareness and adoption of these CSA practices are then explored in order to understand the extent to which information and knowledge gaps contribute to the adoption patterns of female and male farmers. Given that awareness is likely not the only determinant of adoption of CSA practices, we use a Heckman selectivity regression model to examine the correlates of adoption of specific CSA practices, taking into account the endogeneity of awareness. The next section presents the conceptual framework for the study. This section is followed by a description of the survey design and local context using descriptive data from the two survey sites, in Satkhira and Bagerhat districts. The results section presents descriptive results on gender differences in access to different types and sources of information, and findings from the Heckman model on awareness and adoption of CSA practices. The chapter concludes by discussing the implications of the results and the need for increasing awareness and adoption of CSA practices by both women and men.

Conceptual Framework: Understanding Awareness and Adoption

A number of studies emphasize the sequential process of awareness and adoption of agricultural practices, as well as awareness and adaptation to climate change (e.g. Kaliba *et al.*, 2000; McBride and Daberkow, 2003; Deressa *et al.*, 2011; Bernier *et al.*, 2015). Deressa *et al.* (2011) discuss a two-step process in which farmers first perceive a change in climate, and second, act on it through adaptation and apply this to the analysis of the role of perceptions in climate change adaptation in the Nile Basin of Ethiopia. Using data from the CGIAR Research Program on 'Climate Change, Agriculture and Food Security' (CCAFS) for Kenya, Bernier *et al.* (2015) also posit a two-step model in which adoption of CSA practices is conditional on awareness of the practice. They use the Heckman (1979) selection model to account for awareness of a particular CSA practice (at the individual level) and then adoption of the CSA practice (reported at the individual level on a plot managed by the responding individual). Bernier *et et al.* (2015) hypothesize that those who are aware of technologies are not randomly selected, but that there is a self-selection bias that needs to be corrected to obtain an unbiased estimate of adoption. The appropriateness and validity of the Heckman selectivity correction depends on the identification of suitable excluded instruments that affect awareness, but not adoption. In the current study, it is assumed that access to different information sources directly affects awareness but does not directly influence adoption except through the effect on awareness. Therefore, dummy variables are used for sources of information as the excluded instruments to improve statistical identification and precision in this two-step estimation Heckman procedure.

Heckman's sample selection model assumes that there exists an underlying relationship given by:

$$y^*_j = x_j\beta + u_{1j} \tag{10.1}$$

where y^*_j is a latent, binary variable (the adoption of a particular CSA practice), x_j is a k-vector of different variables, which include factors hypothesized to affect adoption, β is the parameter to be estimated, and u_{1j} is an error term.

$y^*_j > 0$ is a dichotomous variable that is observed only if:

$$y^{aware}_j = z_j\delta + u_{2j} > 0 \tag{10.2}$$

$$u_1 \sim N(0, 1)$$

$$u_2 \sim N(0, 1)$$

$$corr(u_1, u_2) = \rho$$

where y^{aware}_j is a binary variable indicating whether the respondent is aware of a CSA practice, z is an m-vector of explanatory variables affecting the awareness of CSA practices, δ is a vector of parameters to be estimated, u_{2j} is an error term, and u_1 and u_2 are normally distributed with mean zero and variance 1. Following Deressa *et al.* (2009), the Heckman two-step procedure is used to estimate the correlates of adoption of CSA practices, conditional on awareness of the practice.

In the adoption equation (Eqn 10.1), x is a vector of individual and household characteristics that affect adoption, including the experience of shocks, women's involvement in agricultural decision making, membership in groups as a measure of social capital and collective agency, farm size, and income as measures of household wealth; z is a vector of variables that affect awareness, including those that affect awareness but not adoption, namely access to various sources of information; β and δ are coefficients to be estimated, and u_1 and u_2 are normally distributed error terms.

Individual characteristics that might affect both awareness and adoption include whether the respondent is female and their age, years of schooling, and asset holdings. It is expected that age, a proxy for experience, is positively correlated with adoption and years of schooling, and increases the likelihood of both awareness and adoption of CSA practices. We expect assets to be positively correlated with both awareness and adoption owing to their potential positive effect on bargaining power within the household. The outcome and selection equations, therefore, control for the value of household and farm assets owned by women, men, and jointly owned assets as a measure of relative bargaining power. Asset ownership (women, men, and joint) is separated because research suggests that women's, men's, and joint assets are often used for different purposes (Rakib and Matz, 2016; Quisumbing *et al.*, 2018). The study also controls for whether the respondent participates in agricultural decisions in both the awareness and adoption models.

Household characteristics include household size, total farmland owned, income, and a district dummy. Larger households are expected to have a larger pool of labor that would facilitate the adoption of labor-intensive CSA practices. Both awareness and adoption are expected to be positively correlated with household wealth. Having a land title is associated with higher land tenure security, which is likely to lead to greater investment in CSA practices, particularly those that require longer time commitments.

Other variables affecting awareness and adoption of CSA practices include whether the respondent has experienced an adverse shock and the domestic time burden of women. A variable for whether the household uses off-farm sources of domestic water is used as a measure of women's time burden. The authors hypothesize that for women, having a heavy domestic time burden is likely to constrain awareness and adoption of CSA practices. The study also includes variables of the respondents' experiences with climate shocks that resulted in crop losses, and their perceptions of future climate risks from drought and flood. The authors hypothesize that respondents who have experienced adverse shocks, or expect them in the future, are more likely to seek out information and adopt CSA.

Access to information sources is the set of variables for individual access to different sources of agricultural and climate information. The authors hypothesize that access to information sources does not have a direct effect on adoption but works through effects on awareness of CSA practices.

The Heckman two-step estimator was chosen to overcome the potential selection bias due to non-random levels of awareness. However, there is also a possible second source of bias (endogeneity bias) that could arise from the inclusion of endogenous explanatory variables. The authors, therefore, estimate the Heckman model with and without the potentially endogenous explanatory variables in what they identify as full and reduced-form models, respectively, as shown in the results section. The results were robust and were not sensitive to inclusion or exclusion of potential endogenous explanatory variables in the estimations, suggesting endogeneity bias is not a serious concern for the estimates. Therefore, the full set of results are presented below.

Survey Design and Context

Description of survey and data collection approach

This chapter uses intra-household survey data collected in 2013 from two sites in Bangladesh, in Bagerhat and Satkhira districts, as part of a larger data collection effort of the CCAFS research. These sites are two of 15 climate-smart villages across West Africa, East Africa, and South Asia that were identified based on a set of criteria developed through consultations with stakeholders, including biophysical and agroecological conditions, socio-economic and demographic factors, risks from climate change, anticipated land use pressures, agricultural potential, and governance and institutional capacity, among other factors. The final sites are each 10 × 10 km blocks that are not representative of national or sub-national administrative boundaries. Within these sites, different research activities were carried out and interventions were implemented to promote uptake of CSA practices. One of these research activities included a comprehensive farm characterization survey called 'IMPACTLite' (Integrated Modeling Platform for Mixed Animal Crop Systems). The intra-household survey data, on which this chapter is based, used the same sampling frame as the IMPACTLite survey, as described in Rufino *et al.* (2013), in order to link the intra-household data to underlying household-level data on livelihood and agricultural production activities. In each site, data were collected from 200 households (~400 individuals), stratified to reflect the different farm production systems that exist within the research grid.[1] Only households engaged in crop and other agricultural activities were interviewed.

The data set contains responses from 373 men and 377 women in 381 households across the two sites (site description below).[2] The survey contained modules on land tenure and ownership; agricultural, livestock, and household decision making; adoption and knowledge of climate-smart practices; access to and use of climate and agricultural information services; access to and use of credit; membership in groups; fuel and water use; coping strategies for dealing with shocks; perception of climate change and its potential impacts; identification of adaptation strategies; and cognitive decision making and personal values questions. The same set of questions was asked of the principal female and male adult decision makers in the household to understand differences in perceptions of climate change, preferences for climate-smart practices, and constraints to adoption of these practices. The survey was implemented by Data Analysis and Technical Assistance Ltd, which has extensive experience in conducting surveys on gender and intra-household issues.

Site description and sample descriptives

Satkhira district (referred to from now on as Satkhira) is located in the south-west coastal area of Bangladesh, and Bagerhat district (henceforth, Bagerhat) is located on the south-central coast. The sites were chosen because they have two different potential water management systems; water sources in Satkhira are highly saline but there is a higher potential for shallow groundwater use due to lower salinity intrusion into the groundwater, while Bagerhat has medium levels of salinity but less groundwater access. Bagerhat is better connected to markets and, therefore, has more options for diversification out of agriculture, whereas Satkhira is less accessible, but is closer to the border of West Bengal. As mentioned above, these sites were chosen to represent farming systems and agroecological zones and are neither nationally nor regionally representative. Cereals are produced in both sites with rice being the dominant crop. In addition to rice, wheat production is also important in Bagerhat, while close to half of households in Satkhira also produce fruits and vegetables. Fish production is an important livelihood activity for about half of households in both sites and more than three-quarters of households in both sites also engage in livestock production. Market sales of rice, fish, fruits and vegetables, and livestock products are higher for farm households in Satkhira compared to those of Bagerhat.

Respondents in Bagerhat are slightly older than those in Satkhira but have roughly the same years of schooling. Household sizes are between 9–10 household members with high dependency ratios, which are not significantly different across districts. Households in Satkhira are significantly wealthier, with higher levels of household assets and land. Agricultural incomes and overall incomes are higher in Satkhira; income from other sources is significantly higher in Bagerhat. Approximately 70% of households produce rice in both sites, although a higher proportion of households in Satkhira sells rice in the market. A higher proportion of households in Bagerhat produces fish, although a smaller percentage sells fish in the market. A higher proportion of households grows vegetables in Satkhira.

Reflecting marriage patterns and gender norms surrounding asset ownership in Bangladesh, husbands are about 9 years older than their wives and own a larger proportion of household assets – 47% compared to 24% for wives. Women, however, own a larger share of household livestock (68%) in terms of the number of livestock, compared to 9% owned solely by men. This does not reflect the value of livestock, however; typically husbands own a dairy cow or buffalo, which is substantially more valuable than the poultry and goats that women usually own (Quisumbing *et al.*, 2013). Consistent with gendered patterns of land ownership throughout Bangladesh (Kieran *et al.*, 2015), men in the study sample are more likely to own land than women, with 60% of male respondents saying that they own land, compared to only 5% of female respondents. Eighty-three percent of men say that they make decisions regarding land use, compared to 38% of women. Responses to questions on whether they use their own, communal, or borrowed or rented land, and the type of land document that they hold, are practically identical between women and men, indicating that they respond about the household's land, rather than their own. This reflects the pattern of joint family farming in Bangladesh, unlike in many parts of sub-Saharan Africa, where men and women farm separate plots and may own and manage land both individually and jointly (Doss *et al.*, 2014; Kieran *et al.*, 2015).

Although women and men work together in family farms, agricultural decisions are taken mostly by men. Among rice producing households, 94% of male respondents say they participate in decisions about what crops to plant, land preparation, input use, planting decisions, and weeding, compared to 47% of women. Similarly, high percentages of male respondents participate in decisions regarding crop management, harvesting, postharvest handling or processing, and crop use. Fewer women participate in agricultural decisions; 43% of female respondents participate in decisions on crop management, for example. Seventy-three percent of men report participating in decisions on use of income from crop sales, while only 39% of women do so. In contrast, women are more engaged in livestock production than are men. For virtually all livestock activities, except for grazing, which involves traveling far from the homestead, a higher proportion of women participate in decision making. This includes decisions regarding livestock production, feeding, veterinary services, housing for livestock, breeding, slaughtering, production, use, and use of income from animal products. Both women and men participate equally in decisions regarding the production of milk products, the sale, disposal or the use of animal products, and income derived from animal products.

Results

Gender differences in access to sources and types of information

Descriptive analysis of the survey data shows significant gender differences in access to information sources and types of information. In general, access to agricultural information from extension agents, service providers, non-governmental organizations (NGO), and farmer groups is low for both women and men. Informal sources of information, such as neighbors and family members, are more common, as are media sources like radio, television, and newspapers/bulletins. Men are more likely to receive agricultural information from almost all sources, including government extension services, agricultural service providers, farmer field days, and community meetings. Women, on the other hand, are more likely to have access to information from informal sources like family members and neighbors, while men are more likely to report having information from traditional sources or their own experience. More men than women report receiving information about agricultural practices from the radio, television, and newspapers. The gender gap in access to agricultural information sources reflects gender gaps in staffing of both the government agricultural extension service, the Department of Agricultural Extension, and the private sector. Except for those working for NGOs, most of the extension workers in the public and private sectors are male (Birner *et al.*, 2010), even though it is widely recognized that female extension workers can serve women beneficiaries better (World Bank *et al.*, 2009; Meinzen-Dick *et al.*, 2010; World Bank and IFPRI, 2010).

Women and men also have access to different types of information regarding climate- and weather-related events. Women are more likely than men to report hearing a forecast about extreme events, such as floods and droughts, as well as the start of rains. Men are more attuned to seasonal and long-term weather forecasts, although both women and men are equally likely to receive short-term weather forecasts. Men are more likely to have information about crop and pest management, but both women and men are equally likely to have information on livestock management and postharvest processing.

Given the differences in gender roles in agriculture and in access to information sources and types, it is not surprising that gaps exist in terms of women's and men's awareness and adoption of CSA practices (Table 10.1). Men are significantly more likely to report being aware of high yielding varieties, composting, efficient fertilizer use, integrated pest management (IPM), and tree planting. Notably, awareness of some key CSA practices, like stress-tolerant varieties, is very low by both women and men, which could reflect the lack of availability of these technologies.

A different pattern of gender differences emerges when examining rates of adoption of CSA practices by women and men, conditional on their awareness of the practice. Among those who are aware of the practice, men are more likely than women to adopt high yielding varieties and irrigation, and to engage in climate-smart methods of maintaining soil fertility, such as manure management and more efficient fertilizer use. Women, on the other hand, are significantly more likely to practice improved feed management for livestock and grain storage. Gender differences in the adoption of other CSA practices are insignificant, suggesting that, conditional on awareness, women and men are equally likely to adopt many CSA practices. That is, while there were significant differences in awareness of composting and tree planting between women and men, when aware, women and men appear equally likely to adopt such practices.

It is interesting to note that the practices both women and men adopt at higher rates are consistent with the gendered roles and responsibilities in Bangladeshi agriculture. Most irrigation is provided to field crops, for which men provide most of the labor, and men are more likely to be concerned about maintaining soil fertility for those crops as well – thus the higher reported practice of soil fertility enhancing techniques by men. On the other hand, women are in charge of feeding livestock and postharvest processing, and are more likely to adopt improved feed management and improved grain storage, once they are aware of these practices.

The descriptive results indicate that gender gaps in awareness are larger than the gender gaps in adoption, conditional on awareness. This finding implies that a first step in encouraging adoption of CSA practices would be to increase women's awareness of those practices. Nevertheless, even among those who are aware of the practice, gender gaps in adoption remain, suggesting that knowledge of the practice does not guarantee that it is adopted; other constraints must also be addressed. Moreover, gendered preferences and needs for information given individuals' roles in the household should be considered when reaching women and men with information on CSA.

Heckman selectivity model results on awareness and adoption of climate-smart agriculture practices

As the descriptive tables do not control for other individual and household characteristics, as well as other factors that may influence awareness and adoption, the results of the Heckman estimation of selectivity-corrected CSA adoption regressions are presented, based on the conceptual model described above. The goal of the econometric analysis in this section is to statistically test the extent to which gender gaps in awareness of CSA practices contribute to gendered patterns of adoption of those practices, while controlling for other factors that influence adoption. The results of this analysis using the Heckman two-step selection model are presented in Tables 10.2 and 10.3. The results in Table 10.2 show the first stage regression results predicting awareness (selection model) and Table 10.3 shows the second stage regression results predicting CSA adoption (outcome model). The inverse Mills ratio is significant for a few practices: irrigation, composting, and IPM. This finding suggests that selection bias is limited in the estimations as it affected very few CSA practices, justifying the use of the Heckman two-step estimator. As the data used for this analysis are cross-sectional and observational, the relationships analyzed here should be interpreted as associative rather than causal.

The selection equation estimates the probability of being aware of the practice as a function of individual and household characteristics, as well as access to information sources. As access to information sources is endogenous (as it is influenced by individual and household characteristics), the authors again caution that their results are only indicative of correlation, not

Table 10.1. Awareness and adoption of climate-smart agricultural (CSA) practices, by sex of respondent, Bangladesh.

	Whether respondent is aware of practice (proportion)						Whether respondent adopted practice (proportion of those who are aware)					
Practices	Men	Obs	Women	Obs	p-value		Men	Obs	Women	Obs	p-value	
Crop choice												
Planting stress-tolerant varieties	0.03	342	0.02	377	0.33		0.33	9	0.17	6	0.51	
Improved, high yielding varieties	0.64	342	0.42	377	0.00	***	0.57	218	0.46	158	0.04	**
Water and soil fertility management												
Irrigation	0.98	342	0.98	377	0.76		0.63	335	0.54	368	0.02	**
Leaving crop residue	0.58	342	0.54	377	0.28		0.43	198	0.41	203	0.61	
Composting	0.81	342	0.71	377	0.00	***	0.37	276	0.40	268	0.42	
Livestock manure management	0.64	342	0.60	377	0.26		0.50	220	0.37	227	0.00	***
More efficient use of fertilizer	0.89	342	0.56	377	0.00	***	0.84	306	0.65	213	0.00	***
No till/minimum tillage	0.31	342	0.28	377	0.45		0.06	105	0.04	106	0.51	
Other practices												
Improved feed management for livestock	0.31	342	0.27	377	0.22		0.54	105	0.71	100	0.01	**
Integrated pest management	0.79	342	0.63	377	0.00	***	0.49	271	0.47	239	0.62	
Grain storage	0.43	342	0.38	377	0.17		0.19	147	0.62	143	0.00	***
Tree planting	0.57	342	0.43	377	0.00	***	0.08	195	0.05	163	0.29	
Improved cookstoves	0.71	342	0.70	377	0.70		0.06	244	0.05	264	0.68	

*$p<0.1$; **$p<0.05$; ***$p<0.01$

Table 10.2. Awareness of climate-smart agriculture (CSA) practices in Bangladesh: selection equation.

	Crop residue	Irrigation	Compost	Integrated pest management	Livestock feed	Improved cookstoves	Grain storage	Fertilizer	Manure	Improved crop varieties	Tree planting
Female	−0.299**	0.526	−0.163	−0.611***	−0.240	0.458***	−0.0833	−0.858***	−0.232	−0.345**	0.00443
	(0.147)	(0.377)	(0.160)	(0.155)	(0.155)	(0.156)	(0.148)	(0.163)	(0.143)	(0.146)	(0.144)
Self-decision maker	−0.220*	0.494	0.0783	0.101	−0.313**	0.268*	−0.174	0.314**	−0.0185	0.408***	0.118
	(0.130)	(0.383)	(0.140)	(0.133)	(0.136)	(0.139)	(0.131)	(0.132)	(0.125)	(0.129)	(0.128)
Membership in groups	0.0715	0.137	0.0327	0.127	0.00670	0.135	0.0194	−0.0633	0.135	0.0180	0.108
	(0.0880)	(0.322)	(0.107)	(0.0979)	(0.0915)	(0.102)	(0.0896)	(0.102)	(0.0875)	(0.0893)	(0.0878)
Experienced crop shock	0.492***	0.231	0.0631	−0.118	0.318**	−0.169	0.469***	0.292*	0.379***	0.0997	0.344***
	(0.136)	(0.508)	(0.153)	(0.139)	(0.128)	(0.137)	(0.128)	(0.154)	(0.129)	(0.129)	(0.127)
Perceives drought risk	0.580***	0.399	0.342***	0.172	0.159	−0.211*	−0.197*	−0.552***	0.321***	−0.165	−0.202*
	(0.120)	(0.341)	(0.132)	(0.125)	(0.119)	(0.125)	(0.118)	(0.128)	(0.113)	(0.117)	(0.115)
Perceives flood risk	−0.278*	0.514	0.131	−0.207	−0.398***	0.0318	−0.218	0.167	−0.137	0.0641	0.0676
	(0.147)	(0.480)	(0.160)	(0.148)	(0.152)	(0.150)	(0.145)	(0.153)	(0.139)	(0.143)	(0.142)
Total farm area (log)	0.0668**	0.237**	0.0528	0.0246	0.0171	0.0420	−0.00901	0.0489	−0.0334	0.0297	−0.0105
	(0.0298)	(0.0987)	(0.0330)	(0.0309)	(0.0306)	(0.0309)	(0.0299)	(0.0325)	(0.0290)	(0.0295)	(0.0292)
Average distance to water sources (minutes)	−0.008**	−0.012	−0.00953***	0.00720**	−0.0136***	−0.00178	−0.0154***	−0.00787**	−0.00341	−0.00932***	−0.0145***
	(0.00313)	(0.00843)	(0.00338)	(0.00353)	(0.00384)	(0.00333)	(0.00340)	(0.00332)	(0.00293)	(0.00312)	(0.00319)
Income (log)	0.0474**	0.0423	0.0241	−0.0149	0.0266	−0.0289	0.0699***	0.0548***	0.0290	−0.00438	−0.00976
	(0.0199)	(0.0430)	(0.0202)	(0.0193)	(0.0195)	(0.0199)	(0.0198)	(0.0203)	(0.0181)	(0.0187)	(0.0183)
Household size (log)	−0.172	−0.739**	−0.189	0.103	−0.112	0.0934	−0.396***	−0.125	0.125	0.0833	−0.00747
	(0.143)	(0.372)	(0.156)	(0.149)	(0.149)	(0.148)	(0.144)	(0.157)	(0.138)	(0.145)	(0.140)
Years of education (log)	−0.0452	0.178	0.154**	0.0851	0.0143	0.144**	0.0849	−0.0533	−0.0133	−0.0523	0.0169
	(0.0633)	(0.167)	(0.0678)	(0.0658)	(0.0653)	(0.0651)	(0.0631)	(0.0711)	(0.0604)	(0.0616)	(0.0609)
Age (log)	0.118	−0.170	−0.471***	−0.0259	−0.192	−0.214	0.0785	0.0497	−0.303*	−0.567***	−0.271*
	(0.164)	(0.366)	(0.181)	(0.164)	(0.162)	(0.166)	(0.160)	(0.176)	(0.158)	(0.160)	(0.157)
Took loan	−0.253**	0.219	0.228*	0.0365	0.181	−0.0573	0.0598	−0.0862	−0.180	0.0953	−0.270**
	(0.124)	(0.335)	(0.131)	(0.123)	(0.125)	(0.126)	(0.120)	(0.133)	(0.119)	(0.120)	(0.119)

Continued

Table 10.2. Continued.

	Crop residue	Irrigation	Compost	Integrated pest management	Livestock feed	Improved cookstoves	Grain storage	Fertilizer	Manure	Improved crop varieties	Tree planting
Has land title (customary or govt.)	0.106	−0.264	0.104	−0.0225	0.0660	0.124	−0.0203	0.154	0.0655	−0.0110	−0.0362
	(0.133)	(0.361)	(0.144)	(0.136)	(0.136)	(0.138)	(0.130)	(0.143)	(0.128)	(0.133)	(0.128)
Bagerhat	−0.295**	−0.111	−0.154	−0.282**	0.499***	−0.726***	0.398***	−0.0880	−0.0370	−0.799***	−0.729***
	(0.130)	(0.376)	(0.144)	(0.136)	(0.136)	(0.137)	(0.131)	(0.143)	(0.127)	(0.131)	(0.129)
Info from govt. extension workers	0.0418	6.351	0.551**	0.0764	−0.205	0.0381	−0.0534	0.276	−0.260*	−0.217	0.247
	(0.160)	(0)	(0.218)	(0.181)	(0.163)	(0.176)	(0.161)	(0.209)	(0.154)	(0.159)	(0.158)
Info from NGOs	−0.164	6.478	0.0216	0.0959	0.190	−0.00332	−0.408**	−0.113	0.173	0.305*	−0.222
	(0.169)	(0)	(0.221)	(0.209)	(0.184)	(0.204)	(0.187)	(0.207)	(0.172)	(0.182)	(0.174)
Info from community meetings	0.875**	−2.870***	−0.296	−0.174	−0.279	−0.703	−1.098**	−1.329***	1.267**	0.0573	0.0749
	(0.420)	(1.002)	(0.476)	(0.448)	(0.457)	(0.452)	(0.531)	(0.426)	(0.560)	(0.431)	(0.438)
Info from religious groups	−0.0182	6.606	−0.145	0.767	0.636*	−0.241	0.561	0.316	0.529	0.245	0.963**
	(0.351)	(0)	(0.425)	(0.550)	(0.366)	(0.399)	(0.423)	(0.554)	(0.378)	(0.370)	(0.431)
Info from agriservice providers	−0.301	5.273	0.0984	0.224	0.0499	−0.179	0.240	0.0523	−0.322*	−0.242	−0.0140
	(0.194)	(0)	(0.264)	(0.246)	(0.200)	(0.225)	(0.199)	(0.270)	(0.195)	(0.207)	(0.196)
Info from family members	0.522***	0.218	0.384***	0.119	0.297**	0.0443	0.652***	0.0206	0.278**	0.156	0.274**
	(0.123)	(0.307)	(0.133)	(0.130)	(0.132)	(0.128)	(0.127)	(0.140)	(0.118)	(0.124)	(0.123)
Info from neighbors	−0.00709	−0.975*	−0.161	0.685***	0.251*	−0.463***	−0.488***	0.221	0.108	−0.214	−0.408***
	(0.145)	(0.560)	(0.160)	(0.143)	(0.152)	(0.153)	(0.142)	(0.160)	(0.138)	(0.141)	(0.141)
Info from radio	0.534***	0.385	0.399***	−0.146	−0.0313	0.259**	0.0813	0.372***	0.310***	0.0942	0.224*
	(0.119)	(0.351)	(0.131)	(0.125)	(0.121)	(0.125)	(0.118)	(0.131)	(0.114)	(0.117)	(0.115)
Info from television	0.139	0.00534	0.0540	−0.153	−0.0148	0.279**	−0.0117	0.0443	0.319**	0.173	0.311**
	(0.131)	(0.320)	(0.140)	(0.133)	(0.135)	(0.134)	(0.130)	(0.134)	(0.126)	(0.129)	(0.129)
Info from newspaper/ bulletin	−0.436**	5.999	−0.422**	0.233	0.248	0.259	−0.204	0.312	−0.519***	0.410**	0.475**
	(0.184)	(0)	(0.207)	(0.212)	(0.194)	(0.213)	(0.193)	(0.243)	(0.178)	(0.193)	(0.188)

Info from traditional sources/own experience	−0.602***	0.384	0.652***	0.717***	0.233**	0.697***	0.175	−0.365***	−0.134	0.439***	0.141
	(0.113)	(0.329)	(0.127)	(0.118)	(0.115)	(0.119)	(0.111)	(0.122)	(0.108)	(0.111)	(0.109)
Info from farmer field days	−0.604**	7.663	0.116	0.186	0.593**	0.671**	0.863***	1.006**	−0.531**	0.337	0.111
	(0.242)	(0)	(0.327)	(0.323)	(0.239)	(0.298)	(0.255)	(0.438)	(0.235)	(0.241)	(0.237)
Inverse Mills Ratio	−0.0591	−0.273	−0.326***	−0.342***	−0.0793	0.00923	−0.113	0.196*	−0.150	0.225*	−0.0678
	(0.0787)	(0.230)	(0.109)	(0.101)	(0.137)	(0.0526)	(0.102)	(0.105)	(0.123)	(0.134)	(0.0700)
Constant	−0.602	2.507	1.405*	−0.115	−0.534	0.877	−0.722	0.353	0.673	2.121***	1.424**
	(0.726)	(1.786)	(0.793)	(0.730)	(0.720)	(0.736)	(0.714)	(0.782)	(0.694)	(0.709)	(0.696)
Observations	719	719	719	719	719	719	719	719	719	719	719

Standard errors in parentheses
*p<0.1; **p<0.05; ***p<0.01

causality. First stage regression estimates on awareness presented in Table 10.2 suggest that, consistent with the descriptive statistics, women are less aware of many CSA practices, including the application of crop residues, integrated pest management, efficient fertilizer use, and improved crop varieties. However, women appear to be more aware of improved cookstoves, when controlling for access to sources of information, and other individual and household characteristics. Other individual characteristics, besides gender, also influence awareness of CSA practices. Older farmers appear to have less awareness of composting, manure management, improved crop varieties, and tree planting as shown by the negative estimates on the log of age. As expected, better educated respondents are more likely to know about composting and improved cookstoves.

Access to different sources of agricultural information also influences awareness of CSA practices, although the results should be treated cautiously given that very few farmers have access to some sources of information that are included in the model. Access to information from government extension workers only increases awareness of composting but is associated with decreased awareness of manure management. This source reaches only 18% of farmers in the sample and among them, most are men. The results highlight the limited reach of formal extension services and also suggest that extension messages are targeted to the interests of male farmers and do not include information on many CSA practices, like livestock feeding, that may be aligned with women's roles. Among other formal sources of information, only farmer field days, where farmers receive demonstrations of agricultural practices, seem effective at increasing awareness of many CSA practices including livestock feeding practices, improved cookstoves, grain storage, and efficient fertilizer use – although only 6% of farmers in the sample participate in field days and most are men. Other, more formal sources of information, such as NGOs, community meetings, and agriservice providers have little influence on farmers' awareness of CSA practices and the results are mixed. Few farmers receive information from community meetings, but those that do are more aware of crop residues and manure management but less aware of irrigation, grain storage, and efficient fertilizer use. Receiving information from NGOs is associated with lower awareness of grain storage but higher awareness of improved crop varieties.

These results suggest that NGOs and community groups deliver only specific messages related to ongoing projects and may not be providing messages about CSA options more broadly. NGOs in Bangladesh have a long history of successfully targeting microfinance and extension messages to women's groups using implementation approaches that address gender-specific constraints to agricultural technology adoption, such as hiring female extension workers (Birner *et al.*, 2010). Through such group-based approaches, women have been able to adopt improved varieties for homestead gardening and polyculture fish technologies (Hallman *et al.*, 2007). Notably, these practices were not included in the list of CSA practices covered by this survey and, therefore, the authors are unable to evaluate whether NGOs have been effective at increasing awareness and adoption of these practices in the study sites. However, the results suggest a missed opportunity for NGOs and other local organizations to share information on CSA practices that may benefit women farmers.

Among informal sources of information, receiving information from family members appears to increase awareness of almost all CSA practices. Receiving information from religious groups also has a positive effect on awareness of some practices, namely improved livestock feeding and tree planting. On the other hand, relying on neighbors for agricultural or climate information has mixed effects on awareness of CSA practices. Those who rely on their own experience or traditional sources of knowledge also were more likely to be aware of some CSA practices like composting, IPM, improved cookstoves, and improved varieties but less aware of crop residues and efficient fertilizer use. Having access to information from radio increases awareness of most practices, while having access to information from television also increases awareness of improved cookstoves and manure management. Newspapers or bulletins tend to have mixed effects on awareness across the range of CSA practices.

Household characteristics also influence awareness of CSA practices. Households with higher income are more likely to be aware of crop residues, improved grain storage, and efficient fertilizer practices. The size of agricultural land is associated with increased awareness of crop residues and irrigation, while having a government right or customary right to land does not influence awareness of any CSA practices. Average distance to domestic water sources appears to be negatively associated with awareness of most CSA practices. This negative association may reflect the remote location of households with limited access to domestic water or the opportunity cost of having to spend significant time collecting domestic water, which may hinder participation in groups or other activities where farmers may have access to information. Borrowing behavior also affects awareness of CSA practices in different directions – households that took a loan in the last year were more aware of composting and less aware of crop residues, and tree planting. Finally, respondents from households that experienced crop losses as a result of climate shocks were more likely to report higher awareness of several CSA practices. This may be because farmers seek out agricultural information when they are faced with crop shocks. Perceiving future climate risks (related to drought and flood) negatively influenced awareness of several CSA practices. This suggests that farmers who are most concerned about climate change impacts may be less able to adopt farming practices to mitigate future climate risks. The exception is in the case of farmers perceiving future risks from drought – these farmers report higher awareness of drought-mitigating practices like using crop residues, irrigation, and manure management.

Table 10.3 presents selectivity-corrected estimates of the probability of adopting CSA practices in the past year. These results show that once awareness is taken into account, women appear to be equally likely to adopt most CSA practices – there are no significant differences in adoption rates for crop residues, irrigation, composting, IPM, improved cookstoves, manure management, and tree planting. Moreover, while women do still appear to be less likely to adopt efficient fertilizer use and improved crop varieties, they are more likely than men to adopt improved livestock feeding practices and grain storage. These results remain consistent with the descriptive results of adoption patterns conditional on awareness, even after controlling for other factors. The results suggest that women prefer to adopt practices that relate to their gendered roles, as livestock feeding and grain storage are typically women's activities.

Other individual characteristics also appear to be correlated with adoption of CSA practices. Being the main decision maker on at least some of the household plots increases the likelihood of adopting many CSA practices. Similarly, membership in groups increases the likelihood of adopting irrigation, improved livestock feed, improved cookstoves, efficient fertilizer use, and improved crop varieties. Receiving a loan in the last year is also associated with higher adoption of crop residues, irrigation, and IPM, suggesting that loans are used for agricultural investments. As with awareness, younger farmers also seem more likely to try new agricultural practices, including crop residues, irrigation, and improved crop varieties, while education is only associated with higher levels of adoption of grain storage.

Household characteristics also appear to influence adoption of CSA practices. Farm size, as a measure of wealth, appears to drive adoption of almost all CSA practices, except for improved cookstoves and tree planting – two practices which are not related to farm size given that trees are typically planted off-farm. Similarly, having a land title is associated with higher adoption of irrigation, composting, IPM, and manure management. Labor availability also appears to be a factor in adoption of CSA practices. Larger households are more likely to adopt crop residues but less likely to use efficient fertilizer practices. Average time to sources of domestic water (a measure of water insecurity and labor burden) increases the likelihood of adopting crop residues and improved cookstoves.

Although experience with climate shocks resulting in crop losses seems to increase awareness of CSA practices, experience with adverse shocks has little impact on adoption of CSA practices, apart from increasing adoption of improved livestock feeding practices. However, perceiving climate risks (more frequent and severe droughts and floods) increases the likelihood of adopting a few practices – households perceiving future drought risks were more likely to adopt crop residues and IPM, while households perceiving flood risk were more likely to plant trees but less likely to adopt improved varieties. This result suggests that farmers may seek to

Table 10.3. Adoption of climate-smart agriculture (CSA) practices in Bangladesh: outcome equation.

	Crop residue	Irrigation	Compost	Integrated pest management	Livestock feed	Improved cookstoves	Grain storage	Fertilizer	Manure	Improved crop varieties	Tree planting
Female	0.0382	0.00654	0.0552	0.0500	0.281***	−0.0200	0.405***	−0.138**	−0.0724	−0.153**	0.00776
	(0.0532)	(0.0419)	(0.0538)	(0.0562)	(0.0808)	(0.0237)	(0.0650)	(0.0565)	(0.0571)	(0.0687)	(0.0338)
Self-decision maker	0.146***	0.219***	−0.0217	−0.0258	−0.00953	−0.0118	−0.154**	0.265***	0.0958	0.152**	0.0537
	(0.0542)	(0.0438)	(0.0542)	(0.0534)	(0.0935)	(0.0246)	(0.0661)	(0.0476)	(0.0602)	(0.0661)	(0.0337)
Membership in groups	0.0264	0.0568**	0.0468	−0.0301	0.110**	0.0332**	0.0588	0.0578**	0.0427	0.109***	0.00132
	(0.0351)	(0.0284)	(0.0338)	(0.0350)	(0.0474)	(0.0161)	(0.0367)	(0.0263)	(0.0371)	(0.0402)	(0.0217)
Experienced crop shock	−0.00225	−0.0116	−0.0603	0.0384	0.137*	−0.0325	0.0591	0.0619	0.00876	0.0278	−0.0298
	(0.0523)	(0.0424)	(0.0503)	(0.0524)	(0.0768)	(0.0247)	(0.0587)	(0.0403)	(0.0586)	(0.0567)	(0.0317)
Perceives drought risk	0.225***	0.0505	0.0398	0.193***	−0.0243	0.0136	−0.0222	−0.0765*	0.00281	0.0197	−0.0155
	(0.0487)	(0.0381)	(0.0474)	(0.0460)	(0.0646)	(0.0224)	(0.0506)	(0.0435)	(0.0528)	(0.0534)	(0.0312)
Perceives flood risk	−0.0426	−0.0386	0.0820	0.0145	−0.153	−0.0199	−0.0941	−0.0690	0.0350	−0.124*	0.0913**
	(0.0597)	(0.0468)	(0.0573)	(0.0609)	(0.103)	(0.0276)	(0.0722)	(0.0458)	(0.0649)	(0.0699)	(0.0389)
Total farm area (log)	0.103***	0.0541***	0.0416***	0.0751***	0.0496***	0.00161	0.0717***	0.0920***	0.0418***	0.0804***	0.00372
	(0.0120)	(0.0103)	(0.0121)	(0.0123)	(0.0180)	(0.00594)	(0.0141)	(0.00984)	(0.0133)	(0.0144)	(0.00784)
Average distance to water sources (minutes)	0.00411***	−0.00131	−0.000563	−0.00105	2.79e−05	0.00165***	0.00176	−0.00157	−0.000917	−0.00307*	−0.000850
	(0.00140)	(0.00101)	(0.00135)	(0.00119)	(0.00292)	(0.000538)	(0.00216)	(0.000983)	(0.00140)	(0.00157)	(0.00106)
Income (log)	0.00950	−0.00379	0.00860	−0.00425	0.0148	−0.00239	−0.0142	0.00482	0.0146	−0.00678	0.00679
	(0.0108)	(0.00648)	(0.00809)	(0.00814)	(0.0115)	(0.00382)	(0.0111)	(0.00716)	(0.00985)	(0.0103)	(0.00529)
Household size (log)	0.0973*	0.0158	0.00772	−0.00531	0.117	0.000731	0.100	−0.103**	−0.0102	−0.0826	−0.0319
	(0.0585)	(0.0480)	(0.0596)	(0.0586)	(0.0882)	(0.0280)	(0.0737)	(0.0462)	(0.0649)	(0.0669)	(0.0378)
Years of education (log)	0.00534	0.00439	−0.00748	−0.0332	0.0298	0.0182	0.0915***	0.0268	0.00226	0.0173	0.0105
	(0.0245)	(0.0205)	(0.0253)	(0.0252)	(0.0369)	(0.0121)	(0.0285)	(0.0189)	(0.0268)	(0.0274)	(0.0160)
Age (log)	−0.109*	−0.109**	0.0203	−0.0668	0.212**	0.0191	0.201***	0.00775	−0.00855	−0.195**	0.00370
	(0.0653)	(0.0548)	(0.0675)	(0.0696)	(0.0921)	(0.0308)	(0.0769)	(0.0540)	(0.0725)	(0.0778)	(0.0429)

Took loan	0.110**	0.0744*	0.00876	0.130**	0.0173	0.0188	−0.126**	0.00690	0.000563	−0.0707	0.0210
	(0.0476)	(0.0397)	(0.0495)	(0.0508)	(0.0818)	(0.0229)	(0.0549)	(0.0384)	(0.0526)	(0.0574)	(0.0325)
Has land title (customary or govt.)	−0.0647	0.101**	0.116**	0.121**	−0.0910	−0.0184	0.0344	−0.0158	0.0953*	−0.0223	0.0362
	(0.0523)	(0.0431)	(0.0535)	(0.0540)	(0.0853)	(0.0254)	(0.0573)	(0.0428)	(0.0573)	(0.0582)	(0.0319)
Bagerhat	0.0873*	0.117***	0.0473	0.214***	0.130	0.0326	0.128**	0.116***	0.121**	−0.296***	0.0265
	(0.0530)	(0.0419)	(0.0520)	(0.0544)	(0.0828)	(0.0306)	(0.0622)	(0.0401)	(0.0555)	(0.0965)	(0.0491)
Constant	−0.0735	0.521**	0.0357	0.401	−0.811*	−0.0594	−0.759**	0.294	0.127	1.117***	−0.0352
	(0.295)	(0.235)	(0.285)	(0.305)	(0.443)	(0.132)	(0.360)	(0.238)	(0.308)	(0.310)	(0.175)
Observations	719	719	719	719	719	719	719	719	719	719	719

Standard errors in parentheses
*p<0.1; **p<0.05; ***p<0.01

avoid the risk of losing the investment in high yielding seeds and/or that farmers are not aware of the benefits of or do not have access to submergence-tolerant varieties.

Conclusions and Policy Implications

This chapter explores the determinants of adoption of CSA practices in Bangladesh, a country characterized both by vulnerability to climate change and a high degree of gender inequality. Based on a specially designed survey administered to women and men within the same households in districts of south-west Bangladesh, the results shed light on possible gender-specific constraints to adoption of CSA practices. Gendered constraints to and patterns of adoption of climate-smart practices are likely to vary across contexts, depending on gender roles in agriculture, the institutional context, the distribution of resources and services, and other factors. While there are likely to be differences across contexts, this chapter highlights the importance of considering gender dimensions when designing and implementing interventions to promote CSA.

The authors find that information access is low all around, for men as well as for women. Consistent with the high degree of gender inequality and their lower levels of human and physical capital, women get less information from formal agricultural information services but are more likely to have access to informal channels of information (family members and neighbors). Extension agents, agriculture service providers, and other formal sources of information do not reach many of their key target beneficiaries and are not always effective at communicating information about CSA practices. Most information comes from the radio and television as well as informal sources, such as neighbors and family members; however, these sources are also inefficient at increasing awareness of CSA practices, with the exception of family members.

Partly as a result of low information access, awareness of many important CSA practices, such as stress-tolerant varieties, minimum tillage, cover cropping, and improved feed management for livestock, is also extremely low. Large gender gaps exist in accessing information on agriculture and climate change and in awareness of practices for CSA. Other individual and household characteristics also influence access to information and awareness of CSA practices. Farmers cannot adopt practices they are not aware of, so improving access to information about these practices and how they can be applied appropriately in the different agroecological contexts of the country is an important step to increase adoption and promote resilience to climate change.

Gender gaps in adoption of CSA practices are smaller among those who are aware of the practice. Although women tend to be at a disadvantage in terms of information access and awareness, they are just as likely as men, and in some cases even more likely, to adopt some CSA practices when aware. The results showed that women are more likely to adopt CSA practices related to livestock feeding and grain storage – practices that are consistent with their traditional roles in the household – while men are more likely to adopt some practices related to field crop production. While lack of information and awareness are found to be key constraints to adoption of CSA practices, other factors also emerge as important determinants of adoption, including household wealth, as measured by size of land holdings, labor availability, access to credit, experience with climate shocks, and perceptions of future climate risk.

Given women's increasing role in agricultural production activities and decisions, and because they are responsible for key activities related to livestock care and postharvest processing, targeting women with extension information, including from government extension services and service providers, makes sense and should be promoted to encourage more widespread adoption of CSA practices. The sectors or commodities in which women are already heavily involved (such as livestock) could be used as an entry point for increasing knowledge about CSA more broadly. Given the success of NGOs at reaching women with services, such as microcredit, in other parts of the country, these organizations may be an important vehicle for further expanding women's access to information about other climate-smart practices. This way, women would learn first about practices related to their spheres of responsibility, and then expand that knowledge in the future through established channels of information. Finally, given the joint nature of family farming in Bangladesh, it makes sense for all family members to become more knowledgeable about CSA. Women and men have different but complementary roles in ensuring environmental sustainability; empowering both women and men with the knowledge and resources to fulfill these roles is an important step to achieve this goal.

Acknowledgments

This work was implemented as part of the CGIAR Research Program on Climate Change, Agriculture and Food Security (CCAFS), which is carried out with support from the CGIAR Trust Fund and through bilateral funding agreements. For details please visit www.ccafs.cgiar.org/donors (accessed 8 February 2021). The views expressed in this document cannot be taken to reflect the official opinions of these organizations. The intra-household survey was developed in the fall of 2012 by researchers from the International Food Policy Research Institute (IFPRI), the International Livestock Research Institute (ILRI), the World Agroforestry Centre (ICRAF), and the International Center for Tropical Agriculture (CIAT), building on a comprehensive farm characterization survey called IMPACTLite that had already been conducted in several sites in Africa, Asia, and Latin America, where CCAFS is promoting CSA practices. The authors would like to thank Agnes Quisumbing and Chiara Kovarik for their contributions to an earlier version of this paper and Claudia Ringler for helpful comments.

Notes

[1] The intra-household survey was implemented in six sites across four countries in 2013 (two sites in Bangladesh, two sites in Kenya, one site in Uganda, and one site in Senegal).

[2] These men and women are not necessarily the household head and spouse: 348 respondents are heads and 347 are spouses, but out of these 348 household heads, 331 are male, and out of the 347 spouses, 336 are female.

References

Ahmad, N. (2012) Gender and climate change in Bangladesh: the role of institutions in reducing gender gaps in adaptation programs. *World Bank Social Development Papers, 126*. World Bank, Washington, DC.

Ahmed, A., Ahmad, K., Chou, V., Hernandez, R., Menon, P. *et al.* (2013) The status of food security in the Feed the Future zone and other regions of Bangladesh: results from the 2011–2012 Bangladesh Integrated Household Survey. Dhaka. Available at: www.ifpri.org/sites/default/files/publications/bihstr.pdf (accessed 3 February 2021).

Alam, G.M.M., Alam, K. and Mushtaq, S. (2017) Climate change perceptions and local adaptation strategies of hazard-prone rural households in Bangladesh. *Climate Risk Management* 17, 52–63.

Ali, A.M.S. (2005) Home gardens in smallholder farming systems: examples from Bangladesh. *Human Ecology* 33, 245–270.

BBS (2018) 2017 Statistical Year Book Bangladesh. Bangladesh Bureau of Statistics, Dhaka. Available at: www.bbs.portal.gov.bd/sites/default/files/files/bbs.portal.gov.bd/page/9ead9eb1_91ac_4998_a1a3_a5caf4ddc4c6/CPI_January18.pdf (accessed 10 June 2019).

Beauman, L. and Dillon, A. (2014) The diffusion of agricultural technologies within social networks: evidence from composting in Mali. In: Ringler, C., Quisumbing, A.R., Bryan, E. and Meinzen-Dick, R. (eds) *Enhancing Women's Assets to Manage Risk under Climate Change: Potential for Group-Based Approaches*. International Food Policy Research Institute (IFPRI), Washington, DC, pp. 47–49.

Bernier, Q., Meinzen-Dick, R., Kristjanson, P., Haglund, E., Kovarik, C. *et al.* (2015) Gender and institutional aspects of climate-smart agricultural practices: evidence from Kenya. *CCAFS Working Paper No. 79*. CGIAR Research Program on Climate Change, Agriculture and Food Security (CCAFS), Copenhagen.

Birner, R., Quisumbing, A.R. and Ahmed, N. (2010) Cross-cutting issues: governance and gender. Paper prepared for Bangladesh Food Security Investment Forum, 26–27 May 2010, Dhaka.

Boserup, E. (1970) *Woman's Role in Economic Development*. St. Martin's Press, New York, 283 pp.

Bryan, E., Theis, S., Choufani, J., De Pinto, A., Meinzen-Dick, R. and Ringler, C. (2017) Gender-sensitive, climate-smart agriculture for improved nutrition in Africa south of the Sahara. In: De Pinto, A. and Ulimwengu, J.M. (eds) *2016 ReSAKSS Annual Trends and Outlook Report*. International Food Policy Research Institute (IFPRI), Washington, DC. Available at: www.resakss.org/sites/default/files/Ch9%20ReSAKSS_AW_ATOR_2016_Final.pdf (accessed 3 February 2021).

Chaudhury, M., Kristjanson, P., Kyagazze, F., Naab, J.B. and Neelormi, S. (2012) Participatory gender-sensitive approaches for addressing key climate change-related research issues: evidence from Bangladesh, Ghana, and Uganda. *CCAFS Working Paper No. 19*. CGIAR Research Program on Climate Change, Agriculture and Food Security (CCAFS), Copenhagen.

Corcoran-Nantes, Y. and Roy, A. (2018) Gender, climate change, and sustainable development in Bangladesh. In: McIntyre-Mills, J., Romm, N. and Corcoran-Nantes, Y. (eds) *Balancing Individualism and Collectivism*. Springer, New York.

De Pinto, A., Seymour, G., Bryan, E. and Bhandari, P. (2019) Women's empowerment and crop diversification in Bangladesh: a possible pathway to climate change adaptation and better nutrition. *IFPRI Discussion Paper No. 1849*. International Food Policy Research Institute (IFPRI), Washington, DC.

Deressa, T.T., Hassan, R.M., Ringler, C., Alemu, T. and Yesuf, M. (2009) Determinants of farmers' choice of adaptation methods to climate change in the Nile Basin of Ethiopia. *Global Environmental Change* 19, 248–255. doi.org/10.1016/j.gloenvcha.2009.01.002.

Deressa, T.T., Hassan, R.M. and Ringler, C. (2011) Perception of and adaptation to climate change by farmers in the Nile Basin of Ethiopia. *The Journal of Agricultural Science* 149, 23–31. doi.org/10.1017/S0021859610000687.

Doss, C.R., Summerfield, G. and Tsikata, D. (2014) Land, gender, and food security. *Feminist Economics* 20, 1–23.

FAO (2010) 'Climate-smart' agriculture: policies, practices and financing for food security, adaptation and mitigation. Food and Agriculture Organization of the United Nations, Rome. Available at: www.fao.org/docrep/013/i1881e/i1881e00.pdf (accessed 3 February 2021).

FAO (2013) Climate-Smart Agriculture Sourcebook. Food and Agriculture Organization of the United Nations, Rome.

Hallman, K., Lewis, D. and Begum, S. (2007) Assessing the impact of vegetable and fishpond technologies on poverty in rural Bangladesh. In: Adato, M. and Meinzen-Dick, R. (eds) *Agricultural Research, Livelihoods, and Poverty: Studies of Economic and Social Impacts in Six Countries*. Johns Hopkins University Press, Baltimore, MD.

Heckman, J.J. (1979) Sample selection bias as a specification error. *Econometrica* 47, 153–161.

Huyer, S. (2016) Closing the gender gap in agriculture. *Gender, Technology and Development* 20, 105–116. doi.org/10.1177/0971852416643872.

Jost, C., Kyazze, F., Naab, J., Neelormi, S., Kinyangi, J. *et al.* (2016) Understanding gender dimensions of agriculture and climate change in smallholder farming communities. *Climate and Development* 8, 133–144.

Kabeer, N. (1994) Women's labor in the Bangladesh garment industry: choices and constraints. In: El-Solh, F. and Mabro, J. (eds) *Muslim Women's Choices: Religious Belief and Social Reality*. Berg Publishers, Oxford, UK, pp. 164–183.

Kabir, M.J., Alauddin, M. and Crimp, S. (2017) Farm-level adaptation to climate change in western Bangladesh: an analysis of adaptation dynamics, profitability and risks. *Land Use Policy* 64, 212–224.

Kaliba, A.R., Verkuijl, H. and Mwangi, W. (2000) Factors affecting adoption of improved maize seed and use of inorganic fertiliser for maize production in the intermediate and lowland zone of Tanzania. *Journal of Agricultural and Applied Economics* 32, 35–47. doi.org/10.1017/S1074070800027802.

Katungi, E., Edmeades, S. and Smale, M. (2008) Gender, social capital and information exchange in rural Uganda. *Journal of International Development* 20, 35–52.

Khalil, M.B., Jacobs, B.C., McKenna, K. and Kuruppu, N. (2020) Female contribution to grassroots innovation for climate change adaptation in Bangladesh. *Climate and Development* 12, 664–676. Advance online publication, doi.org/10.1080/17565529.2019.1676188.

Kieran, C., Sproule, K., Doss, C., Quisumbing, A.R. and Kim, S.M. (2015) Examining gender inequalities in land rights indicators in Asia. *Agricultural Economics* 46, 119–138. doi.org/10.1111/agec.12202.

Kristjanson, P., Bryan, E., Bernier, Q., Twyman, J., Meinzen-Dick, R.S. *et al.* (2017) Addressing gender in agricultural research for development in the face of a changing climate: where are we and where should we be going? *International Journal of Agricultural Sustainability* 15, 482–500. doi.org/10.1080/14735903.2017.1336411.

Lele, U. (1986) Women and structural transformation. *Economic Development and Cultural Change* 34, 195–221.

McBride, W.D. and Daberkow, S.G. (2003) Information and the adoption of precision farming technologies. *Journal of Agribusiness* 21, 1–18.

Meinzen-Dick, R., Quisumbing, A.R., Behrman, J., Biermayr-Jenzano, P., Wilde, V. *et al.* (2010) Engendering agricultural research. *IFPRI Discussion Paper No. 973*. International Food Policy Research Institute (IFPRI), Washington, DC. Available at: www.indiaenvironmentportal.org.in/files/ifpri-may2010.pdf (accessed 3 February 2021).

ND-GAIN (2018) ND-GAIN Country Index. Notre Dame Global Adaptation Initiative, University of Notre Dame, IN. Available at: gain.nd.edu/our-work/country-index/ (accessed 1 February 2021).

Quisumbing, A.R., Roy, S., Njuki, J., Tanvin, K. and Waithanji, E. (2013) Can dairy value-chain projects change gender norms in rural Bangladesh? Impacts on assets, gender norms, and time use. *IFPRI Discussion Paper No. 1311*. International Food Policy Research Institute (IFPRI), Washington, DC.

Quisumbing, A.R., Kumar, N. and Behrman, J.A. (2018) Do shocks affect men's and women's assets differently? Evidence from Bangladesh and Uganda. *Development Policy Review* 36, 3–34. doi.org/10.1111/dpr.12235.

Rahman, A., Alam, M., Mainuddin, K., Ali, L., Alauddin, S.M. *et al.* (2009) The probable impacts of climate change on poverty and economic growth and the options of coping with adverse effects of climate change in Bangladesh. Planning Commission, Government of the People's Republic of Bangladesh and UNDP Bangladesh, Dhaka. Available at: www.climatechange-asiapac.com/resource/probable-impacts-climate-change-poverty-and-economic-growth-and-options-coping-adverse (accessed 1 February 2021).

Rahman, M.R., Lateh, H. and Islam, M.N. (2018) Climate of Bangladesh: temperature and rainfall changes, and impact on agriculture and groundwater – a GIS-based analysis. In: Islam, M. and van Amstel, A. (eds) *Bangladesh I: Climate Change Impacts, Mitigation and Adaptation in Developing Countries*. Climate Change Management series, Springer, New York.

Rahman, S. (2000) Women's employment in Bangladesh agriculture: composition, determinants and scope. *Journal of Rural Studies* 16, 497–507. doi.org/10.1016/S0743-0167(00)00006-1.

Rakib, M. and Matz, J.A. (2016) The impact of shocks on gender-differentiated asset dynamics in Bangladesh. *The Journal of Development Studies* 52, 377–395. doi.org/10.1080/00220388.2015.1093117.

Rasul, G. and Thapa, G.B. (2004) Sustainability of ecological and conventional agricultural systems in Bangladesh: an assessment based on environmental, economic and social perspectives. *Agricultural Systems* 79, 327–351. doi.org/10.1016/S0308-521X(03)00090-8.

Roy, S., Ara, J., Das, N. and Quisumbing, A.R. (2015) 'Flypaper effects' in transfers targeted to women: evidence from BRAC's 'Targeting the Ultra Poor' program in Bangladesh. *Journal of Development Economics* 117, 1–19. doi.org/10.1016/j.jdeveco.2015.06.004.

Ruane, A.C., Major, D.C., Yu, W.H., Alam, M., Hussain, S.G. *et al.* (2013) Multi-factor impact analysis of agricultural production in Bangladesh with climate change. *Global Environmental Change* 23, 338–350. doi.org/10.1016/j.gloenvcha.2012.09.001.

Rufino, M.C., Quiros, C., Boureima, M., Desta, S., Douxchamps, S. *et al.* (2013) Developing generic tools for characterizing agricultural systems for climate and global change studies (IMPACTlite – phase 2). Report of Activities Submitted by ILRI to the CGIAR Research Program on Climate Change, Agriculture and Food Security (CCAFS), Copenhagen. Available at: www.cgspace.cgiar.org/bitstream/handle/10568/42065/Rufino_Report.pdf (accessed 3 February 2021).

Saroar, M.M. and Filho, W.L. (2016) Adaptation through climate smart agriculture: status and determinants in coastal Bangladesh. In: Filho, W.L., Musa, H., Cavan, G., O'Hare, P. and Seixas, J. (eds) *Climate Change Adaptation, Resilience and Hazards*. Climate Change Management series, Springer, New York. doi.org/10.1007/978-3-319-39880-8_10.

Seymour, G. (2017) Women's empowerment in agriculture: implications for technical efficiency in rural Bangladesh. *Agricultural Economics* 48, 513–522.

Shah, T. (2009) Climate change and groundwater: India's opportunities for mitigation and adaptation. *Environmental Research Letters* 4:035005.

Sraboni, E., Malapit, H.J., Quisumbing, A.R. and Ahmed, A.U. (2014) Women's empowerment in agriculture: what role for food security in Bangladesh? *World Development* 61, 11–52. doi.org/10.1016/j.worlddev.2014.03.025.

Tall, A., Kristjanson, P., Chaudhury, M., Mckune, S. and Zougmore, R. (2014) Who gets the information? Gender, power and equity considerations in the design of climate services for farmers. *CCAFS Working Paper*

No. 89. CGIAR Research Program on Climate Change, Agriculture and Food Security (CCAFS), Copenhagen.

Taneja, G., Pal, B.D., Joshi, P.K., Aggarwal, P.K. and Tyagi, N.K. (2019) Farmers' preferences for climate-smart agriculture – an assessment in the Indo-Gangetic plain. In: Pal, B., Kishore, A., Joshi, P. and Tyagi, N. (eds) *Climate Smart Agriculture in South Asia*. Springer, Singapore. doi.org/10.1007/978-981-10-8171-2_5.

Thomas, T.S., Mainuddin, K., Chiang, C., Rahman, A., Haque, A. *et al.* (2013) Agriculture and adaptation in Bangladesh: current and projected impacts of climate change. *IFPRI Discussion Paper No. 1281*. International Food Policy Research Institute (IFPRI), Washington, DC. Available at: www.ifpri.org/sites/default/files/publications/ifpridp01281.pdf (accessed 3 February 2021).

Twyman, J., Green, M., Bernier, Q., Kristjanson, P., Russo, S. *et al.* (2014) *Adaptation Actions in Africa: Evidence that Gender Matters*. International Food Policy Research Institute (IFPRI), Washington, DC.

World Bank (2020) World Development Indicators. Available at: www.databank.worldbank.org/data/views/variableSelection/selectvariables.aspx?source=world-development-indicators (accessed 3 February 2021).

World Bank and IFPRI (2010) *Gender and Governance in Rural Services: Insights from India, Ghana and Ethiopia*, World Bank and International Food Policy Research Institute. The World Bank, Washington, DC.

World Bank, FAO and IFAD (2009) *Gender in Agriculture Sourcebook: Agriculture and Rural Development*. World Bank, Food and Agriculture Organization of the United Nations, International Fund for Agricultural Development. World Bank, Washington, DC.

Yu, W.H., Alam, M., Hassan, A., Khan, A.S., Ruane, A.C. *et al.* (2010) Climate change risks and food security in Bangladesh. Earthscan, London and Washington, DC. doi.org/10.4324/9781849776387.

11 Gender and Climate Change Adaptation in Livestock Production in Tunisia

D. Najjar and B. Baruah

Livestock, Climate Change and Gender

Livestock are an essential component of the livelihood strategies of the rural poor throughout developing countries, as they are the assets, capital, and 'insurance policies' of people without access to banks, and with high vulnerabilities to crop failures, illnesses, and other threats (Njuki and Sanginga, 2013). As such, the economic security that livestock are associated with also represent an important aspect of building resilience (Dumas *et al.*, 2018). Livestock tend to be the primary livelihood activity, especially for rural households and communities located in areas that are dry or have more extreme climates where crop cultivation is not as reliable (Turner and Williams, 2002; Archambault, 2016). In other geographical regions, livestock are integral to crop production as well as livelihood security (Fisher *et al.*, 2000; Debela, 2017). For many communities, livestock can represent a reliable food source (usually in the form of milk), traction for work, a security mechanism which allows households to better absorb shocks and obtain better incomes (Thomas-Slayter and Bhatt, 1994; Curry *et al.*, 1996; Fisher *et al.*, 2000; Turner and Williams, 2002; Njuki and Sanginga, 2013; Archambault, 2016; Dumas *et al.*, 2018). Raising livestock makes up 33% of the global GDP (Galie *et al.*, 2018). However, in developing rural areas, this number can be much higher; for example, 752 million of the world's rural poor keep livestock and it is often the primary means through which they earn livelihoods (FAO, 2013).

As climate change causes conditions for increasing droughts, existing ecological issues become compounded and more dryland areas will be subjected to drought conditions (Fraser *et al.*, 2011). Dryland areas are particularly vulnerable because of their low adaptive capacity and higher sensitivity to change. Climate change will adversely affect wool, meat, and dairy production, mostly because grasslands and rangelands will not be as productive (Calvosa *et al.*, undated, p. 2). In particular, the livelihoods of the 2.5 billion people who inhabit dryland areas are threatened by climate change (Fraser *et al.*, 2011). As women are the most likely to be affected by climate change, identifying their vulnerabilities and increasing their abilities to adapt to changing environmental conditions are paramount (Alexander *et al.*, 2011).

In the dry areas of the world, climatic changes and population growth have led to increased outmigration among rural populations who depend on rain-fed agriculture (Radel *et al.*, 2012; Kristensen and Birch-Thomsen, 2013; Abdelali-Martini and Hamza, 2014; Afifi *et al.*, 2016). According to existing studies and a report compiled from discussions at the International Conference on Food Security in Dry Lands held

DOI: 10.1079/9781789247053.0011

in Qatar in 2012, two-thirds of the world's population live in deserts and dry areas which are most affected by ongoing agrarian feminization trends (Gartaula *et al.*, 2010; Pedrick, 2012; Abdelali-Martini and Hamza, 2014; Pattnaik *et al.*, 2018). Outmigration affects women's roles in agrarian economies with broader implications on productivity and gender equity. The related studies also show that generational, socio-economic, and socio-cultural factors, as well as economic and social remittances affect migration-related agricultural feminization. Little is known about the compounded impacts of male outmigration, climate change, and livestock production in rural Tunisia. The authors of the current study follow up on the ground to understand the role of women and men in a changing rural context in two rural areas in Tunisia.

The findings reveal that both women and men are marginalized from income generation training and loan acquisitions, and are negatively affected by resource degradation and climate change impacts, albeit in different ways. Women are additionally disadvantaged in ownership of land, extension targeting, and drought-mitigation strategies. The findings also reveal that women are more involved in grazing than is generally believed in practitioner and policy circles, and are becoming increasingly involved in livestock and agricultural production. The benefits of the feminization of agrarian labor may be incommensurate with the disadvantages mainly due to the lack of social and economic interventions needed to improve agricultural productivity for women and men in the context of increased climate change impacts and resource degradation.

Methodology

In light of existing vulnerabilities to climate change for men and especially women in the dry areas, along with increased roles for women in agrarian livelihoods, this research was conducted in two communities in Tunisia to understand the coping mechanisms, constraints, and opportunities in rural areas that are dependent on livestock for livelihoods. The study was conducted in three rural areas located in Medenine and Zaghouan with a total of 220 women and men (120 in Medenine and 100 in Zaghouan). A total of 100 interviews were conducted (60 in Medenine and 40 in Zaghouan) along with a total of 12 sex-segregated focus groups with ten respondents in each group. Medenine was of particular interest given its increased vulnerability to climate change due to its relatively higher dependence on rangelands for livestock production. The comparative approach between rangeland and crop production systems allows the authors to identify the key variations in coping mechanisms for livestock production and make their significance more evident. Recognizing a need to integrate gender more effectively within Tunisia's livestock and rangeland projects and policies, the authors brought together multiple government institutions, local communities, and international actors – under the auspices of the CGIAR Research Program on Livestock – to design practical research tools (interviews and focus groups) to guide research activities. The questions are aimed at understanding asset (land and livestock) ownership and effective control, access to training and innovations, gender roles in grazing of livestock, as well as climate change impacts and coping strategies.

Research Findings

This section begins by examining the pattern of women's involvement in livestock livelihoods, and ownership and control of assets (mainly land and livestock) in the study areas. The discussion then moves to changes in gender roles; changes related to climate change; implications and coping mechanisms adopted by women and men; rural services and their role in building resilience; and innovation availability and adoption in the past 5 years.

Women's involvement in livestock livelihoods

The authors found that women's roles are related to all agricultural activities in Medenine, more than in Zaghouan perhaps because there is a higher rate of male outmigration and off-farm work in this area, opportunities which are more available to men than women due to norms of seclusion (Table 11.1). Medenine depends on rangelands for livestock upkeep. Many participants reported a lack of youth engagement in agriculture and a reduction in size of livestock heads over the years due to increase in feed prices and limited availability of labor. Although women did not report grazing as their role in Medenine, perhaps because traditional grazing is a male-related role, men recognized women's roles in grazing. Notably, however, women reported grass collection as an activity for women, which entails collection of alternative feed from rangelands. Both women and men associated women's roles with feeding livestock, milking, processing dairy products, and cleaning of barns. Men were recognized by both genders in both regions as responsible for market concerns, in particular the purchasing and selling of feed, livestock, and inputs. Women in Medenine were also recognized in these roles perhaps due to the dwindling male participation in agriculture, especially the youth. It was found that the average age for male farmers was 60, but for women it was 45.

Asset ownership and decision making power

As land ownership, access, and control are important for livestock production, the current study explored patterns in land ownership and control. The findings reveal that land is mostly owned by men in both areas (Table 11.2). Often the land is not yet divided among the male siblings and other male inheritors. These findings are consistent with other countries in the Middle East and North Africa (MENA) region (e.g. see Najjar *et al.*, 2018, 2020, for work in Egypt). None the less, both women and men in both areas reported being able to use land, regardless of the gender of the owner. 'Use' is defined as the ability to graze livestock on the land or benefit from the land in any way more generally. These findings highlight the role of women in livestock and crop production despite their limited ownership of land.

For the size of livestock flocks in households, in both areas the total number of heads owned per household on average was less than 15. As opposed to land, women owned a significant amount of the livestock (Table 11.3), both alone or with their spouses, with the majority owning sheep (76 of the 100 respondents) and goats (24 of the 100 respondents). The sources of livestock ownership for women were gifts from their husbands due to taking care of livestock, buying from own savings (alone or with family members), inheritance, and/or a development project.

Women also played a significant role in controlling livestock (Table 11.4). We defined control by ability to buy and sell or make other important decisions. Men were more likely to report joint decision making with their spouses with regards to livestock than with regards to land. Men explained that they consult with their spouses because they are the ones who do most of the feeding and tending of livestock, while women explained that they consult with their husbands so that they both agree on the decision: 'I consult with my husband for any big or small issue. Even if I want to get water, before he leaves to work, I tell him that I will be getting water today'; 'We discuss to make decisions, but the final decision belongs to my husband.' These findings suggest that 'jointness' in decision making means different things for the two genders (see also similar findings by Acosta *et al.*, 2019). These differences in perception necessitate asking and probing further as to how and why decisions are made jointly, in order to gain an understanding of decision making power in a particular enterprise. A simple label of joint decision making is misleading to the extent of power a man and especially a woman may hold in an enterprise of a given household.

Changes in gender roles

When asked about the changes seen in roles, both women and men reported that women are now more involved in grazing and in collecting grass as in livestock feed in Medenine; and women are working more in agriculture and livestock due to male outmigration in both areas. Along the same lines, men in both areas reported that men in the community are increasingly leaving for off-farm opportunities

Table 11.1. Gender roles in livestock livelihoods in Medenine and Zaghouan, Tunisia.

Gender roles			Governorate		
			Medenine	Zaghouan	Total
Female respondents	Women's roles	All roles of agriculture (planting, plowing, pruning, harvesting)	70.4%	29.6%	27
		Harvesting	100.0%		2
		Breeding and caring for livestock (barn cleaning, feeding, watering)	38.5%	61.5%	13
		Purchasing feed	100.0%		1
		Grass collection	100.0%		1
		Bringing water		100.0%	2
		Milking and converting milk to yogurt	50.0%	50.0%	2
		Shrinking the role of agriculture and work in factories and construction	100.0%		1
	Men's roles	All roles of agriculture (planting, plowing, pruning, harvesting)	62.5%	37.5%	24
		Breeding and caring for livestock (barn cleaning, feeding, watering)	50.0%	50.0%	2
		Grazing	75.0%	25.0%	4
		Buying and selling livestock	75.0%	25.0%	4
		Purchasing feed	50.0%	50.0%	6
		Bringing water		100.0%	2
		Business of buying and selling in general, financial transactions and product marketing		100.0%	1
Male respondents	Women's roles	All roles of agriculture (planting, plowing, pruning, harvesting)	66.7%	33.3%	24
		Harvesting	33.3%	66.7%	3
		Breeding and caring for livestock (barn cleaning, feeding, watering)	47.4%	52.6%	19
		Grazing	100.0%		3
		Nothing	50.0%	50.0%	2
	Men's roles	All roles of agriculture (planting, plowing, pruning, harvesting)	60.9%	39.1%	23
		Breeding and caring for livestock (barn cleaning, feeding, watering)	25.0%	75.0%	4
		Grazing	62.5%	37.5%	8
		Buying and selling livestock	100.0%		3
		Purchasing feed	80.0%	20.0%	5
		Grass collection	100.0%		1
		Bringing water		100.0%	2
		Business of buying and selling in general, financial transactions and product marketing		100.0%	1
		Supervising workers		100.0%	1
Total respondents	Women's roles	All roles of agriculture (planting, plowing, pruning, harvesting)	68.6%	31.4%	51
		Harvesting	60.0%	40.0%	5
		Breeding and caring for livestock (barn cleaning, feeding, watering)	43.7%	56.3%	32
		Grazing	100.0%		3
		Purchasing feed	100.0%		1
		Grass collection	100.0%		1
		Bringing water		100.0%	2
		Milking and converting milk to yogurt	50.0%	50.0%	2

Continued

Table 11.1. Continued.

Gender roles		Governorate Medenine	Zaghouan	Total
	Shrinking the role of agriculture and work in factories and construction	100.0%		1
	Nothing	50.0%	50.0%	2
Total responses for women's roles		60	40	100
Men's roles	All roles of agriculture (planting, plowing, pruning, harvesting)	61.7%	38.3%	47
	Breeding and caring for livestock (barn cleaning, feeding, watering)	33.3%	66.7%	6
	Grazing	66.7%	33.3%	12
	Buying and selling livestock	85.7%	14.3%	7
	Purchasing feed	63.6%	36.4%	11
	Grass collection	100.0%		1
	Bringing water		100.0%	4
	Business of buying and selling in general, financial transactions and product marketing		100.0%	2
	Supervising workers		100.0%	1
Total responses for men's roles		53	38	91

Table 11.2. Land ownership by sex in Medenine and Zaghouan, Tunisia.

Who owns land		Governorate Medenine	Zaghouan	Total
Respondent	Female		100%	2
	Male	48%	52%	29
Spouse	Female	64%	36%	22
Respondent and spouse	Female	33%	67%	3
	Male	50%	50%	2
Respondent and member(s) of household	Female	75%	25%	4
	Male	100%		3
Spouse and member(s) of household	Female		100%	1
Outsider	Female	33%	67%	3
	Male	82%	18%	11
Respondent and outsider	Male	50%	50%	2
Spouse and member outside family	Female	67%	33%	6
	Male	100%		1
Respondent, spouse and outsider	Female	100%		1
	Male	100%		1
Herdsman	Female	86%	14%	7
	Male		100%	2
Total	Female	61%	39%	49
	Male	59%	41%	51
	Total	60	40	100

in the cities, in manufacturing, for example. Men also reported a shrinking area of rangelands and increased planting of trees. Men reported that tree planting often entailed hiring men over a 20-day duration. In turn, the shrinking size of rangeland limited the availability of feed for livestock and increased pressure on buying feed from the market.

Table 11.3. Livestock ownership by sex in Medenine and Zaghouan, Tunisia.

Who owns livestock		Governorate		Total
		Medenine	Zaghouan	
Respondent	Female	43%	57%	7
	Male	46%	54%	13
Spouse	Female	20%	80%	5
	Male	50%	50%	2
Respondent and spouse	Female	75%	25%	20
	Male	62%	38%	24
Daughter(s)	Female	100%		1
	Male	100%		1
Son(s)	Female		100%	1
Respondent and member(s) of household	Female	67%	33%	9
	Male	71%	29%	7
Outsider	Male	50%	50%	2
Spouse and member outside family	Female		100%	2
	Male		100%	1
Herdsman	Female	100%		4
Total	Female	61%	39%	49
	Male	58%	42%	50
	Total	59	40	99

Table 11.4. Livestock control by sex in Medenine and Zaghouan, Tunisia.

Who controls livestock		Governorate		Total
		Medenine	Zaghouan	
Respondent	Female	73%	27%	33
	Male	69%	31%	16
Spouse	Male	73%	27%	15
Mother-in-law	Female	100%		1
Respondent and daughter	Female	100%		1
Respondent and parent(s)	Male		100%	1
Respondent and sister	Female		100%	1
Total	Female	72%	28%	36
	Male	69%	31%	32
	Total	48	20	68

Changes in climate, implications and coping mechanisms

The Arab world is one of the regions that is most vulnerable to climate change. It is projected to have a 10–30% decrease in precipitation in the coming years, leading to a subsequent decline in groundwater replenishment and seriously overexploited aquifers (Sowers *et al.*, 2011; Schilling *et al.*, 2020; Haddad and Shideed, 2013). In addition, increased temperatures are expected to compound production (Sowers *et al.*, 2011; Haddad and Shideed, 2013). The MENA region in general is among the most water-stressed in the world. More than half of its countries have an average per capita water availability lower than the water scarcity threshold (Sowers *et al.*, 2011). Adaptation to climate change in the MENA region is often framed in terms of water use, where agriculture is the biggest consumer of water (Haddad *et al.*, 2011; Schilling *et al.*, 2020; Waterbury, 2013).

To our knowledge, little is known about the impacts of climate change on livestock production as a coping mechanism in the MENA region. In this section the gendered impacts in this sector are documented. Both women and men reported the following changes in the climate: hotter summers, increased drought incidence, erratic rainfall, and reduced rainfall (Table 11.5). The impacts identified by men included reduced oil yield from olives, degradation of rangelands, reduced profitability of agriculture and higher dependency on purchased feed, and an inability to irrigate as many olive trees, which led to the death of these trees. Men also reported a change in their irrigation timing to night-time in order to avoid high temperatures during daytime. Men reported that as a result of climate change they now have to collect feed from common areas and depend more on install feeding. Both of these activities have increased women's workloads.

When asked who contributes to grazing in the household, respondents reveal the significant contributions of women's participation (Table 11.6). These findings challenge assumptions held in policy circles about the lack of

Table 11.5. Changes in climate as perceived by female and male respondents in Medenine and Zaghouan, Tunisia.

Changes in climate		Governorate		Total
		Medenine	Zaghouan	
Temperature increase	Female	93%	7%	15
	Male	83%	17%	18
Increased cold	Female		100%	1
	Male	100%		1
Drought	Female	36%	64%	25
	Male	37%	63%	27
Trouble with precipitation	Female	100%		5
	Male	100%		2
No change	Female	67%	33%	3
	Male	100%		1
Sirocco (dry winds)	Male		100%	1
There is a change but I don't know how to explain it	Male	100%		1
Total	Female	61%	39%	49
	Male	59%	41%	51
	Total	60	40	100

Table 11.6. Grazing responsibility among female and male respondents in Medenine and Zaghouan, Tunisia.

Who is responsible for grazing		Governorate		Total
		Medenine	Zaghouan	
Respondent	Female	80%	20%	10
	Male	68%	32%	22
Spouse	Female	67%	33%	12
	Male	60%	40%	5
Respondent and spouse	Female	33%	67%	6
	Male	42%	58%	12
Respondent and member(s) of household	Female	45%	55%	11
	Male	43%	57%	7
Spouse and member(s) of household	Female	100%		1
	Male	100%		1
Herdsman	Male	100%		2
Other family member	Female	75%	25%	4
	Male		100%	1
Total	Female	61%	39%	44
	Male	58%	42%	50
	Total	56	38	94

women's participation in grazing activities, which the authors have observed in a stakeholder workshop that preceded the study. This omission has negative implications on providing women with a voice when consulting farmers on how rangelands can be better managed.

Focus groups conducted with women in each of the two regions reveal that women are interested in participating in consultation meetings on how to improve management of rangelands. Women in both areas have also reported performing the following practices to improve the management of rangelands for which they have access: leaving the land fallow; alternating grazing and fallow zones; avoiding grazing during the flowering period; and cultivating feed on farms. Men reported the same efforts in addition to cultivating barley in rangelands for livestock to graze on, as well as planting olives to control for erosion. Men also reported a project, entitled 'Prodifil,' which provides farmers with feed in return for keeping their lands fallow for the season. However, both male and female respondents noted that the beneficiaries are selected based on nepotism and are not the neediest in the community.

Women were notably absent in the Prodifil project. Women in the focus groups explained a need for the following to improve livestock production in their communities: government support for digging wells and subsidizing feed; allowing farmers to graze their livestock in larger areas through expansion in rangeland areas; provision of women with agricultural projects on livestock rearing and dairy production; and limiting fire damages and theft of livestock. Men reported the following practices for improving livestock production: increasing rangeland areas available for grazing; planting of cacti as livestock feed; providing shady areas in rangelands for livestock to rest under; improving access to labor which has significantly decreased with youth outmigration; and provision of feed markets closer to grazing areas.

When asked about climate change impacts on women and men in the two areas, respondents reported that climate change has led to limited production and crop failure; reduced purchasing power due to higher dependence on purchase of feed; reduced grazing time due to high heat and degradation of rangelands; limited job opportunities in the area; having to fetch water from further locations; fatigue; and a lack of interest in farming (Table 11.7).

When women and men were asked how they were coping with the resulting resource degradation, they reported saving water, looking for alternative water sources, storing food, decreasing herd sizes, and decreasing areas cultivated (Table 11.8).

Rural services and their role in increasing resilience

The research also aimed at examining whether services are helping women and men cope with stresses and improving their livelihoods; the authors asked about access to credit, extension advice and training, as well as the adoption of innovations. The findings revealed that very few households take loans (26%). It is worth noting that the majority of respondents knew about the availability of credit from lending institutions, yet many have opted for not acquiring credit (Table 11.9).

When asked why credit was not obtained the following reasons were provided by female and male respondents: lack of land title, religious reasons, lack of guarantees to repay loans and fear of failing to repay, lack of stable income, and lack of project ideas (Table 11.10). Women are especially affected by lack of title deed to land, and as revealed earlier, women seldom own land: 'Opportunities to acquire bank loans for private projects are non-existent because women do not own land.'

Innovations and extension can be sources of practices and technologies which alleviate the impacts of climate change and strengthen the resilience of men and women in affected rural areas. The majority of the respondents (83%) reported lack of visits from extension agents in the past year, with a larger proportion being women (Table 11.11).

Given the strong sex segregation in Tunisian culture, respondents were asked about preferences for the gender of extension agent. The majority (81%, more men than women at 96% vs 66%) felt indifferent to the gender of the extension agent, while 28% of the women preferred a female extension agent. The findings further suggest that it may be more effective for

Table 11.7. Climate change impacts on female and male respondents in Medenine and Zaghouan, Tunisia.

Climate change impacts		Governorate: Medenine	Governorate: Zaghouan	Total
Drought of crops and lack of production	Female	67%	33%	9
	Male	73%	27%	11
Lack of interest in agricultural work	Female	100%		1
	Male		100%	1
Lack of water and the need to go further to collect it	Male	50%	50%	2
Affecting purchasing power negatively due to the high price of feed	Female	50%	50%	8
	Male	50%	50%	14
Fatigue and difficulty in working	Female	50%	50%	4
	Male	17%	83%	6
Farmers get sick frequently	Female	30%	70%	10
	Male	100%		2
Livestock and crop diseases	Male	50%	50%	4
Decreased grazing time	Female	100%		2
	Male	67%	33%	3
Optimism for new season	Female	100%		2
	Male	100%		1
Going to nearby areas for showers	Female		100%	1
No impact	Female	100%		5
	Male	100%		1
Disturbance in agricultural and household activities and lack of work	Female	75%	25%	4
	Male	100%		3
Total	Female	61%	39%	46
	Male	58%	42%	48
	Total	56	38	94

Table 11.8. Coping mechanisms adopted by female and male respondents in Medenine and Zaghouan, Tunisia.

Coping mechanism		Governorate: Medenine	Governorate: Zaghouan	Total
Saving water	Male		100%	2
Storage of food to face any crisis and changing diet	Female	67%	33%	3
Changing domestic and agricultural work program to adapt to new situation	Female	64%	36%	14
	Male	50%	50%	8
Collecting, storing and drying herbs	Female	100%		2
	Male	100%		2
Increasing amount of food and buying it	Female		100%	2
	Male	58%	42%	12
Nothing special	Female	67%	33%	3
Decreased herd size	Female	100%		1
	Male	100%		1
Searching for other sources of drinking water and allocating budget for purchasing water	Female		100%	4
Lack of work in agriculture and material loss	Female	86%	14%	7
	Male	57%	43%	7
Lack of hunting and more problems protecting livestock from predators	Male		100%	1
Degradation of vegetation and shrinkage of cultivated areas	Female		100%	3
	Male	50%	50%	6
Total	Female	56%	44%	39
	Male	54%	46%	39
	Total	43	35	78

Table 11.9. Sources of credit available to female and male respondents in Medenine and Zaghouan, Tunisia.

Source of credit		Governorate		Total
		Medenine	Zaghouan	
Agricultural bank	Female	63%	37%	38
	Male	51%	49%	37
Solidarity bank	Female	100%		4
	Male	77%	23%	13
Development association for financing	Female		100%	2
Don't know	Female	40%	60%	5
	Male	100%		1
Total	Female	61%	39%	49
	Male	59%	41%	51
	Total	60	40	100

Table 11.10. Reasons for not obtaining credit in Medenine and Zaghouan, Tunisia.

Reason for not obtaining credit		Governorate		Total
		Medenine	Zaghouan	
Loans are prohibited by religion due to presence of interest	Female	100%		1
	Male	60%	40%	5
I don't like loans	Female	38%	62%	8
	Male	43%	57%	7
Absence of guarantees and inability to pay	Female	71%	29%	14
	Male	67%	33%	15
Good situation, I don't need a loan	Female	100%		4
	Male	83%	17%	6
No goal or project to need to take a loan	Female	100%		6
	Male	100%		1
Bureaucracy and no land ownership contract	Female	100%		1
	Male	100%		1
Intermediate	Female	67%	33%	3
	Male	33%	67%	3
Inability to work	Male	50%	50%	2
Total	Female	73%	27%	37
	Male	62%	38%	40
	Total	52	25	77

women to have a female extension agent providing advice and training because those who preferred women had very strong reasons: 'I prefer a woman extension agent because I feel I can be more comfortable and express myself better with a woman extension agent.' There were generational differences in preferences, with young unmarried women more likely to prefer a woman extension agent due to cultural norms: 'My parents will not accept that I meet with a man. They would prefer that a woman gives advice and trainings.'

When asked about the topics of training desired, these overlapped with training offered, with notable exceptions, such as one for women dealing with drought (Table 11.12). Such training was directed to men only. Men also asked for training not offered to them and not offered more generally; these included training on acquiring and repaying loans and on tree production. Men asked for training in cheese making which is only available to women. Both men and women asked for training in marketing, which was offered to neither.

Some women explained that the reason for asking for training on distillation was due to availability of medicinal and aromatic plants in

Table 11.11. Visits by extension agents during the past year in Medenine and Zaghouan, Tunisia.

		Governorate		
Number of visits		Medenine	Zaghouan	Total
0 visits	Female	60%	40%	45
	Male	55%	45%	38
1 visit	Male	75%	25%	4
2 visits	Female	100%		1
	Male	50%	50%	2
3 visits	Female	100%		1
	Male	100%		4
12 visits	Male	100%		1
Several visits	Female		100%	1
Unable to remember	Male		100%	2
Total	Female	60%	40%	48
	Male	59%	41%	51
	Total	59	40	99

Table 11.12. Topics desired for training through extension programs in Medenine and Zaghouan, Tunisia.

		Governorate		
Topic		Medenine	Zaghouan	Total
All about livestock and feeding	Female	65%	35%	17
	Male	67%	33%	12
Beekeeping	Male	25%	75%	4
Water solutions and coping with drought	Male	33%	67%	3
Marketing and promotion of agricultural	Female	100%		2
products	Male		100%	1
Techniques for organic farming	Male	100%		2
Livestock diseases, methods of treatment	Female	50%	50%	2
and protection	Male	100%		1
Tree pruning	Male	67%	33%	3
Tree diseases	Male		100%	2
Getting loans	Male	33%	67%	3
All about agriculture and its production	Female	75%	25%	4
	Male	100%		3
Nothing	Female	57%	43%	14
	Male	67%	33%	12
Educational development projects	Female	33%	67%	3
	Male	50%	50%	4
Cheese making	Male	100%		1
Solutions to high cost of production	Female	100%		1
Poultry and rabbit keeping	Female	67%	33%	3
Medicinal herbs distillation	Female	50%	50%	2
Total	Female	62%	38%	48
	Male	59%	41%	51
	Total	60	39	99

their environments, as well as to reduce pressure on rangelands through diversification of livelihood options and income. Others explained wanting the training in order to open their own business, while some felt that their information could be wrong and wanted to acquire the right information. One woman explained that 'all the information I have are second hand from my husband or from my neighbors. I would like to obtain scientific information from the experts.' Men justified their choices for the training themes to the subjects being a lucrative skill (e.g. tree pruning), wanting to start a commercial enterprise for the related subject (e.g. beekeeping), and achieving self-sufficiency, for example in olive oil consumption. These reasons also provide justification for the objectives and nature of future training.

When asked about the last time each of the respondents attended training, the majority explained (55%) that they had never done so. Surprisingly, a larger proportion of men said this than women (Table 11.13). This finding could be due to the purposive sampling of the current study whereby the authors have selected for respondents who are women and have participated in a development project. Men attended training related to feeding of small ruminants, rearing cows and small ruminants, beekeeping, shearing livestock wool with machines, forming and benefits of SMSAs (Mutual Agricultural Services Companies) and GDAs (Agricultural Development Groups), dealing with water scarcity, organic farming, and dairy cows. Some of the women's trainings overlapped with those of men while others were unique to their traditional roles. Women received training about rearing of small ruminants, milking cows, forming farmer groups with a focus on food processing techniques, poultry production, dairy processing, and distillation of aromatic and medicinal plants. One woman who attended training on the formation and managing of farmers' cooperatives explained that as a result some women in attendance formed their own SMSA cooperative. One respondent stated the following:

> The cooperative trainings were a useful experience. The location was close to our house whereby we spent about 2 hours daily for a month, which is not a lot of time to affect our workloads negatively. We learned a lot about communication techniques and food processing methods and its storage, as well as forming cooperative and managing it. As a result, we gained confidence and we opened our own SMSA cooperative after the training.

Innovation availability and adoption in the past 5 years

When asked about the most important innovations that have entered the community in the past 5 years, both women and men reported machines, fallow practice to sustain rangelands, new goat

Table 11.13. Training last attended by female and male respondents in Medenine and Zaghouan, Tunisia.

Training attended		Governorate		Total
		Medenine	Zaghouan	
Never attended training	Female	50%	50%	20
	Male	54%	46%	35
1997	Male		100%	1
1998	Male		100%	1
2003	Male		100%	1
2009	Female		100%	1
	Male		100%	1
2009–2011	Female		100%	1
2011	Female		100%	1
	Male	100%		1
2014	Female	100%		1
2015	Female	50%	50%	2
	Male	100%		1
2016	Female	67%	33%	6
	Male	100%		5
2017	Female	67%	33%	9
	Male	80%	20%	5
Six months ago	Female	100%		1
Past month	Female	100%		1
Total	Female	56%	44%	43
	Male	59%	41%	51
	Total	54	40	94

and sheep breeds, organic farming, vaccination, distillation of medicinal and aromatic plants, forage crop introduction, cooperative structures, and concentrated feed (Table 11.14). Only men reported irrigation technologies, soil analysis, improved cow breeds, fertilizers, and hydroponics. This finding could be related to lack of targeting of women in irrigation technologies as well as men's better connections to public organizations, such as soil analysis centers. Only women reported agricultural extension training as something new for them, as well as machines for milking and olive harvesting machines, both of which are largely used in their roles.

For both women and men, the justifications for the importance of these innovations ranged between improved production, increased income, and reducing workloads. Creating job opportunities was an additional justification provided by men.

When asked whether innovations were adopted, 52% of the respondents reported adopting at least one of the most important innovations entering their communities in the past 5 years (Table 11.15). The 48% who did not adopt

Table 11.14. Most important innovations introduced in the past 5 years in Medenine and Zaghouan, Tunisia.

		Governorate		
Innovation		Medenine	Zaghouan	Total
Machines	Female	60%	40%	5
	Male	70%	30%	10
All types of irrigation	Male	100%		3
Deferring grazing and pasture use	Female	100%		3
	Male	100%		3
Raising goats (Dimachki race)	Female	100%		5
	Male	100%		6
Vaccination and medications	Female	50%	50%	2
	Male	33%	67%	3
Fertilizers	Male		100%	1
Medicinal herbs distillation	Female		100%	2
	Male		100%	1
Olive seedlings and other new types of trees	Female		100%	2
	Male		100%	2
Improving race of cows	Male		100%	1
Raising sheep (Tibar race)	Female		100%	1
	Male		100%	3
Biological agriculture	Female		100%	1
	Male	50%	50%	2
To cultivate forage crops and olive trees	Female		100%	3
	Male		100%	2
Special machines for collecting olives and milking	Female	100%		2
No innovation	Female	38%	62%	8
	Male	50%	50%	6
To be organized in professional structure	Female	100%		2
	Male	100%		1
Plant in water/off-ground cultivation	Male	100%		1
Concentrated feed	Female	50%	50%	2
	Male	100%		1
Soil analysis	Male		100%	1
Agricultural extension and training	Female	100%		1
Don't know	Female	100%		2
Total	Female	56%	44%	41
	Male	57%	43%	47
	Total	50	38	88

Table 11.15. Innovations adoption from the most important innovations introduced in the past 5 years in Medenine and Zaghouan, Tunisia.

Innovation		Governorate		Total
		Medenine	Zaghouan	
Machines	Female	67%	33%	3
	Male	54%	46%	13
All types of irrigation	Male	100%		2
Deferring grazing and pasture use	Female	100%		2
	Male	100%		2
Raising goats (Dimachki race)	Female	100%		2
	Male	100%		3
Vaccination and medications	Female	100%		2
	Male	33%	67%	3
Fertilizers	Female		100%	1
	Male		100%	1
Medicinal herbs distillation	Female		100%	2
	Male		100%	1
Olive seedlings and other new types of trees	Female		100%	1
Raising sheep (Tibar race)	Female		100%	2
	Male	50%	50%	2
To cultivate forage crops and olive trees	Female		100%	1
Special machines for collecting olives and milking	Female	100%		1
No innovation adoption	Female	55%	45%	11
	Male	50%	50%	14
To be organized in professional structure	Female	100%		2
	Male	100%		1
Concentrated feed	Female	50%	50%	2
	Male	100%		1
Agricultural extension and training	Female	100%		1
Reconstruction of stables with bricks	Female		100%	1
Total	Female	56%	44%	34
	Male	58%	42%	43
	Total	44	33	77

comprised of a similar number of female and male respondents. Most of the innovation adoptions centered around new machinery, followed by new goat breeds, and leaving rangelands fallow. The differences between adoption reported between women and men were similar to differences in innovations identified as most important. For example, men exclusively reported adoption of irrigation mechanisms while women exclusively reported mechanized olive harvesting and attending extension programs.

Discussion and Conclusion

In their study on the myths around women's participation in farming in China, de Brauw *et al.* (2008) identify two types of feminization: labor feminization and managerial feminization. The authors define 'labor feminization' as the situation where women in a household perform an increasing share of on-farm work whereas 'managerial feminization' refers to the increasing participation of women in agricultural decision making along with greater access to agricultural resources. Both de Brauw *et al.* (2008) and Abdelali-Martini and Dey de Pryck (2015), however, assert that more women are engaging in labor feminization rather than managerial feminization. Most of these studies are based on crop production. Through the focus on livestock in the current study, the findings reveal an increase in managerial feminization along with a simultaneous decrease in number of livestock heads reared by a given household. A decline in grazing area is especially serious because it impacts the production of fodder and feed. Fodder production is important to rural people that are dependent on rangelands because it allows livestock keepers to mitigate the risks of food shortages and maintain the productivity of livestock (Ayantunde *et al.*, 2017). Additionally, given the important role that livestock plays in the empowerment of women through strengthening their decision making power, it is important that feed options are strengthened with attention paid to women's workloads.

Numerous studies have found that women and children are among the world's most impoverished and vulnerable to climate change shocks, and could, in theory, benefit most from risk management tools (Chanamuto and Hall, 2015; Bageant and Barrett, 2017). At the same time, because of their existing vulnerability, women are often unable to access innovations which may absorb the effects of shocks. This study echoes these findings, as women were largely marginalized from skills and trainings related to drought and irrigation. As such, it is paramount that women's as well as men's capacities are strengthened in drought adaptation mechanisms. Even for men, most of the training was related to supplementary irrigation and irrigation approaches more generally. Other techniques like drought tolerant crops, such as cactus, and animal breeds remain absent from the extension and training programs, as well as available innovations. An essential first step is the validation of women in perceived masculine roles (e.g. irrigation, rangeland use, purchase and selling in the market). This will enable women to participate in rangeland and agricultural programs on an equal footing with men and voice their concerns in policy dialogues.

References

Abdelali-Martini, M. and Dey de Pryck, J. (2015) Does the feminisation of agricultural labor empower women? Insights from female labor contractors and workers in northwest Syria. *Journal of International Development* 27, 898–916. doi.org/10.1002/jid.3007.

Abdelali-Martini, M. and Hamza, R. (2014) How do migration remittances affect rural livelihoods in drylands? *Journal of International Development* 26, 454–470. doi.org/10.1002/jid.2896.

Acosta, M., van Wessel, M., van Bommel, S., Ampaire, E.L., Twyman, J., Jassogne, L. and Feindt, P.H. (2019) What does it mean to make a 'joint' decision? Unpacking intra-household decision making in agriculture: implications for policy and practice. *The Journal of Development Studies* 56, 1–20.

Afifi, T., Milan, A., Etzold, B., Schraven, B., Rademacher-Schulz, C. *et al.* (2016) Human mobility in response to rainfall variability: opportunities for migration as a successful adaptation strategy in eight case studies. *Migration and Development* 5, 254–274. doi.org/10.1080/21632324.2015.1022974.

Alexander, P., Nabalamba, A. and Mubila, M. (2011) The link between climate change, gender and development in Africa. *African Statistical Journal* 12, 119–140.

Archambault, C.S. (2016) Re-creating the commons and re-configuring Maasai women's roles on the rangelands in the face of fragmentation. *The International Journal of the Commons* 10, 728–746.

Ayantunde, A., Yameogo, V., Traore, R., Kansaye, O., Kpoda, C. *et al.* (2017) Improving livestock fodder production through greater inclusion of women and youth: results from the 'realizing the full biomass potential of mixed crop-livestock systems in rapidly changing Sahelian agro-ecological landscapes' project. *Water, Land and Ecosystems Briefing Series, No. 10*. CGIAR Research Program on Water, Land and Ecosystems, International Water Management Institute, Colombo, Sri Lanka.

Bageant, E.R. and Barrett, C.B. (2017) Are there gender differences in demand for index-based livestock insurance? *The Journal of Development Studies* 53, 932–952.

Calvosa, C., Chuluunbaatar, D. and Fara, K. (undated) Livestock and climate change. *IFAD Thematic Paper*. International Fund for Agricultural Development, Rome, pp. 1–19.

Chanamuto, N.J.C. and Hall, J.G. (2015) Gender equality, resilience to climate change, and the design of livestock projects for rural livelihoods. *Gender and Development* 23, 515–530.

Curry, J., Huss-Ashmore, R., Perry, B. and Mukhebi, A. (1996) A framework for the analysis of gender, intra-household dynamics, and livestock disease control with examples from Uasin Gishu District, Kenya. *Human Ecology* 24, 161–189.

de Brauw, A., Li, Q., Liu, C., Rozelle, S. and Zhang, L. (2008) Feminization of agriculture in China? Myths surrounding women's participation in farming. *The China Quarterly* 194, 327–348. doi.org/10.1017/S0305741008000404.

Debela, B.L. (2017) Factors affecting differences in livestock asset ownership between male and female-headed households in northern Ethiopia. *European Journal of Development Research* 29, 328–347.

Dumas, S.E., Maranga, A., Mbullo, P., Collins, S., Wekesa, P., Onono, M. and Young, S.L. (2018) 'Men are in front at eating time, but not when it comes to rearing the chicken': unpacking the gendered benefits and costs of livestock ownership in Kenya. *Food and Nutrition Bulletin* 39, 3–27.

FAO (2013) Tackling climate change through livestock: a global assessment of emissions and mitigation opportunities. Food and Agriculture Organization of the United Nations, Rome.

Fisher, M.G., Warner, R.L. and Masters, W.A. (2000) Gender and agricultural change: crop-livestock integration in Senegal. *Society & Natural Resources* 13, 203–222. doi.org/10.1080/089419200279063.

Fraser, E.D.G., Dougill, A.J., Hubacek, K., Quinn, C.H., Sendzimir, J. and Termansen, M. (2011) Assessing vulnerability to climate change in dryland livelihood systems: conceptual challenges and interdisciplinary solutions. *Ecology and Society* 16:3.

Galie, A., Teufel, N., Korir, L., Baltenweck, I., Webb Girard, A., Dominguez-Salas, P. and Yount, K.M. (2018) The women's empowerment in livestock index. *Social Indicators Research* 142, 799–825. doi.org/10.1007/s11205-018-1934-z.

Gartaula, H.N., Niehof, A. and Visser, L. (2010) Feminisation of agriculture as an effect of male out-migration: unexpected outcomes from Jhapa district, eastern Nepal. *The International Journal of Interdisciplinary Social Sciences: Annual Review* 5, 565–578.

Haddad, N. and Shideed, K. (2013) Mainstreaming adaptation to climate change into the development agenda. In: Sivakumar, M.V.K., Lal, R., Selvaraju, R. and Hamdan, R. (eds) *Climate Change and Food Security in West Asia and North Africa*. Springer, Amsterdam, The Netherlands, pp. 301–315.

Haddad, N., Duwayri, M., Oweis, T., Bishaw, Z., Rischkowsky, B., Aw Hassan, A. and Grando, S. (2011) The potential of small scale rainfed agriculture to strengthen food security in Arab countries. *Food Security* 3(Suppl. 1), S163–S173.

Kristensen, S.B.P. and Birch-Thomsen, T. (2013) 'Should I stay or should I go?' Rural youth employment in Uganda and Zambia. *International Development Planning Review* 35, 175–201.

Najjar, D., Baruah, B., Aw-Hassan, A., Bentaibi, A. and Kassie, G.T. (2018) Women, work, and wage equity in agricultural labor in Saiss, Morocco. *Development in Practice* 28, 525–540.

Najjar, D., Baruah, B. and El Garhi, A. (2020) Gender and asset ownership in the old and new lands of Egypt. *Feminist Economics* 26, 119–143.

Njuki, J. and Sanginga, P.C. (2013) Gender and livestock: key issues and opportunities. In: Njuki, J. and Sanginga, P.C. (eds) *Women, Livestock Ownership and Markets: Bridging the Gender Gap in Eastern and Southern Africa*. Routledge, New York, pp. 1–8.

Pattnaik, I., Lahiri-Dutt, K., Lockie, S. and Pritchard, B. (2018) The feminization of agriculture or the feminization of agrarian distress? Tracking the trajectory of women in agriculture in India. *Journal of the Asia Pacific Economy* 23, 138–155. doi.org/10.1080/13547860.2017.1394569.

Pedrick, C. (2012) *Strategies for combating climate change in drylands agriculture: synthesis of dialogues and evidence presented at the International Conference on Food Security in Dry Lands, Doha, Qatar, November 2012*. The International Center for Agricultural Research in the Dry Areas (ICARDA), Aleppo, Syria and CGIAR Research Program on Climate Change, Agriculture and Food Security (CCAFS), Copenhagen, 36 pp.

Radel, C., Schmook, B., Mcevoy, J., Méndez, C. and Petrzelka, P. (2012) Labour migration and gendered agricultural relations: the feminization of agriculture in the Ejidal sector of Calakmul, Mexico. *Journal of Agrarian Change* 12, 98–119.

Schilling, J., Hertig, E., Tramblay, Y. and Scheffran, J. (2020) Climate change vulnerability, water resources and social implications in North Africa. *Regional Environmental Change* 20:15.

Sowers, J., Vengosh, A. and Weinthal, E. (2011) Climate change, water resources, and the politics of adaptation in the Middle East and North Africa. *Climatic Change* 104, 599–627.

Thomas-Slayter, B. and Bhatt, N. (1994) Land, livestock, and livelihoods: changing dynamics of gender, caste, and ethnicity in a Nepalese village. *Human Ecology* 22, 467–494.

Turner, M. and Williams, T. (2002) Livestock market dynamics and local vulnerabilities in the Sahel. *World Development* 30, 683–705.

Waterbury, J. (2013) The political economy of climate change in the Arab Region. *Arab Human Development Report Research Paper Series*. United Nations Development Programme, New York.

12 The Nexus Between Climate Change, Migration and Gender

D. Hummel

Introduction

In the coming decades, the world population will have to deal with adverse impacts of global climate change such as sea level rise, floods, droughts, land degradation, and loss of biodiversity with possibly devastating effects on agricultural production, economic development, and human health. Countless publications deal with actual and potential increases in human mobility as a response to these events and their negative impacts on people's livelihoods, particularly in the Global South. Interactions between climate change, environmental degradation, and the mobility of people is a topic of increasing attention in the scientific as well as in the public debate. Although there has been remarkable progress in reflecting on gender at the level of international climate policy, as well as considerable advances in the fields of climate and migration research, it is surprising that gender dimensions of climate-induced migration have only recently received more attention in the scientific domain. Fortunately, gender dimensions are becoming more and more important in conceptual and empirical research on the topic. The purpose of this chapter is to explore gender-specific dimensions of so-called 'climate-induced migration' and to elucidate their relevance for research on climate change, gender, and livelihoods in the Global South. To this end, the chapter starts with a brief overview on the state of the scientific debate on the interlinkages of climate driven environmental change and migration, followed by a discussion of conceptual approaches applicable for the inclusion of the gender dimension. The chapter provides empirical evidence for gender relevance using a case study on the West African Sahel and ends with some considerations on policy issues and further research. In doing so, the chapter illustrates in which way a 'gender lens' adds to more robust knowledge and in what way the perspective of social ecology is of particular value.

Interlinkages of Climate, Environment and Migration

For more than two decades there has been an increasing interest in interlinkages between environmental changes and migration in the scientific as well as in the public debate. With better knowledge about the mechanisms and impacts of anthropogenic climate change, the academic discourse has developed considerably in recent years (Hunter *et al.*, 2015; Bettini and Gioli, 2016; Kaczan and Orgill-Meyer, 2020; Hoffmann *et al.*, 2020). 'How many people could be displaced as a result of climate change?' is a frequently asked question in this debate (Adger *et al.*, 2014). However, it is difficult to give a well defined, unequivocal answer. Estimated numbers of current and potential future migrant populations that might emerge as a result of climatic and

 Gender, Climate Change and Livelihoods: Vulnerabilities and Adaptations (eds J. Eastin and K. Dupuy)
DOI: 10.1079/9781789247053.0012

environmental changes vary considerably. Depending on the different temporal and spatial scales, data, and methodology used, the numbers range from the current 10–30 million people displaced as a result of environmental changes to 200 million or even 700 million people by 2050. All these numbers are controversial and have been criticized as unreliable or speculative and, therefore, confidence in these estimates is low. For example, a population 'at risk' from sea level rise or drought does not necessarily entail that people will actually move (Kaczan and Orgill-Meyer, 2020; Cattaneo *et al.*, 2019). As Kelman stresses, 'labelling, counting and calculating climate change migrants and non-migrants depend on political choices, so any numbers reached might not be scientifically robust' (Kelman, 2019, p. 2). Moreover, the increasing political and scientific concerns about the consequences of climate change for population mobility have not led to an answer on what exactly constitutes an environmentally or climate driven move. The according literature offers different concepts and terminologies such as 'environmental refugees,' 'climate refugees,' or 'climate-induced migration,' all of which are problematic, since they assume direct causal relationships. The Intergovernmental Panel on Climate Change (IPCC) therefore demanded caution when using such terms, and states that 'it is difficult to categorise any individual as a climate migrant' (Adger *et al.*, 2014, p. 768). Still, the mechanisms and effects of (climate driven) environmental changes affecting population movements are barely understood. Research studies are usually context-specific and cannot be generalized. Even case studies that explore issues of the same geographical region vary in the subject, methods, concepts, and theoretical framework (if any) used and are therefore difficult to compare. The available studies refer to different types of environmental changes associated with climate change (e.g. rainfall variability, desertification, changes in vegetation, or flood events), and to different migration types (e.g. internal, international, short- or long-term, and seasonal), and they differ in spatial scales ranging from village up to continental level.

Despite these challenges, research on the interlinkages between climate, environment, and migration has made considerable progress in recent years and a growing consensus is emerging on several important issues. The academic literature mentions various types of environmental hazards that are associated with climate change and which can be assigned to different patterns of migration. In addition to the distinction between direct and indirect links it is reasonable to differentiate between fast-onset events (e.g. hurricanes, floods, landslides, etc.) which are mostly associated with temporary migration, and slow-onset events (e.g. sea level rise, desertification, land degradation) which are mostly associated with progressive population movements (Hummel *et al.*, 2012; Cattaneo *et al.*, 2019). In most cases, climate change does not directly cause people to move, but is accompanied by very specific environmental effects that can exacerbate existing vulnerabilities. Research on migration and slow-onset environmental changes in particular has shifted towards a more positive and proactive understanding of migration as an adaptation strategy to diversify income sources and to reduce people's vulnerability to environmental and non-environmental stress. Here, we have a view on migration as a continuum from voluntary to forced movement, while most migrations include both elements, forced and voluntary (van der Land, 2018, p. 3). Earlier studies considered migrants as the most vulnerable to environmental stress, while recent literature regards those who are not able to migrate due to poverty as 'trapped populations' (Adger *et al.*, 2014). Recently, this field of research has taken the insights of current migration theory into account and has moved beyond linear, simplistic 'push' and 'pull' approaches towards a greater integration of context, including micro-, meso-, and macro-level interactions (Hunter *et al.*, 2015; van der Land *et al.*, 2018). These changes in discourse can also be observed for the debate on the nexus of environmental and climate change, migration, security, and conflict (Scheffran *et al.*, 2012; Adger *et al.*, 2015; Veronis *et al.*, 2018). These discursive shifts are frequently associated with the argument of environmentally and climate driven migration as multicausal phenomena (Black *et al.*, 2011; McLemann and Gemenne, 2018; Cattaneo *et al.*, 2019). Since ecological, social, economic, cultural, and political factors interact, it is virtually impossible to isolate single factors from the bundle of manifold drivers. Migration in the context of climate change constitutes a complex phenomenon that cannot be addressed by monocausal explanatory approaches but requires a nuanced perspective on the interplay of social, economic, political, and cultural factors – including gender issues.

Pointing the 'Gender Lens' at Migration in the Context of Climate Change

Gender constitutes an essential principle of societal structuring that positions women and men, as well as gender-diverse people (i.e. people who do not want to or cannot assign themselves to either category 'female' or 'male'), in gender relations and associated inequalities, hierarchies, and power relations. While 'sex' refers to the biological attributes that distinguish people as female, male or intersex, 'gender' refers to psychological, social, and cultural factors that shape attitudes, behaviors, and knowledge (Tannenbaum *et al.*, 2019, p. 138). For gender studies in the field of climate and migration, it is useful to look at gender in terms of relations, structures, and processes (Gioli *et al.*, 2014; Hummel and Stieß, 2017; Gioli and Milan, 2018). 'Gender relations' refer to characteristics socially ascribed to femininity and masculinity, such as socially constructed roles, opportunities, and responsibilities ascribed to women, men, and gender-diverse people; they are context-specific, fluid, and changeable over time. Gender relations are structured by specific social inequalities, different access to and control over resources, and decision making. Conceptualizing 'gender as a process' allows us to study gendered discourses and practices that reproduce or oppose existing power hierarchies (Pessar and Mahler, 2003, p. 813). In social interactions and discourses in everyday life, gender and the societal norm of a binary gender order (femininity vs masculinity) are constantly and actively produced and reproduced in a manner of 'doing gender' (West and Zimmerman, 1987). Conceptualizing 'gender as a structure' means to take power relations into consideration that are beyond the individual with regard to gendered dimensions of institutions, policies, and governance structures (Denton, 2002; Pessar and Mahler, 2003; Gioli *et al.*, 2014; Gioli and Milan, 2018, p. 135).

An examination of gender impacts of migration in the context of climate change must therefore consider that gender cannot be equated with 'woman.' Furthermore, genders are not homogeneous. Factors of social inequality or social marginalization must also be considered. This consideration is reflected in the approach of intersectionality, which can be defined as 'the interaction between gender, race and other categories of difference in individual lives, social practices, institutional arrangements, and cultural ideologies and the outcomes of these interactions in terms of power' (Davis; cited in Kaijser and Kronsell, 2014, p. 418f). Interactions between gender and other characteristics of social inequality such as age, income, education, physical condition, ethnic background, etc. are then explored according to the specific problem context.

Integrating gender into the picture: conceptual approaches

Considering the different dimensions of gender – gender as relation, process, and structure – the context dependency and complexity of the climate–migration–gender nexus needs to be reflected in this field of research. In order to investigate empirical issues of the nexus in more detail, a theoretical framework is required that adequately addresses these challenges and ideally makes a generalization possible beyond the specific case. However, the scientific discourse on climate change, environment, and migration builds on very heterogeneous epistemological approaches and theoretical premises. Given the high degree of complexity resulting from different temporal, spatial, and social scales and their interactions, a theoretical–conceptual pluralism is required for the study of the climate–migration–gender nexus. The following approaches are particularly valuable and suitable for the analysis of gender aspects of climate migration with a focus on livelihoods in the Global South.

Sustainable livelihoods

Research on the climate–environment–migration nexus has been largely inspired by the 'livelihoods framework' and the 'sustainable livelihoods approach' (SLA), respectively (Chambers and

Conway, 1992; Carney, 1998). Accordingly, the relative availability of different 'capital assets' shapes the livelihood options of individuals and households. These assets comprise human capital such as labor and education, financial capital such as savings, physical capital such as agricultural equipment, social capital (e.g. kin networks), and natural capital such as wild foods from communal land. The relative availability of these assets is shaped by individual and household actions, as well as broader socio-economic and political structures and processes (Hunter and David, 2011). SLA assumes that people have access to different assets. These assets can be translated into different livelihood strategies and can determine the capacity or incapacity of people to make a living under conditions of stress. Therefore, some studies combine the SLA with the concept of social vulnerability. It refers to 'the ability or inability of individuals and social groupings to respond to, in the sense of cope with, recover from or adapt to, any external stress placed on their livelihoods and well-being' (Kelly and Adger, 2000, p. 328).

Both approaches, or rather 'concepts,' are useful analytical tools for the analysis of gender aspects from an intersectional point of view, since they emphasize the mediating function of social factors within environment–migration interactions. Focusing on the household, the SLA acknowledges the multicausality of migration and considers migration as a household strategy. It thus offers two explanatory dimensions with respect to migration. The first dimension is migration as a strategy to minimizing risks and vulnerability; the second dimension is migration as a strategy to improving the livelihoods and well-being of the household and hence contributing to poverty reduction (van der Land, 2018, p. 29). A great benefit of the SLA is that the livelihood situation of households is comprehensively analyzed and not restricted to environmental stressors. Furthermore, it highlights poor people's agency (or lack thereof) to improve their livelihoods in special structural circumstances. However, there is criticism that the SLA's focus on the household as a unit of analysis involves the risk of overlooking intra-household dynamics, inequalities, and unequal power dynamics among household members, particularly in terms of gender, age, and generations (Gioli and Milan, 2018, p. 143; van der Land, 2018, p. 31).

Capability and aspirations

Given the challenges of the SLA illustrated above, van der Land (2018) introduced a 'capability and aspirations approach' that is based on a critical combination of SLA and Amartya Sen's 'capability approach' (Sen, 2000). It considers people's capabilities to migrate (or to stay), as well as their individual aspirations and motivations.

> If people do have the capability to choose freely between different options, the selection of a certain action from the capability set occurs according to people's needs, preferences and aspirations. People's assets and their socio-economic position – often depending on individual characteristics – strongly influence whether people have the capability to choose between migrating and staying.
>
> (van der Land, 2018, p. 35)

This approach is conducive for the analysis of gender dimensions since it supports a combined analysis of structural conditions and individual agency. The emphasis on people's aspirations can contribute to a better understanding of why some people migrate while others stay, despite living under similar conditions and facing similar environmental risks. Moreover, it facilitates the analysis of existing gender-specific norms, values, perceptions, expectations, and related social practices.

Social–ecological systems

Social ecology is dedicated to the exploration of the relations between society and nature (Görg *et al.*, 2017; Hummel *et al.*, 2017). From the perspective of this emerging inter- and transdisciplinary scientific field, migration takes place under specific social–ecological conditions. Different actions and strategies taken by individuals, groups, and societies to cope with changes of the natural environment depend on specific societal relations to nature. This concept refers to patterns of relations between humans, society, and nature that emerge from the culturally specific and historically conditioned forms in which people shape and regulate their relations to nature, both materially (e.g. cultivation of the soil, collecting firewood) and symbolically (e.g. norms, values, categorizations, and meanings of 'female'/'male' responsibilities). Hence, changes in ecosystems such as land degradation or loss of biodiversity are not only a result of biophysical dynamics but also of human decisions, actions, or omissions which in turn are influenced by social, cultural, political, and economic settings. For the empirical analysis, complex interactions between climate-induced environmental changes, social transformations, and migration can be explored as 'social–ecological systems' (SES). They consist of interactions between societal actors and ecosystems, which are mediated by technology, practices, institutions, and knowledge (Becker, 2012; Hummel, 2016). This approach facilitates transdisciplinary analysis that integrates natural scientific and social–scientific knowledge, as well as practical knowledge of relevant societal stakeholders. Furthermore, it allows an analysis of migration as a socially differentiated – and thus gendered – practice performed by different actors, which is influenced by institutional settings (including cultural norms and traditions), as well as situated knowledge (Hummel, 2016; Liehr *et al.*, 2016.). That way, gender-specific aspects of migration can be studied with a focus on gender relations and gendered societal structures.

Empirical evidence: gender matters

Migration in the context of climate and environmental change is far from being gender-neutral. Given their specific experiences, needs, and priorities, women and men perceive and act upon climate change and migration quite differently. However, in the academic debate on climate-induced migration, gender dimensions represent a relatively novel topic; hence, scientific knowledge is still limited and dispersed (Hunter and David, 2011; Chindarkar, 2012; van der Land and Hummel, 2013; Gioli and Milan, 2018). This is probably also related to the challenge that a 'gender lens' is required to connect rather scattered and heterogeneous knowledge: feminist and gender studies, research on gender and migration, gender and climate change, gender and development, gender, environment, and conflict, etc. Yet, as this section of the chapter shows, the significance of gender is increasingly admitted in the debate on the climate–migration nexus and research has made considerable progress. Without claiming to be a systematic review, the state of debate is sketched below.

Different pathways

As a general rule, studies dealing with gender dimensions of migration in the context of climate change focus on different aspects: (i) gender-specific vulnerability due to the consequences of climate change; (ii) gender-specific preferred adaptation strategies to climate change; (iii) gender-specific burdens of climate adaptation; and (iv) gender-specific issues of decision making and institutions.

> As an example of gendered dimensions of livelihoods and livelihood decision making, socio-culturally defined gender roles shape both perceived value of 'assets' (e.g. differential value placed on men's and women's human capital), as well as actual household and individual use of the assets themselves (e.g. natural resource collection perceived as 'women's work').
>
> (Hunter and David, 2011, p. 307f)

Following Hunter and David (2011) two pathways can be distinguished through which gendered migration impacts of climate change may manifest: first, shifts in proximate natural resources and agricultural potential, and second, increases in extreme weather events. Both refer to the distinction between slow-onset and fast-onset environmental changes illustrated above.

Shifts in proximate natural resources and availability of natural capital

With regard to the first pathway, changes in the availability of natural capital that impact resource-based activities (e.g. collecting wild herbs or firewood) can result in a diversification of households and livelihood strategies including migration. These strategies are influenced by environmental factors interacting with gendered labor processes and the household division of labor (Hunter and David, 2011, p. 311). Gendered impacts of migration are illustrated in various case studies. For example, studies conducted in the Chitwan Valley of Nepal show evidence for increasing outmigration from rural areas due to declining agricultural productivity. The Himalaya region is known as a region highly affected by climate and environmental changes such as rising temperatures, variations in rainfall, and water scarcity. Bohra-Mishra and Massey (2011) revealed that the links between environmental degradation and migration are an important factor for both women and men, but result in gendered migration patterns due to gendered working tasks. Female migration is happening primarily on the local level, while international migration is largely a male phenomenon. For the residents of Chitwan Valley, increasing collection times for fodder (a traditionally female task) as well as firewood (traditionally a male task) due to environmental degradation have both caused a rising probability of local migration. The work burden due to environmental change clearly increased at the expense of women. The study found that increasing time to gather fodder had no significant effect on male migration, 'but an increase in the time required to collect firewood is significant in predicting both local and distant moves by men' (Bohra-Mishra and Massey, 2011, p. 97). Lamicchane Dhimal (2018) illustrates that migration can also put particular stress on men who have to find sources of income to make up for the loss of or inadequate livelihood. In her comparative study about gender dimensions of health impacts of climate change in two regions of Nepal – the highlands Gorkha district and lowlands Morang district – Dhimal revealed an increasing trend of migration from rural to urban areas as well as from the highlands to riverbanks. However, she found that migration also depends on respective prosperity, i.e. socio-economic factors. People who were better-off migrated to safer and more suitable places while poorer people were still struggling with the environmental changes such as variations in rainfall and water scarcity. Considered as a whole, migration patterns were revealed unambiguously as gender-specific. In the context of climate-induced population movements, international outmigration appears to be predominantly a male phenomenon but can be accompanied by the wives' and children's internal migration to nearby suburban and urban centers. These examples highlight gender-specific vulnerabilities, adaptation strategies, impositions, and burdens, as well as agency, all of which are embedded in broader socio-cultural orders and power structures.

Gendered migration in the context of disaster and extreme weather-related events

Regarding the second pathway illustrated by Hunter and David (2011) – extremes in weather events – gender analyses on migration patterns largely refer to research on disaster events and vulnerabilities that expose different social groups to different levels of risk. For example, several studies about the impacts of hurricane Katrina in New Orleans, USA, in 2005 illustrated the intersections of gender, 'race', class, and age leading to different migration outcomes (Tuana, 2008; Weber and Hilfinger Messias, 2012). Adger *et al.* (2014, p. 767) emphasize that gendered vulnerabilities are particularly significant when women lose their social networks and their social capital. With respect to gender-diverse people, a study in Pakistan showed that transsexual, transgender, and intersex people are less likely to be reached by early warning systems or compensated for arising losses (Gioli and Milan, 2018, p. 137). As an exemplary case for the Global South, Bangladesh is one of the countries that is considerably affected by extreme events such as floods with devastating effects on people's life, health, and livelihoods. Empirical studies revealed notable gender differences in terms of risk perception, needs, and strategies for action. For example, women and girls learn to swim less often than boys and men and do not find refuge in time during evacuations because patriarchal norms prevent them from leaving their homes without male company (Ashraf and Azad, 2015; Rezwana, 2016; Ferdous and Mallick, 2019, p. 91). According to a recent study on water and migration (Nagabhatla *et al.*, 2020), water-related environmental changes have a disproportionate impact on gender. As a rule, men tend to migrate more often than women in crises related to water, food, and climate change. 'As women stay behind, they take on the responsibilities of the men; yet, they are faced with unequal access to resources (financial, social, etc.); thus, gaining an increase in responsibilities with fewer rights and accessibility' (Nagabhatla *et al.*, 2020, p. 18). Similarly, a study from the West Karakoram region in the Upper Indus basin in Pakistan illustrated that migration adopted by the population as a core response to environmental pressure is highly gendered. People's movement in this case constitutes either an *ex ante* form of risk mitigation of households in the face of uncertain agricultural yields, or an *ex post* coping mechanism in the wake of environmental shocks such as landslides and floods. However, only men participated in circular labor migration while women were left behind in the care economy, i.e. subsistence agriculture, household chores, and care for children and sick and elderly family members (Gioli *et al.*, 2014). Thus, it may be concluded that not only climate driven environmental changes, but also migration adds to strains and can exacerbate existing gender hierarchies and power relations. Several studies, particularly on the subjects of gender and disaster, but also conceptual works, address issues of conflict/security (e.g. Arora-Jonnson, 2011; Brzoska and Fröhlich, 2016; Rothe, 2017), or even the nexus of climate–gender–conflict migration (e.g. Raleigh *et al.*, 2008; Myrttinen, 2017). Yet, a more detailed literature review is required for elaborating the differences and similarities between the different topics and theoretical perspectives, and their potential for research on gender and livelihoods in the context of climate-induced migration.

Nevertheless, there is increasing consensus that gender impacts of migration in the context of climate change go beyond differential vulnerabilities and livelihood strategies, as the case study on the West African Sahel will illustrate in the following section.

Case Study: Climate Change, Migration and Gender in the West African Sahel

This section draws on empirical works of the research project 'Migration, climate and environmental change' (MICLE), which investigated social–ecological conditions of population mobility in two regions – Bandiagara in Mali and Linguère in Senegal (Hummel *et al.*, 2012; van der Land and Hummel, 2013; Hummel, 2015, 2016; Liehr *et al.*, 2016; van der Land, 2018; van der Land *et al.*, 2018). Both study regions (Fig. 12.1) are highly affected by land degradation and an increasing rainfall variability, and the causes and consequences of environmental changes are both climatic and anthropogenic (Brandt *et al.*, 2014). Owing to the high dependence on agriculture and thus on the natural environment, for the majority of the rural population, climate change can pose a serious threat to their livelihoods. Therefore, the Sahel region is frequently regarded as a hotspot of environmental displacement. Indeed, the rural areas of both study regions experience considerable out-migration. Mobility, however, has a long tradition in West Africa and is deeply rooted in the culture.

The empirical research and data collection were conducted in close cooperation between the natural and the social scientists involved in the project. The social–empirical research was conducted at two empirical levels: with villagers in selected rural study areas in the Bandiagara region in Mali, and in the Linguère region in Senegal, and it addressed migrants originating from these regions but now living in the capitals Bamako and Dakar, respectively, as destinations (see Fig. 12.1). The empirical data collection was divided into two phases combining different qualitative and quantitative methods. The explorative phase in 2011 included qualitative interviews and expert interviews to prepare for the main

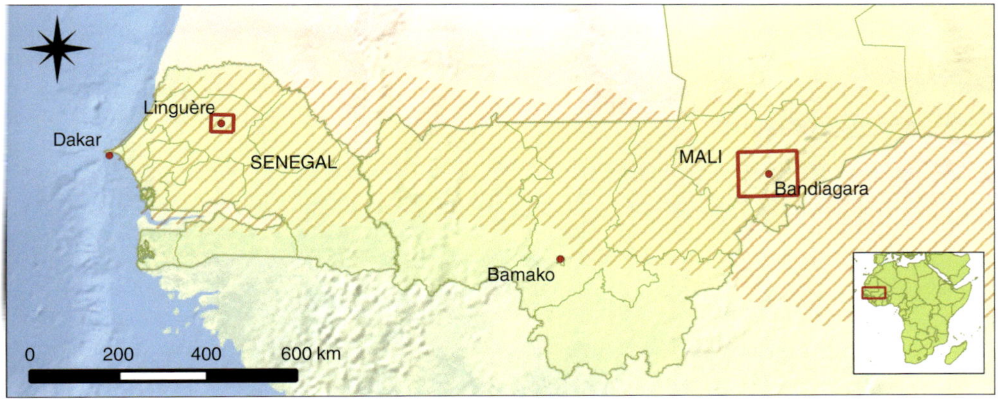

Fig. 12.1. Location of study areas in Mali and Senegal for 'Migration, climate and environmental change' (MICLE) research project. Study areas located in red rectangles. Shaded area indicates the Sahel's extent, delimited by the 250 mm/a isohyet in the north and 900mm/a in the south (based on the definition of the MEA, 2005). From DIVA-GIS, Natural Earth Data; cartography by Lukas Drees.

fieldwork. In addition, natural scientific data on temperature, precipitation, vegetation cover indicators for land degradation, etc. were collected. In the second and main phase in 2012, a total of 905 individuals in the regions of origin as well as in the two capitals were questioned by means of a standardized survey. In addition, 60 semi-structured and narrative interviews with migrants and non-migrants, and eight focus group discussions with villagers and migrants were conducted over several months, in addition to participant observation. The survey included 661 questionnaires with individuals in the rural study areas and 244 questionnaires with the migrants in Bamako and Dakar, both with and without migration experience. Where possible, interviews were conducted in French, but more often in local languages with the help of local field assistants. Interviewees and survey participants were selected randomly and differed in terms of age, formal education, ethnicity, sex/gender, and migration experience. The survey was confined to people at minimum age of 18 years and older, whereas the qualitative interviews also addressed younger people. The sampling of the survey participants was based on a quota sample with an equal distribution of women and men in three different age groups (18–30 years, 31–50 years, 51 years and older). In the two capitals Bamako and Dakar, migrants from the study areas were identified by means of a snowball sampling method. The qualitative interviews were analyzed by applying a content analysis with different categories. In the standardized analysis, cross-tabulations were applied to identify relations between different attributes and variables and to test the hypotheses and assumptions. Statistical significance was tested by applying chi-square (χ^2), with the significance level set at $p \leq 0.05$ (van der Land and Hummel, 2013; Hummel, 2016; van der Land, 2018).

The study adopted both the SES approach for the interdisciplinary analysis of climate–environment–migration interlinkages, and the 'capability and aspirations approach' for the study of migration patterns and livelihoods as described above (see section on 'Integrating gender into the picture' of this chapter).

Gendered migration patterns

As already demonstrated by numerous studies on the West African Sahel (Findley, 1994; Scheffran *et al.*, 2012; Morrissey, 2014), migration also proved to be common in the rural study areas in Mali and Senegal. The majority of survey participants (81%) had experienced migration personally at least once in their lives. Surprisingly, the extent of migration experience does not differ significantly between women and men from Senegal, while in Mali, migration experience is more common among men than among women (Fig. 12.2).

In some towns in the Bandiagara study area, migration of unmarried women is prohibited by the community or the village elders. For men, on the other hand, temporarily leaving their home village is the norm. Especially in Bandiagara, the so-called *aller en aventure* is part of growing up for adolescent men and a *rite de passage* of the Dogon culture. Having no migration experience is rather the exception among men (van der Land, 2018). In terms of temporal and spatial migration patterns, migration is usually temporary and circular, either within the country of origin or to neighboring countries. Here, significantly gendered migration patterns are found, particularly in Mali. Women in Mali migrate more often for longer periods of time while, generally speaking, men migrate for shorter periods of time, depending on the reasons for migration, the type of labor performed during migration, the financial gains, and the number of people of working age in the village (van der Land, 2018, p. 21).

According to the literature, due to traditional gender roles, men in West Africa usually migrate for economic reasons while women primarily migrate for marriage or other family reasons (Findley, 1994; Hunter and David, 2011). The results of the MICLE study revealed significant differences in the distribution of motives concerning reasons for migration, providing a more nuanced picture. Both female and male survey respondents in Mali and Senegal mentioned 'search for money/job' as the most important reason for migration, although it was more frequently mentioned by male respondents. Interestingly, the motive 'food security' was only slightly more often mentioned by men. This indicates the relevance of the motive also for women. However, there are some migration motives that are clearly gender-specific. For instance, men migrate more often for educational reasons or vocational training. On the other hand, female survey participants more often mentioned 'family

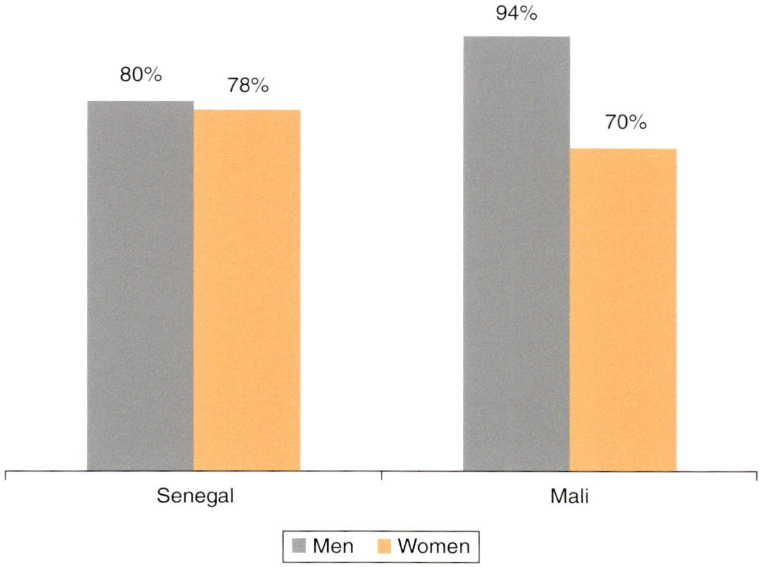

Fig. 12.2. Migration experience of men and women in Mali and Senegal. From 'Migration, climate and environmental change' (MICLE) survey 2012 (Senegal n = 460; Mali n = 445).

reasons,' 'visit,' and 'buying clothes/dowry' (see Figs 12.3 and 12.4).

Women specified in the qualitative interviews that the 'search for money/job opportunities' comprises the aim to prepare weddings and to be able to afford kitchen equipment as well as fashionable clothing or jewelry, whereas for men 'the search for money/job opportunities' in seasonal or temporary migration refers to economic security and investments. Due to the lack of work and employment opportunities during the dry season, migration can, particularly for young men, constitute a necessity to support their families with money and staple food. Thus, men usually migrate even if the home crop yields have been satisfying, in order to improve financial bases, e.g. to build or maintain houses or to buy agricultural equipment, livestock, or other goods such as motorcycles, clothes, or mobile phones (van der Land, 2018). Here, clear indications are shown that migration motives go beyond securing livelihoods.

Since motives for migration may alter over time, the study investigated such changes during the period between 1970 and 2012. This could be realized with the data provided by the standardized survey and due to the wide age span of survey participants. The questionnaire included items about the individual migration biography (e.g. first and last time of migration, duration, and respective motives for the migration). Surprisingly, the results show that the men's motives hardly changed at all whereas those of women changed substantially. While family reasons have become less important for women of the younger age groups during the last decade, economic motives have clearly gained in importance. On the one hand, the study results confirm the still existing traditional role and responsibility of men as 'breadwinners,' while women did not indicate using the money earned in urban areas to support the household's livelihoods in the rural home region. On the other hand, the findings show that migration from areas affected by environmental degradation is not solely aimed at improving the economic situation. Particularly for young people, both women and men, individual migration decisions can also indicate a rebellion against traditional norms and hierarchies (Hunter and David, 2011, p. 313; van der Land, 2018, p. 55). Hence, gendered traditional social norms influence the decision to move or stay very differently for women and men. Existing male migration culture facilitates men's mobility, while the decision to stay might even have to be justified. The

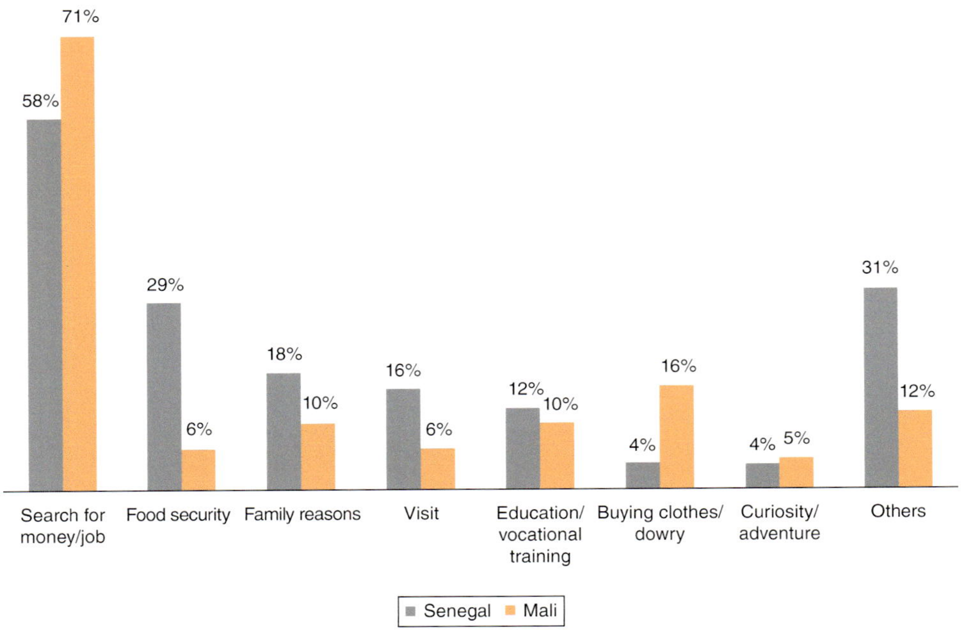

Fig. 12.3. Migration motives per country (multiple answers possible). From 'Migration, climate and environmental change' (MICLE) survey 2012 (Senegal n = 388; Mali n = 398).

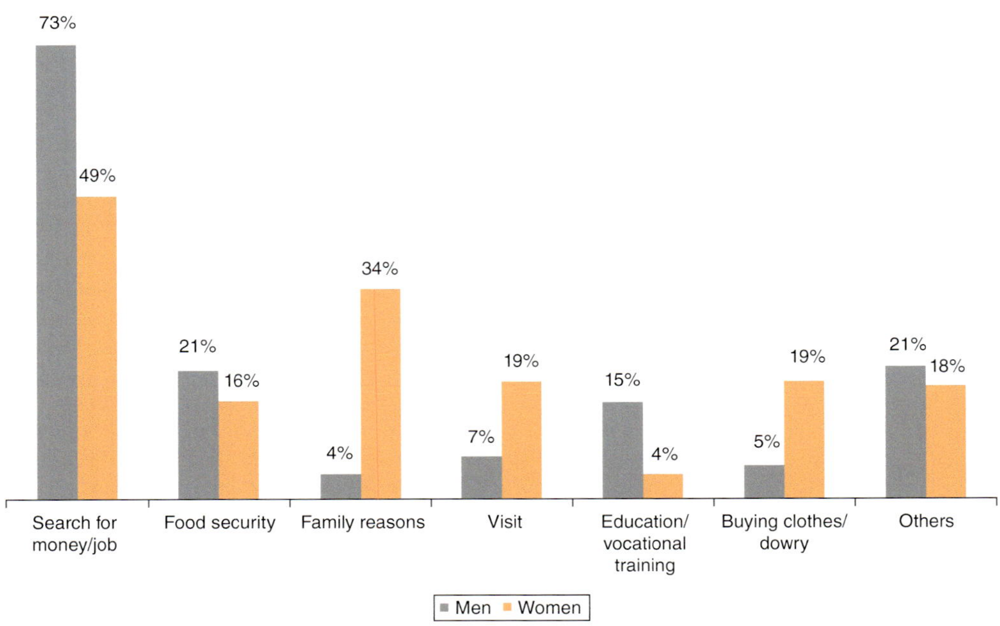

Fig. 12.4. Migration motives according to gender (multiple answers possible). From 'Migration, climate and environmental change' (MICLE) survey 2012 (Senegal n = 388; Mali n = 398).

opposite applies for women; for them, migrating rather than staying has to be justified since the independent migration of women has not been customary until recently. Age and gender are therefore important factors to explain migration decisions:

> In the Sahel societies, men and parents are attributed with more power in the household, compared to women and children, and thus decide over their own migration and the mobility of other family members. [...] Household members with less power due to their age and gender might either be put under pressure to migrate or be excluded from migration.
>
> (van der Land, 2018, p. 31)

These findings illustrate the relevance of gender as relation but also as structure, found in sociocultural norms and institutions.

Intersections of gender, education and age

Although the MICLE study did not pursue a systematic intersectional methodology, it univocally revealed interactions of gender with other forms of social differentiation, particularly age and education (van der Land and Hummel, 2013). Overall, the older generation migrates primarily to generate income for the family, while for the younger generation economic independence and the purchase of consumer goods are essential. Food security plays an increasingly minor role as the level of education rises. Interestingly, the MICLE results show no significant relationship between people's migration experience and their level of education, neither for women nor for men. Nevertheless, people's formal education is one of several important social aspects influencing migration decisions under fragile social–ecological conditions. The overall level of education is very low in both study regions: 76% of the survey respondents have not received any formal education and the illiteracy rate is high (69%). Altogether, the level of education is much higher among men, but school enrolment and higher education of girls have increased in recent years as a result of government education programs, particularly in Senegal.

> Girls and boys with a secondary education or higher have usually far better employment and vocational prospects in the urban centers than in the structurally weak rural areas. Therefore, it is not surprising that the better educated more often migrate for education or vocational reasons than for economic or environmental reasons.
>
> (van der Land, 2018, p. 96f)

Generally speaking, the lower the level of formal education, the more likely people are engaged in small-scale farming as their main source of income. The study revealed that people are particularly vulnerable to environmental changes, such as increasing rainfall variability, if they have only small financial means and social capital to compensate income losses as a result of bad harvests. Under these circumstances, migration may be the only way to ensure food security.

Migration as an adaptation or livelihood strategy and aspirations for a better life

A considerable share of the population in both rural study areas live under precarious economic conditions. For these segments of the population, more diversified and multilocal sources of income through migration have become highly important to sustain their livelihoods. The significance of climatic change and land degradation for migration decisions clearly increases in conjunction with the people's dependence on agriculture. The lower the opportunities of a rural household to diversify its income and the fewer alternative job offers available, the stronger the impacts of environmental change on the mobility of the population. Given these social–ecological conditions, migration constitutes an adaptation strategy to deal with harmful impacts of climate change since it minimizes risks of crop failures and contributes to a diversification of income and better food security (Hummel, 2016). However, as Fig. 12.5 illustrates, it is just one of several other coping strategies to compensate for unfavorable environmental conditions in both Sahel regions.

The results from the quantitative and qualitative data indicate that many young people do not see their future in agriculture, and instead

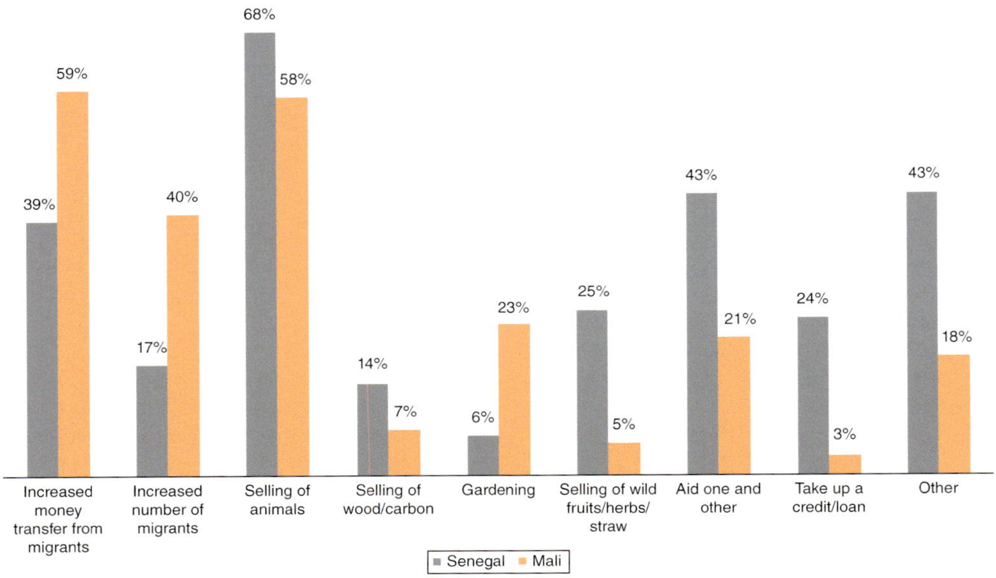

Fig. 12.5. Distribution of strategies to compensate for bad harvests (multiple answers possible). From 'Migration, climate and environmental change' (MICLE) survey 2012 (Senegal n = 411; Mali n = 443).

aspire to alternative occupations. This can be attributed to increased school attendance but also to increasingly unfavorable conditions for agriculture and decreasing yields due to highly variable or scarce rainfall and decreasing soil fertility. Thus, instead of struggling with these unfavorable circumstances, many young people migrate to urban areas to pursue business and other income generating activities that promise both a higher income and better living standards. 'For many, migration appears to be the easiest, most effective, and probably the only solution to move out of poverty and more generally to succeed in life and go ahead. As such, migration is often an expression of personal success and progress that goes beyond fulfilling basic needs' (van der Land, 2018, p. 75).

As the case study illustrates, migration is prevalent in both study areas regardless of socio-demographic characteristics such as gender and formal education level. For many people, migration is a proactive choice and part of their agency. Social ties and migrant networks can stimulate follow-up migration, which in turn facilitates working and living conditions at the respective destinations. Experiences and skills gained during migration as well as remittances and investments can enhance the diversification of the livelihoods in the rural communities (Scheffran *et al.*, 2012). However, as the results of the MICLE study illustrate, migration cannot be reduced to its function as an adaptation and livelihood strategy. For a large number of interview participants it is 'a way of life.' The case study illustrates the very context-specific social–ecological conditions of migration and the relevance of gendered role attributions, norms, and practices, as well as underlying structural power hierarchies. However, generalizations for the nexus between climate change–migration–gender can only be derived to a limited extent, but the case illustrates the context-specific gendered patterns of societal relations.

Conclusion

While the scientific debate on the interdependencies of climate change, environmental change, and mobility has progressed considerably, research on gender impacts of the climate–migration–environment nexus has been addressed only sparsely for a long time. This probably has to do with the challenge that a 'gender lens' requires connecting very scattered and heterogeneous scientific knowledge. As this chapter illustrates, considerable progress can be seen in bringing gender issues into the debate. Current literature stresses the relevance of intersectional or relational approaches, which contribute to a deeper understanding of complex interplays between gender and other categories of social difference within societal power structures. As Gioli and Milan (2018) argue, the next generation of research on the climate–environment–migration nexus should take advantage of interdisciplinary research approaches and 'must integrate a strong relational gender perspective in order to bridge the gaps' (Gioli and Milan, 2018, p. 144). Similar to other fields of research such as gender and development and gender and climate change, research on gender impacts of climate-induced migration has largely focused on 'women' or gender differences, which usually means comparing impacts for women and men. Arguing from a feminist perspective, such data disaggregation by sex (including disaggregation by the sex of the household head in sustainable livelihoods research) is an important step for the integration of a gender perspective. However, this should only be a first step since it does not allow the capture of the social dynamics and power constellations behind the data (Gioli and Milan, 2018; see also Tannenbaum *et al.*, 2019). What is more, a sole perspective on gender *differences* may unintentionally contribute to traditional gender stereotypes and dichotomies. Intersectionality promises to meet these challenges by providing a more comprehensive picture of entangled power constellations, although its empirical applicability still constitutes a methodological challenge (Kaijser and Kronsell, 2014). However, its adoption for the climate migration issue seems appealing and very necessary.

Notwithstanding existing research gaps, there is considerable evidence for migration in the context of climate change being a highly gendered issue. Research, however, that seeks solutions to social–ecological problems for fostering livelihoods for all people, must complement system knowledge (i.e. for better explanation of the nexus' complexity) by problem- and policy-oriented transformation knowledge. Research in the field of climate, migration, and gender must therefore address to a greater extent normative issues of equity, human rights, and justice (including gender justice!) as prerequisites for sustainability transformations. This requires not least, more transdisciplinary cooperation between science and different societal actors and stakeholders.

References

Adger, W.N., Pulhin, J.M., Barnett, J., Dabelko, G.D., Hovelsrud, G.K. *et al.* (2014) Human security. In: Field, C.B., Barros, V.R., Dokken, D.J., Mach, K.J., Mastrandrea, M.D. *et al.* (eds) *Climate Change 2014: Impacts, Adaptation, and Vulnerability. Part A: Global and Sectoral Aspects. Contribution of Working Group II to the Fifth Assessment Report of the Intergovernmental Panel on Climate Change*. Cambridge University Press, Cambridge, UK and New York, pp. 755–791.

Adger, W.N., Arnell, N.W., Black, R., Dercon, R., Geddes, A. and Thomas, D.S.G. (2015) Focus on environmental risks and migration: causes and consequences. *Environmental Research Letters* 10:060201.

Arora-Jonsson, S. (2011) Virtue and vulnerability: discourses on women, gender and climate change. *Global Environmental Change* 21, 744–751.

Ashraf, M.A. and Azad, M.A.K. (2015) Gender issues in disaster: understanding the relationships of vulnerability, preparedness and capacity. *Environmental and Ecology Research* 3, 136–142.

Becker, E. (2012) Social-ecological systems as epistimic objects. In: Glaser, M., Krause, G., Ratter, B. and Welp, M. (eds) *Human-Nature Interactions in the Anthropocene: Potentials of Social-Ecological Systems Analysis*. Routledge, New York, pp. 37–59.

Bettini, G. and Gioli, G. (2016) Waltz with development: insights on the developmentalization of climate-induced migration. *Migration and Development* 5, 171–189. doi.org/10.1080/21632324.2015.1096143.

Black, R., Adger, W.N., Arnell, N.W., Dercon, S., Geddes, A. and Thomas, D. (2011) The effect of environmental change on human migration. *Global Environmental Change* 21(Suppl. 1), S3–S11. doi.org/10.1016/j.gloenvcha.2011.10.001.

Bohra-Mishra, P. and Massey, D. (2011) Environmental degradation and out-migration: evidence from Nepal. In: Piguet, E., Pécoud, A. and Guchteneire, P. (eds) *Migration and Climate Change*. Cambridge University Press, Cambridge, UK, pp. 74–101.

Brandt, M., Romankiewicz, C., Spiekermann, R. and Samimi, C. (2014) Environmental change in time series: an interdisciplinary study in the Sahel of Mali and Senegal. *Journal of Arid Environments* 105, 52–63. doi.org/10.1016/j.jaridenv.2014.02.019.

Brzoska, M. and Fröhlich, C. (2016) Climate change, migration and violent conflict: vulnerabilities, pathways and adaptation strategies. *Migration and Development* 5, 190–210.

Carney, D. (ed.) (1998) *Sustainable Rural Livelihoods: What Contribution Can We Make?* Department for International Development (DFID), London.

Cattaneo, C., Beine, M., Fröhlich, C.J., Kniveton, D., Martinez-Zarzoso, I. *et al.* (2019) Human migration in the era of climate change. *Review of Environmental Economics and Policy* 13, 189–206.

Chambers, R. and Conway, G.R. (1992) Sustainable rural livelihoods: practical concepts for the 21st century. *IDS Discussion Paper 296*. Institute of Development Studies (IDS), Brighton, UK.

Chindarkar, N. (2012) Gender and climate change-induced migration: proposing a framework for analysis. *Environmental Research Letters* 7:025601.

Denton, F. (2002) Climate change vulnerability, impacts, and adaptation: why does gender matter? *Gender and Development* 10, 10–20.

Ferdous, J. and Mallick, D. (2019) Norms, practices, and gendered vulnerabilities in the lower Teesta basin, Bangladesh. *Environmental Development* 31, 88–96.

Findley, S.E. (1994) Does drought increase migration? A study of migration from rural Mali during the 1983–1985 drought. *The International Migration Review* 28, 539–553.

Gioli, G. and Milan, A. (2018) Gender, migration and (global) environmental change. In: McLemann, R. and Gemenne, F. (eds) *Routledge Handbook of Environmental Displacement and Migration*. Routledge, London and New York, pp. 135–150.

Gioli, G., Khan, T., Bisht, S. and Scheffran, J. (2014) Migration as an adaptation strategy and its gendered implications: a case study from the upper Indus basin. *Mountain Research and Development* 34, 255–265.

Görg, C., Brand, U., Haberl, H., Hummel, D., Jahn, Th. and Liehr, S. (2017) Challenges for social-ecological transformations: contributions from social and political ecology. *Sustainability* 9:1045.

Hoffmann, R., Dimitrova, A., Muttarak, R., Crespo Cuaresma, J. and Peisker, J. (2020) A meta-analysis of country-level studies on environmental change and migration. *Nature Climate Change* 10, 904–912.

Hummel, D. (2015) Climate change, environment and migration in the Sahel. *Rural 21 – The International Journal for Rural Development* 1, 40–43.

Hummel, D. (2016) Climate change, land degradation and migration in Mali and Senegal: some policy implications. *Migration and Development* 5, 211–233. doi.org/10.1080/21632324.2015.1022972.

Hummel, D. and Stieß, I. (2017) Social ecology: a transdisciplinary approach to gender and environment research. In: MacGregor, S. (ed.) *Routledge Handbook of Gender and Environment*. Routledge, London and New York, pp. 186–201.

Hummel, D., Doevenspeck, D. and Samimi, C. (2012) Climate change, environment and migration in the Sahel: selected issues with a focus on Mali and Senegal. *MICLE working paper No. 1*. Institute for Social–Ecological Research, Frankfurt am Main, Germany.

Hummel, D., Jahn, Th., Keil, F., Liehr, S. and Stieß, I. (2017) Social ecology as critical transdisciplinary science: conceptualizing, analyzing and shaping societal relations to nature. *Sustainability* 9:1050.

Hunter, L.M. and David, E. (2011) Displacement, climate change and gender. In: Piguet, E., Pécoud, A. and Guchteneire, P. (eds) *Migration and Climate Change*. Cambridge University Press, Cambridge, UK, pp. 306–330.

Hunter, L.M., Luna, J.K. and Norton, R.M. (2015) The environmental dimensions of migration. *Annual Review of Sociology* 41, 377–397.

Kaczan, D.H. and Orgill-Meyer, J. (2020) The impact of climate change on migration: a synthesis of recent empirical insights. *Climatic Change* 158, 281–300. Advance online publication, doi.org/10.1007/s10584-019-02560-0.

Kaijser, A. and Kronsell, A. (2014) Climate change through the lens of intersectionality. *Environmental Politics* 23, 417–433. doi.org/10.1080/09644016.2013.835203.

Kelly, P.M. and Adger, N.W. (2000) Theory and practice in assessing vulnerability to climate change and facilitating adaptation. *Climatic Change* 47, 325–352.

Kelman, I. (2019) Imaginary numbers of climate change migrants? *Social Sciences* 8:131. doi.org/10.3390/socsci8050131.

Lamicchane Dhimal, M. (2018) *Gender Dimensions of Health Impact of Climate Change in Nepal: a Comprehensive Empirical Study*. Nomos, Baden-Baden, Germany.

Liehr, S., Drees, L. and Hummel, D. (2016) Migration as a societal response to climate change and land degradation in Mali and Senegal. In: Yaro, J.A. and Hesselberg, J. (eds) *Adaptation to Climate Change and Variability in Rural West Africa*. Springer, Cham, Switzerland, pp. 147–169.

McLemann, R. and Gemenne, F. (2018) Environmental migration research: evolution and current state of the science. In: McLemann, R. and Gemenne, F. (eds) *Routledge Handbook of Environmental Displacement and Migration*. Routledge, London and New York, pp. 3–16.

MEA (2005) *Ecosystems and Human Well-being: Synthesis*. Millennium Ecosystem Assessment, Island Press, Washington, DC.

Morrissey, J. (2014) Environmental change and human migration in Sub-Saharan Africa. In: Piguet, E. and Laczko, F. (eds) *People on the Move in a Changing Climate: Global Migration Issues 2*. International Organization for Migration, Geneva, Switzerland, pp. 81–109.

Myrttinen, H. (2017) The complex ties that bind: gendered agency and expectations in conflict and climate change-related migration. *Global Policy* 8(Suppl. 1), 48–54.

Nagabhatla, N., Pouramin, P., Brahmbhatt, R., Fioret, C., Glickman, T., Newbold, K.B. and Smakhtin, V. (2020) Water and migration: a global overview. *UNU-INWEH Report Series 10*. United Nations University Institute for Water, Environment and Health, Hamilton, ON, Canada. Available at: www.reliefweb.int/sites/reliefweb.int/files/resources/Migration-and-Water-A-Global-Overview.pdf (accessed 3 February 2021).

Pessar, P.R. and Mahler, S.J. (2003) Transnational migration: bringing gender in. *The International Migration Review* 37, 812–846.

Raleigh, C., Jordan, L. and Salehyan, I. (2008) Assessing the impact of climate change on migration and conflict. *Paper commissioned by the World Bank Group for the Social Dimensions of Climate Change workshop*. World Bank,Washington, DC.

Rezwana, N. (2016) Disasters and access to healthcare in the coastal region of Bangladesh: a gendered analysis. Doctoral thesis, Durham University, UK. Available at: www.etheses.dur.ac.uk/11432/ (accessed 3 February 2021).

Rothe, D. (2017) Myths and stereotypes in the discourse on climate-induced migration. *Global Policy* 8, 40–47.

Scheffran, J., Marmer, E. and Sow, P. (2012) Migration as a contribution to resilience and innovation in climate adaptation: social networks and co-development in northwest Africa. *Applied Geography* 33, 119–127.

Sen, A. (2000) *Development as Freedom*. Oxford University Press, New Delhi.

Tannenbaum, C., Ellis, R.P., Eyssel, F., Zou, J. and Schiebinger, L. (2019) Sex and gender analysis improves science and engineering. *Nature* 575, 137–146.

Tuana, N. (2008) Viscous porosity: witnessing Katrina. In: Alaimo, S. and Hekman, S. (eds) *Material Feminisms*. Indiana University Press, Bloomington, IN, pp. 188–212.

van der Land, V. (2018) *Migration and Environmental Change in the Sahel. Why Capabilities and Aspirations Matter*. Earthscan, London and New York.

van der Land, V. and Hummel, D. (2013) Vulnerability and the role of education in environmentally induced migration in Mali and Senegal. *Ecology and Society* 18:14. doi.org/10.5751/ES-05830-180414.

van der Land, V., Romankiewicz, C. and van der Geest, K. (2018) Environmental change and migration: a review of West African case studies. In: McLemann, R. and Gemenne, F. (eds) *Routledge Handbook of Environmental Displacement and Migration*. Routledge, London and New York, pp. 163–177.

Veronis, L., Boyd, B., Obokata, R. and Main, B. (2018) Environmental change and migration: a review. In: McLemann, R. and Gemenne, F. (eds) *Routledge Handbook of Environmental Displacement and Migration*. Routledge, London and New York, pp. 42–70.

Weber, L. and Hilfinger Messias, D.K. (2012) Mississippi front-line recovery work after Hurricane Katrina: an analysis of the intersections of gender, race, and class in advocacy, power relations, and health. *Social Science & Medicine* 74, 1833–1841.

West, C. and Zimmerman, D.H. (1987) Doing gender. *Gender & Society* 1, 125–151.

13 Gendered Livelihood Adjustments in the Context of Climate-Induced Disasters

M. Alston

Introduction

The increasing frequency and intensity of climate-induced disasters indicates that the argument about whether 'climate change is real' has become an historical artifact of a more innocent time. Across the globe, humanity is facing an uncertain future typified by major climate-induced environmental disasters. The catastrophic dimensions of these events have been evident in the author's country, Australia, in 2020 where summer bushfires burned through 12.6 million ha, killing 34 people and over 1 billion animals and affecting 11.6 million people through the thick smoke haze that lingered over the cities (Werner and Lyons, 2020). The arrival of the COVID-19 pandemic has added significant anxiety to an already burdened populace. None the less, our lack of preparedness for the scale of climate change disaster events, and the necessity to adapt, both in Australia and elsewhere, is evident despite the fact that these events have been forecast for some time (e.g. see the regular reports from the Intergovernmental Panel on Climate Change, including IPCC, 2001; IPCC, 2014).

Meanwhile, many of the world's politicians and policy makers are delaying, or downplaying, actions to address climate change and the threats inherent in climate-induced disasters. As a result, and with minimal policy and practice guidance and support, people are dealing with the major social changes caused by unpredictable weather events and their assault on livelihoods at a local level. As Schofield and Gubbels (2019, p. 93) note, 'the everyday reality of adaptation in these localities may bear little resemblance to the technical, depoliticized notion of adaptation that is predominant in policy and planning.' Further, such developing policies notably and rarely resemble those most affected in terms of gender or levels of poverty (Kaijser and Kronsell, 2014). It is little wonder that the reality of disasters has moved beyond the slow lane of political ineptitude, with the burden falling on communities and families forced to adapt as best they can.

This chapter focuses on the impact of climate-induced disasters on gendered livelihood adaptations in three contrasting countries where the author has been involved in research projects over the last 8 years – Lao PDR, Bangladesh, and Australia. At first glance there appears little commonality between these countries and their responses. Yet, while undertaking research, the author and her team witnessed broad similarities in adaptive responses, particularly in relation to outmigration. The focus of this chapter is on the women in agricultural communities in these countries, where families are now adjusting to the reality of ongoing climate-induced disasters by seeking new ways of supporting themselves. Where disasters have eroded communities and critical local resources (such as agricultural land, animals, and productive capacity), families face the need to restructure income generation.

DOI: 10.1079/9781789247053.0013

Universally these agricultural livelihood adaptations involve complex gender rearrangements as people reconfigure how best to adapt. The case studies presented here illustrate the significant restructuring in both developed and developing countries and the remarkably similar although complex gendered livelihood adaptations under way. Three critical points are raised in this chapter:

- While both women and men undertake paid labor as a critical response to disaster, the outmigration of men is resulting in the increased involvement of women in agricultural labor, food production, and food security.
- Despite this trend, women have inequitable access to the resources, training, education, and other supports they need to ensure ongoing food security (Alston and Whittenbury, 2016).
- This trend to increasing women's involvement in agriculture has facilitated the opening up of restrictions on women's movements and *purdah* practices in some countries.

What Do We Mean by Climate-Induced Disasters, Gender and Livelihood Strategies?

'Climate-induced disasters' refer to extreme weather events and other climatic changes that arise because of the build-up of greenhouse gases (GHG) in the atmosphere. Human activities have been deemed largely responsible (IPCC, 2014), and hence the appeal to countries by the IPCC and other United Nations (UN) bodies to reduce emissions is focused on the survival of the planet as we know it. The major forums where this is discussed include the annual Conference of the Parties (COP) meetings attended by nations who are signatories to the UN Framework Convention on Climate Change (UNFCCC). At these meetings regular attempts are made to enact binding agreements to reduce GHGs and to address climate-induced disasters. Yet, these attempts are not necessarily focused on the complex outcomes for people and communities affected, nor on the gendered arrangements being reshaped across the world.

In the politically overcharged environments where economic matters appear to be prioritized over humanity, there are ongoing delays in global cooperation aimed at addressing emissions. At the same time there is a lack of attention to the complexity of livelihood adaptations and the potential impact on food security. It is little wonder that, in the absence of a supportive international conversation, families and communities are adopting their own strategies and adaptations to deal with climate-related impacts on their agricultural livelihoods (Ogra and Badola, 2015). As we shall see, the gendered complexities inherent in reshaped livelihoods are very evident.

'Livelihoods' refer to the way individuals and families source the income needed to maintain and support themselves. In the agricultural context this includes food production as well as control of assets such as land, livestock, and labor. When disasters occur, there are significant challenges to livelihoods in agricultural areas affected by the loss of assets and broken production cycles. These impacts can mean the loss of a whole year's income and the result can have potentially catastrophic outcomes, requiring significant adjustments. This chapter focuses on outmigration as one such strategy and the gendered implications of this.

In the context of this chapter, the author adopts the definition of 'gender' as 'what is expected, allowed and valued in a woman or a man in a given context. It determines opportunities, responsibilities and resources as well as power associated with being female or male' (UNDP Bureau for Crisis Prevention and Recovery, 2010, p. 1; quoted in Kinnvall and Rydstrom, 2019, p. 5). These culturally shaped gendered patterns of behavior that are the basis of women's vulnerability, are arguably heavily policed within diverse cultural contexts, and are central to the 'status order' (Fraser, 1996; noted in Kinnvall and Rydstrom, 2019). Yet, culturally sanctioned gender-based power relations are not easily eroded despite the changes wrought by disasters. Thus, as Kinnvall and Rydstrom (2019, p. 5) note, 'a catastrophe does not land in a socio-economic and political void, but rather in a place where the autonomy and rights of women and girls might be severely limited. A disaster event interlocks with gender specific inequalities... and can exacerbate gendered susceptibility.' In a disaster context where events move rapidly, established gender inequalities can be reinforced and indeed solidified around supposedly gender-neutral actions and gender-blind responses, further disadvantaging women (e.g. see UNDP, 2014).

In an agricultural context, gender-blind policy responses and resource distribution can ignore the significant changes in reshaped gender roles in agricultural production. In examining the complexities of gender rearrangements in agriculture, the author is conscious of the advice of scholars including Mollett (2017, p. 150), Djoudi *et al.* (2013), and Arora-Jonsson (2011) to avoid the tendency to define a unidimensional 'third world woman.' In the context of this chapter the author also wishes to avoid creating a unidimensional, vulnerable 'woman in agriculture' devoid of context, feelings, complexities, and resilience. It is important to note that intersectional factors are critical to the vulnerability experienced by women in disaster sites and to their capacity to adapt positively.

Outmigration

Among the most common forms of gender-based diversification and livelihood adjustment following climate-induced disasters in rural locations is the outmigration of one or more family members seeking work and income (Alston, 2015; Djoudi *et al.*, 2016). Outmigration in the context of climate events refers to the voluntary or forced movement of people away from their homes, either temporarily or permanently. There are suggestions that approximately 200 million people will be forced to leave their homes permanently by 2050, joining the wave of climate refugees across the globe (Brown, 2008). In response to devastating, and more permanent, climate events such as sea level rises, the migration of climate refugees has been under way for some time. In other areas where catastrophic climate-induced disasters wreak havoc on agricultural production, this can be a semi-permanent or temporary measure requiring one or more family members to move away in search of work. Outmigration is a livelihood strategy that can quickly generate remittance income to be sent back to families. However, such a system is fraught with problems including the loneliness and stress caused to the remitter living precariously in a city, as well as the hazards for the family left behind waiting for remittance funds, and the sometime erosion of the family's security if the remitter, over time, chooses to set up a second home/family (Alston, 2015). However, in the context of this chapter, and in the immediate response to climate change disasters, outmigration is a significant option for families previously reliant on agricultural production.

Outmigration of one or more family members in response to climate-induced disasters is a significant trend in several countries. The push factors for outmigration resulting from climate-induced disasters encompass the destruction of livelihoods including farms and crops; the destruction of homes and other assets; poverty; a lack of relief services; and ongoing threats to life (Kartiki, 2011; Alston, 2015). The pull factors include work opportunities and wages; friends and family moving as well; knowledge of the destination; support systems at the destination; and a lack of incentives to stay (Kartiki, 2011).

The outmigration of men is a common factor in many areas across South Asia. This trend is reshaping the agricultural workforce and having a significant impact on its composition and, ultimately, on food production and food security in the region. Estimates suggest that, in response to male outmigration, approximately 43% of the global agricultural workforce is now female (World Bank, 2012), and in some countries this is closer to 50% (World Bank, 2017).

In two of the case studies presented here, Bangladesh and Lao PDR, outmigration from agricultural areas also includes young women moving to work in garment factories, as well as women undertaking domestic labor. In these countries men might move to cities and other countries to work in the construction industry or as street vendors, *tuk tuk* drivers or farm laborers. When men leave, this leads to a very visible phenomenon that Saha *et al.* (2018) describe as 'the left-behind women' – women who now play an invisible, largely undocumented, but critical role in agriculture, adding to the worldwide feminization of agriculture evident in developing countries. Meanwhile, in the third case study site, Australia, and by contrast, in agricultural families and communities it is more likely that women will be working away from home to earn the income needed for the family to survive in agriculture (Alston *et al.*, 2017). This trend is not necessarily viewed as outmigration in the developed world context. None the less, closer scrutiny indicates that it is indeed also the earning of income away from the farm and used for the purpose of supporting the family; in this sense it is outmigration to earn income.

Despite this expanding role for women in agricultural industries, they are often not recognized through ownership and control of land and assets. Women own less than 20% of the world's land (World Economic Forum, 2020). Eastin (2018) reminds us that following the outmigration predominantly of men, women's ability to achieve economic independence, or to take part in decision making forums, is hampered by their lack of official ownership of land and problematic access to health and education services, as well as critical issues relating to food, fuel, and water access (Bell, 2011).

Case Study Methodology

The three case studies presented in this chapter are drawn from research undertaken in Australia, Lao PDR, and Bangladesh (Alston *et al.*, 2014; Alston, 2015; Whittenbury and Alston, 2016; Clarke *et al.*, 2017). In these studies, the author and her colleagues were examining the impacts of climate-induced disasters on women engaged in agricultural industries.

In Lao PDR in 2016, a scoping study of women's engagement in agribusiness value chains was undertaken, funded by the Australian Centre for International Agricultural Research (ACIAR). As part of this research a dozen interviews with key informants (representatives of government and non-governmental organizations, NGO) were completed in Vientiane, as well as a small number of interviews with women farmers on their farms and at various points of the value chain. The research in Bangladesh was a much larger 3-year study focused on the gendered impacts of climate changes and funded by the Oxfam–Monash Alliance. As part of this project 31 focus groups (mixed, men only, and women only) and 63 interviews were undertaken with women in nine villages, located in three separate areas of Bangladesh. In addition, a survey of 617 people (298 women and 319 men) was conducted in these villages. The Australian research funded by the Australian Research Council and the Gardiner Foundation focused on the impacts on Australian dairy families resulting from the withdrawal of irrigation water entitlements by the government as a result of drought. As part of this research 12 key informant interviews, and interviews and focus groups with dairy officials and women and men in dairy families (92 participants) were completed over 2014–2015. This was followed up with a survey in 2016 completed by 128 dairy family members.

It is important to note that these three studies had different aims and objectives resulting from the funding body specifications. None the less, the three studies focused on gender factors associated with climate changes in agricultural communities. Of note is the similarity of factors shaping gendered livelihood adjustments. This suggests that climate-induced disasters are reshaping agricultural livelihoods in highly gendered ways. In particular for this chapter, the author focuses on outmigration and the increasing role of women in agriculture as common trends that cross borders and cultural boundaries.

Lao PDR

Lao PDR is an elongated, landlocked, fertile country in Asia with a population of a little more than seven million, one-quarter of whom live in poverty (UN Human Rights Office of the High Commissioner, 2019). Lao PDR is particularly vulnerable to climate change impacts including floods and droughts (FAO, 2020) and there are significant issues relating to clean water access and sanitation evident in the country, particularly in disaster-affected areas (ADB and World Bank, 2012).

Men tend to control ownership of land and resources in Lao PDR, and yet are increasingly likely to outmigrate at least for part of each year, while women now form a significant part of the agricultural workforce. As a consequence of climate challenges, outmigration is now a significant strategy adopted in agricultural areas of the country, as men in particular move to the capital city, Vientiane, or to neighboring countries including especially Thailand. Meanwhile, young women are moving to the garment factories and some older women may move to work as domestic laborers in bordering countries or the Middle East.

Despite women's increased role in agriculture, traveling to district centers for training and extension services is limited for women because of household responsibilities, mobility issues, literacy levels, and not being recognized as the official landowners (ADB, 2013). Where women do have a claim to land, there is some evidence of land grabbing and resettlement of ethnic minorities in areas where women are particularly vulnerable to predatory behavior (Cornfold and Matthews, 2007). Key informants noted that there are emerging issues resulting from the incorrect use of chemicals and pesticides because of a lack of training, and women themselves noted the continued presence of unexploded ordinance (UXO) left over from the Vietnam war, as an ongoing source of critical danger for those engaged in agriculture.

The research revealed that outmigration is a critical aspect of income generation for Lao PDR families engaged in agriculture. Key informants from NGOs in Vientiane noted that outmigration is now such a significant trend that: 'sometimes when we are active in the village we find only the elderly, the old people and the children' (Key informant, NGO).

Outmigration is an important source of remittance income for left-behind family members and is provided by those who have moved to the city. In describing outmigration, one government official notes that:

> My [relative] works in the garment factory and she usually sends money to her family. When the family needs money, somebody is sick or they want to repair the house or buy new machines. Many people who cannot go back to their home town will send money to help their parents back home.
>
> (Government representative)

As a result of the increase in outmigration it appears that agriculture is becoming increasingly reliant on left-behind women undertaking agricultural production (Phouxay and Tollefsen, 2011). The research revealed the need for more accurate data on women's increasing agricultural workforce input and land ownership to adequately address the factors shaping women's vulnerability (Kelly and Adger, 2000). It was also found that a lack of roads and infrastructure have a significant impact on women being able to get their produce to market, and this has led to subsistence farming now dominating in more remote communities. Women's advocates noted that women's capacity to produce commercial quantities of food is hampered by distance and poor infrastructure. This limits the capacity of women to achieve livelihood stability.

> In [province] where the villages are very far from the city and there is no road access – only walking. The closest village in 12 kilometers... We saw people, especially women, carrying products to sell in the town, they can carry only one basket... They walk about two or three hours then arrive in town to sell.
>
> (Key informant, NGO, Vientiane)

In these more remote villages, girls receive less formal education and this impacts on their ability to become financially and technologically literate. None the less, and indicating the complexity of livelihood arrangements, women in Lao PDR communities are involved in the rich and highly skilled handicrafts development. In agricultural farms visited, intricately designed, exquisite products under development were part of the paraphernalia of household life, providing an extra source of income.

Bangladesh

Bangladesh is a low-lying rich agricultural delta country with a population of approximately 166 million, making it one of the most densely populated countries in the world. Bangladesh has been heavily affected by climate challenges including rising temperatures; increasing rainfall causing flooding; cyclones; drought; and storm surges. Climate change in Bangladesh has led to significant loss of life, impacts on health and well-being, contaminated water supplies, major issues of food security, and particular impacts on agricultural employment and livelihoods (Alston, 2015; Alston and Akhter, 2016). In 2011–2014 a large-scale study of three regions of Bangladesh, each impacted by a different climate challenge, was undertaken. The results indicate that outmigration is a significant response to climate challenges and includes men moving to construction sites and laboring work in the cities, either seasonally or permanently; young women going to the garment factories; and women working as domestic labor.

The research reveals that women now form a significant part of the agricultural workforce.

> It's like the feminisation of poverty – the feminisation of the impacts of climate change – because women, to begin with, are dispossessed and marginalized. And then outmigration leaves them even more vulnerable... Men have to go and find livelihoods... women are still expected to remain where they are and look after the children... [any remaining] cattle or food or crop.
>
> (Key informant, NGO, Dhaka)

In response to the challenges of climate events and subsequent livelihood adaptations, women have emerged in recent years from strict *purdah* (or home-bound isolation and the need for a male escort when out of home). Women who took part in the current study noted that, as well as looking after their own plots, they may also be undertaking paid labor close to home on farms or in road construction, and they viewed this as a major development in their lives. This livelihood adaptation has enabled a modicum of economic independence.

> We are needy; we need something to do to earn. If we do this job covering our body and head according to the custom of Islam will it do any harm? Poor Muslim women need to eat also. We work hard to earn.
>
> (Women's focus group, Barguna)

A small but emerging trend relates to men outmigrating and then ceasing to remit funds, further exacerbating women's poverty.

> We have waterlogging and females are having miserable lives... nothing can now be produced there... the males have all left their areas and more than 60% of the households are led by women, not really led... they are burdened by the responsibilities... managed by women but under unacceptable circumstances... the males have all left either temporarily or permanently. Maybe 15–20% of these households they don't remit anymore, they don't assume any responsibility for their village family.
>
> (Focus group participant, Dhaka)

Of the Bangladeshi survey respondents, 21% of women and 26% of men noted they had increased their waged labor; and 12% of women and 17.5% of men had undertaken seasonal migration. In addition, 50% of women and 43% of men noted the lack of available paid work locally as a critical issue in their discussions about whether to move away. When asked what information they most wanted, 23% of women and 27% of men noted that information on alternative income generation was a critical need. Meanwhile, 64% of women and 43% of men reported the lack of food as a significant issue in their lives which impacts their capacity to undertake agricultural labor (Alston *et al.*, 2014).

Australia

Australia is an island continent, home to 25 million people. In recent times climate-induced disasters in Australia have included droughts, fires, and floods. Like their Asian counterparts, women involved in agricultural production are increasingly being drawn into agricultural labor to supplant hired labor. Australian agricultural women are more likely to be the family member who outmigrates for paid work, earning the much-needed remittance income required by the family to stay in farming. Official statistics suggest that women constitute 32% of Australia's agricultural workforce and earn 80% of off-farm income, indicating their increased significance to agricultural production (Department of Agriculture, Water and the Environment, 2020). The author's most recent study with Australian agricultural women (Alston *et al.*, 2017) encompassed the dairying communities in the Murray–Darling basin where farms are reliant on irrigation water drawn from the river systems to support their enterprises. The drought had a major impact on the water available in the rivers for irrigation and led to reduced allocations, including for the dairy farms under study, thereby critically affecting production.

The author's research revealed the significant hardship experienced by dairy farm families as water levels were so low that their capacity to produce irrigated pastures for their herds was compromised, as was their entire dairy enterprise. This resulted in significant hardships including the need to buy in feed and the need to lay off hired dairy employees. The predictable result was that women have been drawn into the dairy sheds replacing hired labor, to assist or indeed to take over the twice-daily milking. Many had also taken up off-farm jobs. As one female dairy farmer noted:

> I milk the cows, I'm the main milker so I milk seven days a week, twice a day… I do the most milking, [husband] helps me and we also have two others, we have [husband's] mum that milks as well so I work with her predominantly and then we have people on the weekends to give her a spell; she wants to do it and she's [in her 70s] and she wants to stay in it as long as she can so that's not us telling her to. So, I milk with her mostly and do the bookwork, and when I'm needed I do *whatever else he tells me to do.*
>
> (Woman on dairy farm impacted by drought; full-time farm worker)

Another woman commented on her contributions which include full-time work off-farm and a major role on-farm.

> We've been on the farm two years [husband full-time on-farm, wife on-farm plus full-time off-farm]. I milk every morning and at night and I do anything involved with the cows, picking bulls and drying them off… I sync our programs so anything to do with cows is my area, anything to do with cows is what I manage… farming is absolutely essential to me… [but] we need the stability of my [off-farm] income.
>
> (Female dairy farmer and full-time off-farm worker)

As livelihoods are adjusted around agricultural labor and remittance income, women are undertaking more physical farm labor, and increasing their hours of work off-farm. Women on smaller farms are doing more of the twice-daily milking and are more engaged in attending to calves and livestock. Many women noted the toll that increased physical labor had taken on their health.

> It hasn't been easy. [I was working on and off the farm] we were both absolutely exhausted… we've never had enough cash to pay all the bills each month, it's a huge pressure, a huge stress… when we first started I milked seven days a week and then did the farm work in the middle of the day and then did all the bookwork at night, and now I only fill in the gaps… I don't milk anymore because I have [an injury]…
>
> (Female partner on dairy farm)

As well as the interviews with farm family members, the author and her colleague's collaboration with a national farmer survey (Regional Well-being Survey operated out of the University of Canberra) assisted them to seek the wider, more comprehensive views of dairy farmers. As a result, 274 women and men in the dairy communities under study were surveyed to assess how they felt about changes in their community and on-farm. When the study focuses on the responses to livelihood issues that formed part of the survey, it is clear that women are feeling the pressure to increase their off-farm remittance, income generating workloads, meaning that they must undertake a commute or move away (outmigrate) to a regional center. Their capacity to work from home is severely limited by unreliable (63%) and costly internet access (65%), and poor mobile coverage (63%). In addition, 14.8% of women noted that their off-farm work had increased over the previous 12 months, and 46.4% of women noted that they would be seeking additional off-farm income over the next 5 years. Women's on-farm work had also increased with 44.8% of women noting that their workload on-farm was increasing.

Reserve Army of Labor

What our research reveals is that women in both the developing and developed countries are co-opted during and after climate crises into a greater role in agricultural production – a role that is largely invisible and yet expanding rapidly as climate crises unfold. Because women are less likely to own land and are often shut out of decision making forums, technology transfer opportunities, credit and extension services, and agricultural aid, the UN's Food and Agricultural Organization (FAO, 2013) estimates that agricultural productivity in some areas is down 10–20% and may reduce further.

In all countries explored in the current research, the expanding work roles of women in agricultural areas are added to their household and caring work with sometimes significant impacts on health and well-being (Alston, 2015; Alston *et al.*, 2018). Further, following a disaster, household and care work may be expanded by the need to care for the sick and frail, clearing and rebuilding infrastructure, rehabilitating land, and building livestock numbers. In countries such as Bangladesh and Lao PDR, women's expanding roles are made more complex by contaminated water and fuel sources resulting from disasters, and therefore by longer distances they might now need to walk to source new supplies.

Gendered Analysis

Outmigration and the sourcing of remittance income is a critical livelihood adaptation in response to climate events adopted by families engaged in agriculture across the world. This chapter highlights the commonality of this response, noting that the outmigration of men is a major trend across the two Asian countries highlighted in this chapter, and that the outmigration of women is an emerging factor in Australia. The impacts of climate-induced disasters, consequent livelihood diversification in agriculturally based communities, and the sourcing of remittance income have resulted in a significant rise in outmigration. While not neglecting the fact that in Australia it may be women who are the main outmigraters, what does appear as an outcome of outmigration from agricultural communities is the similarity of women's increased agricultural work in each of these case studies. While women have always been engaged in agricultural production, their input to agriculture is increasing significantly in the context of climate-induced disasters. This supports the UNFCCC opinion (UNDP, 2014) that 'women are at the center of the climate change challenge.' Yet, this does not go far enough in calling for a gendered analysis of livelihood adaptations following disasters (Holmgren and Arora-Jonsson, 2015).

The reshaping of agricultural livelihoods around outmigration and women's engagement in response to climate-induced disasters challenges long-held gender–power relations. In countries like Lao PDR and Bangladesh, access to education, training, credit services, and decision making power continue to be circumscribed for women. Thus, in areas where gendered norms and power relations have traditionally and more visibly contained women's roles and opportunities, their increasing agricultural contributions take place within rearranged structures that continue to support male privilege. This also occurs in Australia and, as noted in the Australian research:

> As communities contract and labor requirements expand, families are organizing their labor in different and highly gendered ways. The dominant hegemonic masculinity which underpins gender and power relations in rural areas shapes the way family labor is allocated. What appears evident is that male farmers direct the labor of women, proving that even though labor requirements are changing, what has not changed are the power relations.
> (Alston *et al.*, 2017, pp. 262–263)

Arguably this outcome represents the re-establishment and reshaping of what Sachs and Patel-Campillo (2014, p. 5) refer to as 'inequality regimes' – the systems of power that act to support masculine privilege. It appears that as women are increasingly drawn into agriculture, gendered livelihood reconfigurations are redefined in ways that support existing gender–power relations.

However, we must tread carefully to avoid setting up a 'feminisation of victimisation' (Djoudi *et al.*, 2016, p. 248) in relation to women undertaking agricultural work. There is no intention to decry women's increased roles as somehow incapacitating. Rather, we need to acknowledge women's increasing centrality in agriculture and call for policy and practice responses at local, national, and international levels that will better support women's significant contribution to food production and food security (Alston and Whittenbury, 2016). For example, Djoudi *et al.* (2016) note that the greater economic freedom accorded to women through these changes may be significantly empowering and emancipatory, and indeed, this is supported by the current work with Bangladeshi women where there was some evidence of a relaxation of *purdah*. In the complex aftermath of disasters there is room not only for challenge to existing positions, but for empowering practices to be adopted and women's increasing involvement to be celebrated. Women have significant local agricultural and environmental knowledge and experience, are critical to community and family cohesion, and are deeply implicated in livelihood adaptations. Acknowledging this and providing support, recognition, and rights-based resourcing is critical.

Yet, gender equality measures should be introduced in ways that enhance and support women. For example, Ampaire *et al.* (2020) note that while governments in East Africa are taking into account gender impacts in climate change policy developments, the effectiveness of these efforts is reduced as governments treat gendered differences in climate experiences as women's issues rather than as mainstream. This lack of attention to structural inequalities effectively reduces the potential outcomes.

The current work reveals that outmigration is a major livelihood adaptation in response to climate-induced disasters across the world. In undertaking a gendered analysis, the work also reveals the significant livelihood adaptations across communities and cultures that result from outmigration. The lack of global attention to these adaptations – especially on outmigration – will have significant repercussions for food security (Alston and Akhter, 2016; Alston and Whittenbury, 2016) and community resilience into the future. What is required is recognition of the impacts of outmigration and the gendering of agriculture. Such recognition must begin with the introduction of policies and practices that facilitate 'equality regimes.' Policy makers and planners working on climate actions at UNFCCC and other levels must understand community level adaptations and introduce policies that support and acknowledge gendered livelihood adaptations. As Dominelli (2020) notes, we must also rethink masculinity and gender–power relations that are holding back our attention to the critical environmental issues of our time.

References

ADB (2013) Gender equality and food security: women's empowerment as a tool against hunger. Asian Development Bank, Mandaluyong City, Philippines.

ADB and World Bank (2012) Country gender assessment for LAO PDR: reducing vulnerability and increasing opportunity. Asian Development Bank, Manila, Philippines and Washington, DC.

Alston, M. (2015) Women and climate change in Bangladesh. *Women in Asia Series*. Routledge, Abingdon, UK.

Alston, M. and Akhter, B. (2016) Gender and food security in Bangladesh: the impact of climate change. *Gender, Place and Culture* 23, 1450–1464.

Alston, M. and Whittenbury, K. (2016) Women and food security in South Asia. In: Kubik, W. and Fletcher, A. (eds) *Women in Agriculture Worldwide: Key Issues and Practical Approaches*. Gower Publishers, Canada.

Alston, M., Whittenbury, K. and Haynes, A. (2014) Gender and Climate Change in Bangladesh. *Report to Monash-Oxfam*. Monash University, Department of Social Work, Melbourne, Australia.

Alston, M., Clarke, J. and Whittenbury, K. (2017) Gender relations, livelihood strategies, water policies and structural adjustment in the Australian dairy industry. *Sociologia Ruralis* 57(Number S1), 752–768.

Alston, M., Clarke, J. and Whittenbury, K. (2018) Limits to adaptation: reducing irrigation water in the Murray-Darling Basin dairy communities. *Journal of Rural Studies* 58, 93–102.

Ampaire, E.L., Acosta, M., Huyer, S., Kigonya, R., Muchunguzi, P., Muna, R. and Jassogne, L. (2020) Gender in climate change, agriculture, and natural resource policies: insights from East Africa. *Climatic Change* 158, 43–60.

Arora-Jonsson, S. (2011) Virtue and vulnerability: discourses on women, gender and climate change. *Global Environmental Change* 21, 744–751.

Bell, K.C. (2011) Lessons from the reconstruction of post-tsunami Aceh. International Finance Corporation. Available at: www.openknowledge.worldbank.org/handle/10986/10442 (accessed 19 August 2018).

Brown, O. (2008) Migration and climate change. *Migration Research Series 31*. International Organisation for Migration, Geneva, Switzerland. Available at: www.publications.iom.int/books/mrs-ndeg31-migration-and-climate-change (accessed 30 August 2020).

Clarke, J., Alston, M. and Whittenbury, K. (2017) Social sustainability in dairy communities affected by the Murray-Darling Basin water plan. Qualitative Report. Monash University, Melbourne, Australia.

Cornfold, J. and Matthews, N. (2007) *Hidden Costs: the Underside of Economic Transformation in the Greater Mekong Subregion*. Oxfam Australia, Carlton, Australia.

Department of Agriculture, Water and the Environment (2020) Young people and women boost Australia's agricultural workforce. Available at: www.agriculture.gov.au/abares/news/media-releases/2018/young-people-women-boost-aus-ag-workforce (accessed 3 February 2021).

Djoudi, H., Brockhaus, M. and Locatelli, B. (2013) Is adaptation to climate change gender neutral? Lessons from communities dependent on livestock and forests in northern Mali. *Regional Environmental Change* 13, 493–508.

Djoudi, H., Locatelli, B., Vaast, C., Asher, K., Brockhaus, M. and Basnett Sijapati, B. (2016) Beyond dichotomies: gender and intersecting inequalities in climate change studies. *Ambio* 45, 248–262. doi.org/10.1007/s13280-016-0825-2.

Dominelli, L. (2020) Rethinking masculinity in disaster situations: men's reflections of the 2004 tsunami in southern Sri Lanka. *International Journal of Disaster Risk Reduction* 48:101594. Advance online publication, doi.org/10.1016/j.ijdrr.2020.101594.

Eastin, J. (2018) Climate change and gender equality in developing states. *World Development* 107, 289–305.

FAO (2013) FAO policy on gender equality: attaining food security goals in agriculture and rural development. *No. I3205E/1/02.13*. Food and Agriculture Organization of the United Nations, Rome.

FAO (2020) Building resilience to climate change in Laos. Food and Agriculture Organization of the United Nations, Rome. Available at: www.fao.org/in-action/building-resilience-to-climate-change-in-laos/en/ (accessed 3 February 2021).

Holmgren, S. and Arora-Jonsson, S. (2015) The Forest Kingdom – with what values for the world? Climate change and gender equality in a contested forest policy context. *Scandinavian Journal of Forest Research* 30, 235–245.

IPCC (2001) Climate change: third assessment report. Intergovernmental Panel on Climate Change, Geneva, Switzerland. Available at: www.ipcc.ch/publications_and_data/publications_and_data_reports.shtml#.T8BtkVFpu8U (accessed 26 May 2012).

IPCC (2014) Climate change 2014: synthesis report. In: Pachauri, R.K. and Meyer, L.A. (eds) *Contribution of Working Groups I, II and III to the Fifth Assessment Report of the IPCC.* Intergovernmental Panel on Climate Change, Geneva, Switzerland. Available at: www.ipcc.ch/report/ar5/syr/ (accessed 3 February 2021).

Kaijser, A. and Kronsell, A. (2014) Climate change through the lens of intersectionality. *Environmental Politics* 23, 417–433. doi.org/10.1080/09644016.2013.835203.

Kartiki, K. (2011) Climate change and migration: a case study from rural Bangladesh. *Gender and Development* 19, 23–38.

Kelly, P.M. and Adger, W.N. (2000) Theory and practice in assessing vulnerability to climate change and facilitating adaptation. *Climatic Change* 47, 325–352.

Kinnvall, C. and Rydstrom, H. (2019) Introduction. In: Kinnvall, C. and Rydstrom, H. (eds) *Climate Hazards, Disasters and Gender Ramifications*. Routledge, London, pp. 1–25.

Mollett, S. (2017) Gender's critical edge: feminist political ecology, postcolonial intersectionality and the coupling of race and gender. In: MacGregor, S. (ed.) *Routledge Handbook of Gender and the Environment*. Routledge, London, pp. 146–158.

Ogra, M.V. and Badola, R. (2015) Gender and climate change in the Indian Himalayas: global threats, local vulnerabilities, and livelihood diversification at the Nanda Devi Biosphere Reserve. *Earth System Dynamics*, 6:505. Gale Academic OneFile, available at: www.link.gale.com/apps/doc/A481516590/AONE?u=monash&sid=AONE&xid=da2b3147 (accessed 28 August 2020).

Phouxay, K. and Tollefsen, A. (2011) Rural–urban migration, economic transition, and status of female industrial workers in Lao PDR. *Population, Place and Space* 17, 421–434.

Sachs, C. and Patel-Campillo, A.P. (2014) Feminist food justice: crafting a new vision. *Feminist Studies* 40, 1–14.

Saha, S., Goswami, R. and Paul, S.K. (2018) Recursive male out-migration and the consequences at source: a systematic review with special reference to the left-behind women. *Space and Culture, India* 5, 30–53. doi.org/10.20896/saci.v5i3.289.

Schofield, D. and Gubbels, F. (2019) Informing notions of climate change adaptation: a case study of everyday gendered realities of climate change adaptation in an informal settlement in Dar es Salaam. *Environment and Urbanisation, Thousand Oaks* 31, 93–114.

UN Human Rights Office of the High Commissioner (2019) UN expert calls for reform of poverty-driving policies in Lao PDR. United Nations, New York. Available at: www.ohchr.org/en/NewsEvents/Pages/DisplayNews.aspx?NewsID=24710&LangID=E#:~:text=Almost%20a%20quarter%20of%20the,less%20than%20%242.50%20per%20day.&text=Lao%20PDR%20does%20not%20have,adequately%20in%20health%20and%20education (accessed 3 February 2021).

UNDP (2014) Human Development Report – Sustaining human progress: reducing vulnerabilities and building resilience. United Nations Development Program, New York. Available at: www.hdr.undp.org/en/content/human-development-report-2014 (accessed 5 September 2020).

UNDP Bureau for Crisis Prevention and Recovery (2010) 2010 Annual Report. United Nations Development Program, New York. Available at: www.undp.org/content/undp/en/home/librarypage/crisis-prevention-and-recovery/preventing-crisis-enabling-recovery-2010-annual-report.html (accessed 3 February 2021).

Werner, J. and Lyons, S. (2020) The size of Australia's bushfire crisis captured in five big numbers. *ABC News, March 5*. Available at: www.abc.net.au/news/science/2020-03-05/bushfire-crisis-five-big-numbers/12007716 (accessed 3 February 2021).

Whittenbury, K. and Alston, M. (2016) The constraints to / opportunities for agribusiness / market engagement by rural women in Lao PDR. Report to ACIAR.

World Bank (2012) Advancing gender equity for food security. Available at: www.worldbank.org/en/news/feature/2012/10/03/advancing-gender-equity-food-security (accessed 9 November 2015).

World Bank (2017) Help women farmers get to equal. April 18. Available at: www.worldbank.org/en/topic/agriculture/brief/women-farmers-getting-to-equal (accessed 3 February 2021).

World Economic Forum (2020) Women own less than 20% of the world's land. It is time to give them equal property rights. Available at: www.weforum.org/agenda/2017/01/women-own-less-than-20-of-the-worlds-land-its-time-to-give-them-equal-property-rights/ (accessed 3 February 2021).

14 Climate-Induced Migration, Women and Decision Making Power in the Agricultural Wage Sector in Saiss, Morocco

D. Najjar, B. Dhehibi, B. Baruah, A. Aw-Hassan and A. Bentaibi

Introduction

The Middle East and North Africa (MENA) region is highly vulnerable to climate change and is undergoing rapid social and political transformation (Sowers *et al.*, 2011; Haddad and Shideed, 2013; Waterbury, 2013). This region is projected to have rising temperatures (2–3°C) and declining precipitation (10–30%) in the coming years (IPCC, 2014). Climate change is affecting rural livelihoods in multiple ways in rural Morocco, and little is known about the gender dynamics of these impacts and related adaptation measures (de Haas, 2006; de Haas and Van Rooij, 2010; Najjar, 2015). This study examines the gendered effects of drought-induced migration in rural Morocco for settler migrants and farmers who stay behind in sending communities. Due to state investments in irrigation, the Saiss plains of Morocco are experiencing rural–rural migration as an adaptive strategy for many who are escaping climate change and unemployment, to take advantage of labor opportunities in agricultural sectors elsewhere. The well-being and decision making power of male and female migrants in receiving communities (Betit and Sidi Slimane) and women staying behind in sending communities (Ain Jemaa) are examined (Fig. 14.1). The current study is timely, as worsening climate change effects and increasing intensification and globalization make migration the most viable option for livelihood improvement, particularly in the dry areas of the world.

In Saiss, migration comes with prospects of not only more work opportunities in irrigated fruit and vegetable sectors, but also with the promise of land and housing in informal settlements. As such, we explore the extent and determinants (age, location, migration, and marital status) of decision making power for female and male wage workers in host communities with regards to work participation and income and asset control. We also examine the decision making power dynamics for women who are left behind as men migrate from the rain-fed area of Ain Jemaa. Very little research has been conducted on the nature of household decision making and its gendered meanings (Acosta *et al.*, 2019; Bernard *et al.*, 2020). This study attempts to explore the factors influencing this type of decision making as well as the underlying meanings.

The chapter begins with a literature review on decision making power, gender, migration, and work in rural areas. Following this, the case study characteristics are presented, which detail how climate change is fueling migration, gender norms in host and sending communities, as well as the gender dynamics in accessing economic opportunities and decision making power. The chapter ends with recommendations to strengthen the women's decision making power as migration continues, with a focus on strengthening landed property ownership for women.

 Gender, Climate Change and Livelihoods: Vulnerabilities and Adaptations (eds J. Eastin and K. Dupuy)
DOI: 10.1079/9781789247053.0014

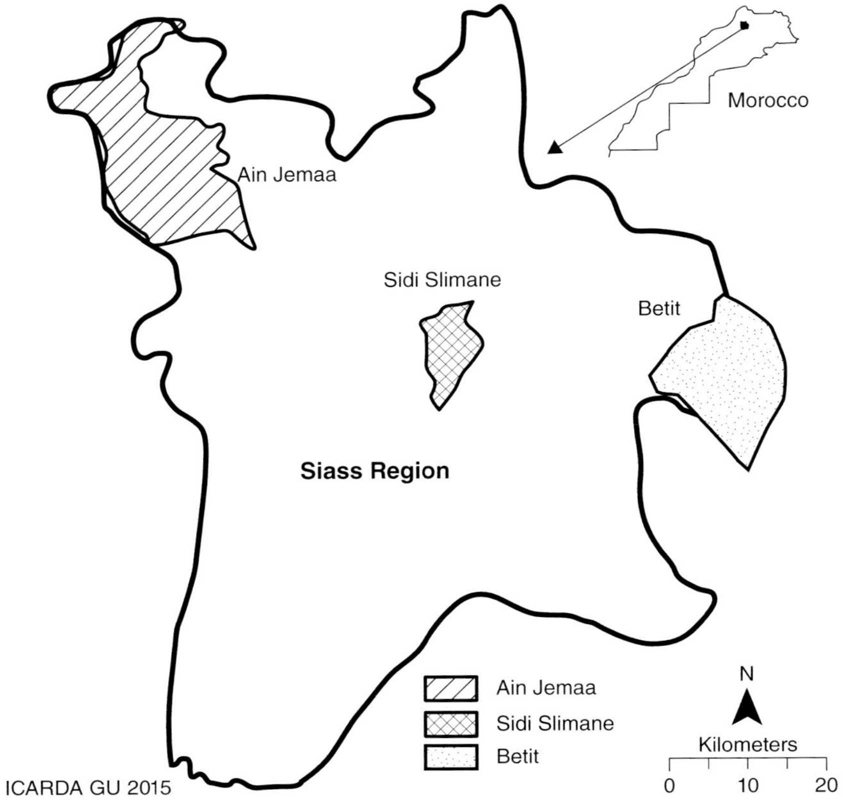

Fig. 14.1. Map of case study area for Saiss region and three research communities of Betit, Ain Jemaa, and Sidi Slimane. With permission from ICARDA Geo-informatics Unit.

Decision Making, Migration and Gender

Accelerating outmigration due to climate change is a determining feature of the drylands (Gartaula *et al.*, 2010; Pedrick, 2012; Abdelali-Martini and Hamza, 2014; Pattnaik *et al.*, 2018). In their study on women's participation in farming in rural China, de Brauw *et al.* (2008) identify two types of agriculture feminization: labor feminization and managerial feminization. 'Labor feminization' refers to women taking on an increased amount of farm work, often to compensate for the outmigration of men in the household or for men's increased participation in non-farming economic activities. 'Managerial feminization' refers to women playing a more prominent and visible role in agricultural decision making while gaining greater access to financial and social resources to optimize agricultural productivity. Based on the findings from their study, de Brauw *et al.* (2008) arrived at the conclusion that rural China was experiencing more labor feminization than managerial feminization. In other words, women were contributing increasing amounts of labor to agriculture without experiencing a commensurate increase in access to resources or authority to make decisions about farming. Abdelali-Martini and Dey de Pryck (2015) arrive at a similar conclusion about women's participation in farming in north-western Syria.

Studies related to agricultural intensification and investments in irrigation in the MENA region note that in Syria, Morocco, and Egypt intensification of agriculture, particularly in the irrigated areas, is leading to increased demand for female labor (Abdelali-Martini *et al.*, 2003; Najjar *et al.*, 2018). This outcome falls within a broader global pattern of increased dependence on women's cheap labor in efforts to commercialize agriculture (Razavi, 2009). These opportunities come alongside a move to fill employment positions on a seasonal or more temporary basis through relocation. Less studied is the internal rural–rural migration, especially from less to more resource endowed areas. Morocco is particularly vulnerable to climate change due to its increased dependence on rain-fed agriculture (Schilling *et al.*, 2012). With increased investments in irrigation, studies have yet to examine the implications for gendered livelihoods of rural–rural migration into these irrigated areas. Researchers cite employment opportunities as a rationale for migration decisions (Schmalzbauer, 2004; Yeoh and Ramdas, 2014). Women often take advantage of these opportunities because work for them is more acceptable or available than in their community of origin. In host communities, women might confront less rigid norms on mobility and greater potential for increased decision making power through income generation. In other words, migration can weaken pre-existing patriarchal gender norms and relax social controls for women (Schmalzbauer, 2004; Yeoh and Ramdas, 2014).

Case Study

The Moroccan government has been attentive to agricultural concerns. The new governmental strategy for agricultural development, the 'Green Morocco Plan' (GMP, 2016), subsidizes drip irrigation and well digging for farmers. With state support for access, significant changes have occurred in the types of crops grown, labor demands, and in the profitability of farming in areas where groundwater is available or accessible. An increase in the demand for labor has created job opportunities in resource endowed areas such as Betit and Sidi Slimane. Due to the availability of water in communities like these and the demand for more water-intensive crops such as onions, potatoes, and fruit trees, there is a greater need for hired agricultural labor. Onions, potatoes, and fruit trees are particularly laborious crops and require additional labor, especially during planting and harvesting. In contrast to Betit and Sidi Slimane, Ain Jemaa is rain-fed and fairs considerably worse in terms of natural resource endowments and labor markets where food crops such as wheat, chickpeas, fava beans, and olives are grown. Agricultural wages are scarce and significantly lower than in Betit and Sidi Slimane. For these reasons, laborers from rain-fed areas such as Ain Jemaa, including women, migrate in large numbers to the irrigated areas.

Gender norms in Ain Jemaa, Betit, and Sidi Slimane are determined, in part, by economic dynamism. In places with lower access to natural resources, such as Ain Jemaa, women enjoy limited mobility and are not allowed to work outside or far away from their homes. These patriarchal gender norms can be also attributed to limited job opportunities (low economic dynamism), as men are more likely to carry out paid work. The findings of the current study reflect these points as well, indicating that greater economic dynamism corresponds to more progressive gender norms. In Betit and Sidi Slimane, women predominantly participated in the wage sector due to the greater prevalence of year-round wage work. However, women are also experiencing an increased role in farming their household plots as men migrate to irrigated areas, cities, and outside of Morocco to Europe and elsewhere.

Methods and Data Analysis

The empirical data for the current research are drawn from a survey administered during November and December 2014, to 400 laborers (179 women and 221 men) in the Saiss region of Morocco. Laborers constitute a significant portion of first- or second-generation landless migrants. The number of laborers surveyed is proportional to the size of labor community present in each of the three areas. Sidi Slimane had the largest number of wage workers as the farms are larger and depend more on hired labor. We attempted to understand the reasons for migration, especially for first-generation migrants, as well as the nature and extent of decision making power over working, income, and accumulated assets. In the new settlement, workers were able to stake a claim to a piece of land and build a house on it. Many of the laborers and participating researchers told us that the government plans to legalize the ownership to these informal homes and land parcels.

In Ain Jemaa (the sending community in the current study), a qualitative study was also conducted. This portion comprised part of the global study: 'Innovation and Development Through Transformation of Gender Norms in Agriculture and Natural Resource Management – a comparative case study' (GENNOVATE), which included a series of interviews with community leaders, innovators in wheat farming, and individuals with varying levels of wealth. Focus group discussions were undertaken with middle-class participants at varying age levels, in order to understand the linkages among gender norms, agency, agricultural innovation, and development outcomes. The research sought to understand in greater depth the issues women farmers face in this rain-fed community in context of high male outmigration.

In this chapter, the authors report findings generated from the survey, focusing on variations in decision making within the household. The analysis reveals the factors that influence women's decision making power in migrant sending and receiving communities in the Saiss region. Decisions concern work participation, income expenditures, and control over assets, especially the house, as it was often the most important outcome of the migration process. A detailed description of regression analytics for this research can be found in Najjar *et al.* (2017). In this piece, broader patterns evident in the survey results are discussed, with a closer focus on qualitative data generated from interviews and focus groups with survey respondents, to paint a fuller picture of key causal processes and outcomes (Bernard, 2011).

Gendered decision making and work participation

The studies were held in Betit, Ain Jemaa, and Sidi Slimane because these communities differ significantly in their gender norms, levels of economic development, and biophysical dynamics. While Betit and Sidi Slimane are flourishing due to onion and fruit production, Ain Jemaa is inhibited in its economic development due to its dependence on rainfall for irrigation. Women in Betit and Sidi Slimane have more freedom in their mobility, and their employment in the agricultural wage sector is prominent and more likely to be socially accepted. While women in Ain Jemaa are increasingly taking on agricultural roles due to male outmigration, they are not yet recognized in their roles as farmers by themselves, local community members, and development organizations alike.

The findings reveal that control over decision making on work is influenced by the respondents' migration history. Regardless of gender, it is found that migrants (seasonal or permanent) have a greater likelihood of controlling their decisions regarding labor force participation. This finding suggests that either migration has a positive impact on decision making power with regards to work, or that the migrants already have higher decision making power, which might also explain why they were able to migrate in the first place. Figure 14.2 further reveals that the percentage of migrants, both female and male, who have decision making power over their work is higher than the percentage of non-migrants. There is also a narrower gender gap for migrants with regards to decision making power over work.

Men in Ain Jemaa reported that worsening droughts and a lack of agricultural subsidies mean that '... farmers in the community can't live only by agricultural income, so they seek off-farm jobs even out of the village.' A focus group with young men in Ain Jemaa revealed that 'it is easy to decide to migrate in this community, and many young men decide to leave.' A different scenario presents itself for the women left behind in the community of Ain Jemaa who have explained that their role is to 'replace her husband as head of the household during his absence.' However, these women have limited opportunities or decision making power in adopting new innovations, which is crucial for adapting to a changing climate and sustaining a livelihood, as compared to men. Women farmers in Ain Jemaa explained that '... There's a big difference between the man and the woman. The man has the right to decide, but the woman doesn't. Even if she wants to do something new, she doesn't have the full freedom to do it.'

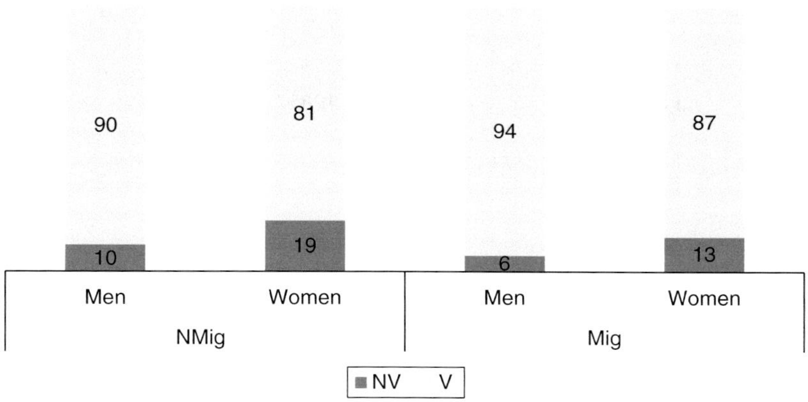

Fig. 14.2. Overview of gendered decision making power over work in the three research communities of Betit, Ain Jemaa, and Sidi Slimane. NV refers to having no voice (or decision) and V refers to having a voice. NMig refers to non-migrant and Mig refers to migrant.

In 2006, state-led land distribution efforts in Saiss provided men, for the most part, with land. Women's access to land was highly limited, available to a few who were heads of household (HH) or upon their husbands' deaths (Bossenbroek and Zwarteveen, 2015). As landowners are often targeted for extension support and varietal demonstrations, female farmers were largely marginalized from these endeavors. To illustrate, men reported to mainly learning from Farmer Field Schools, on-farm trials, and extension services about new varieties and agronomic practices. Women, on the other hand, emphasized the benefits of farmer-to-farmer learning (from neighbors, family, and friends). Among all respondents, women lacked knowledge about results related to ongoing varietal trials and innovations related to wheat, such as conservation agriculture, and legume and wheat rotations: 'They [local research institute] did an experiment on a variety of wheat called '*rihan*' and chickpeas last year and this year and we [the women] don't know... it's the men who know these things.'

Women working in the agricultural wage sector are more likely to have agency over their decision to work. Many of the women working in the wage sector mentioned that their husbands' work is not stable, and hence they chose to participate in wage labor to increase the financial security of the household. Social norms prohibit men from asking their wives to work as men are considered the breadwinners in their families, which suggests that women are making these decisions and by doing so transgressing gender norms (Kandiyoti, 1988). This outcome indicates that migration has had an emancipating effect on women, possibly also due to the increased availability of jobs in the host communities. Furthermore, many of the women who work in the wage sector were HHs, some of whom were migrants who explicitly mentioned that they had migrated because women are allowed to work in the migrant areas of Betit and Sidi Slimane. All of these factors could have contributed to the result of women being more likely to decide over their own participation in work. In contrast, women in the sending community of Ain Jemaa felt that they did not have decision making power over working outside their homes and that this was due to their needing men's approval: 'If the project is out of the house area and not in front of his eyes he would forbid it.'

It was found that younger people (>35 years) are more likely to make decisions alone, without other members of the household. A similar scenario presents itself when the respondent is the HH. In these cases, she or he were more likely to make their decision alone. This situation could be attributed to the youth's rite of passage and their fulfilment of independence from parental guardianship. The finding for young people also suggests that migration has an emancipatory effect, similar to that for women, as they are capitalizing on greater work opportunities.

The region plays a significant role in control over work-related decisions, specifically whether decisions are made jointly or alone. Betit wage work respondents, regardless of their gender, were more likely to make the decision for participating in work alone. The qualitative findings of the study complement these results. Social norms in Betit were the most progressive in that they were the least prohibitive for women's participation. By contrast, women in Ain Jemaa enjoyed minimal freedom to work outside their households, and this community was also the least progressive in its gender norms. Here, women were cloistered while men earned income. There were also fewer agricultural job opportunities due to lack of water and increasing drought. Women in Ain Jemaa were restricted to seed cleaning and sieving, largely in their own homes or in nearby houses. Furthermore, almost all women laborers in Ain Jemaa were HHs, which suggests that agricultural wage work is an option of last resort. Women farmers in Ain Jemaa reported that 'women of course consult with the men in almost everything and anything.' This suggests the prevalence of sole decision making authority for men in agriculture in Ain Jemaa when men are present. Women replace their husbands when they are away, but only on a temporary basis, and men are likely to regain this control when they return (de Haas and Fokkema, 2010). In Betit and Sidi Slimane, on the other hand, wives, who constituted almost 60% of the female survey participants in these two areas, were largely employed in the wage sector.

Gendered decision making and income expenditures

For wage workers, control over income derived from agricultural work is lower for married respondents and greater for single respondents, regardless of gender. Family size also influences the control over agricultural income decision making. The greater the number of the dependents, the lower the control over spending income. This decrease in solo decision making for married people could be attributed to having to consult with other members of the household. Survey findings (see Fig. 14.3) confirm that laborers consulted with their children and other family members (mother, father, brothers), who can also be dependents, when making decisions regarding how to spend income.

Not surprisingly, men are more likely to have decision making power over expenditures than women, especially when they are the HH. In contrast, even when women are the ones generating income, they were more likely to consult with their husbands than the men were with their wives. Many of the answers to who makes the decisions on how to spend the income were justified on the basis of HH status, indicating that such decisions were part of the HH responsibilities.

The results indicate that control over income expenditure is joint if the respondent is either the HH or if the respondent is a woman; as indicated earlier, women were more likely to make joint decisions with their spouses over the control of income they generated in agriculture. The community also played a role in decision making dynamics related to income. Ain Jemaa had the least incidence of men consulting with their wives over spending of agricultural income, while Sidi Slimane had the greatest. Men in Ain Jemaa asserted that 'wives work only in the house' and it is 'men's responsibility to decide on spending.' The fact that the women rarely worked or earned income in Ain Jemaa could contribute to this trend.

On average, women were far more likely to consult with their husbands on spending agricultural income (Tables 14.1 and 14.2). Sharing responsibility with husbands was often the most cited reason for women's joint decision making of income generated through agriculture. Unsurprisingly, when women were the HHs, they made the spending decisions for income they generated through agriculture either alone, or in a few cases, with their sons.

In addition to regional differences, migration status played a role in shaping decision making over agricultural income. Both migrant women and men were more likely to share decisions related to income spending with their spouses than non-migrant women and men. More broadly, migrants were more likely than non-migrants to share decisions. These findings might occur because in settler communities there are far more work opportunities, and as such more household members are likely to work and pool resources, and so collectively have greater say.

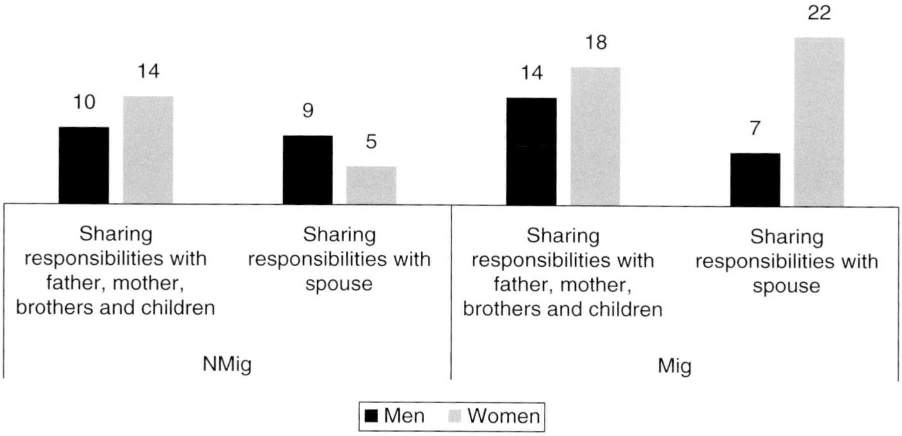

Fig. 14.3. Overview of gendered joint decision making in the three research communities of Betit, Ain Jemaa, and Sidi Slimane. NMig refers to non-migrant and Mig refers to migrant.

Table 14.1. Men's reported decision making control over income they generated from agriculture in the three communities of Betit, Ain Jemaa, and Sidi Slimane.

	Men					
	Ain Jemaa		Betit		Sidi Slimane	
	N	%	N	%	N	%
Men	11	100%	43	69%	40	56%
Women					1	1%
Joint			12	19%	16	22%
Men and a member or more from the family			4	6%	13	18%
Other(s) from the family			3	5%	2	3%
Total	11		62		72	

Table 14.2. Women's reported decision making control over income they generated from agriculture in the three communities of Betit, Ain Jemaa, and Sidi Slimane.

	Women in male-headed households						Female-headed households					
	Ain Jemaa		Betit		Sidi Slimane		Ain Jemaa		Betit		Sidi Slimane	
	N	%	N	%	N	%	N	%	N	%	N	%
Men	1	33%			3	5%						
Women	1	33%	15	68%	38	58%	1	100%	20	95%	31	97%
Joint	1	33%	2	9%	20	31%					1	3%
Women and a member or more from the family			1	5%	3	5%			1	5%		
Other(s) from the family			4	18%	1	2%						
Total	3		22		65		1		21		32	

Women in Ain Jemaa reported limited ability to control income from agriculture, as explained by a group of women in a focus group in this area: '... We are afraid of the rain; we do not even dare planting these days... The land needs water and the rain is much less than before.' They were more likely to generate and control profit related to livestock production (cows and small ruminants). It is customary for women to accumulate assets in the form of livestock, and to own and control this asset. To illustrate, women in Ain Jemaa explained that when they give birth they often turned '*Zroura*' money into livestock purchases which then they sold in times of crises and built on for their sustenance.

Gendered decision making and assets

The authors' analysis reveals that wage working women are more likely to control houses than men. This finding could be attributed to the fact that the women at issue are earning and this gives them decision making power in the assets purchased or rented in the household. In addition, when the respondent is the HH there is great probability of asset control.[1]

On the other hand, regardless of gender, migration and age both contributed to less control over assets such as a house. Young people were less likely to control decision making than older people. The decreased decision making power for migrants over houses could be due to the insecure tenure of housing in the settler communities. Migrants are often located in informal settlements. Settlers acquire an area of land on which they can build a structure of tin and plastic, locally known as a '*koukh*.' They do not have a title deed to the houses, and many were distressed in anticipations that the government would demolish these structures. Despite the lack of tenure security, these houses are sold, bought, and highly valued by the settlers, and fuelled their migration. Settlers mentioned that housing in host communities is affordable and one of the main reasons why they migrated, in addition to abundance of wage work opportunities.

In Ain Jemaa, women rarely acquired land titles to their own names. This outcome marginalized them from accessing agricultural training and knowledge to new, climate-friendly innovations and/or credit opportunities. However, women were more likely due to both biophysical and institutional limitations to depend on livestock as an asset for which they could both own and control. This exclusion of women from property ownership, none the less, has negative consequences on food security and climate change adaptation at large in the community, and possibly more broadly in other rain-fed sending communities.

Gendered decision making and asset-related income

Contrary to the findings for income expenditures, the larger the size of family, the smaller the likelihood of joint decision making in controlling asset-based income.[2] While individual members in the household are more likely to generate income when the income is pooled and a house is purchased, it is often in the name of one person, usually the male HH. Migration status, on the other hand, revealed a greater impact over decisions related to the house, in comparison to those who did not migrate. Again, similar to participating in decision making over work, migration is also attributed to increased decision making power with regards to houses.

It was found that the more education a respondent has (high school, college, or university levels) the greater the likelihood for making solo decisions over assets. This finding could be attributed to the social norms which stipulate that an educated person knows best and as such is more likely to make the major decisions and is expected to do so. Survey findings confirm these results. Some respondents attributed their decision making power over income spending to their education. Finally, respondents from Betit were more likely to take their decision on control over their houses alone in comparison to the respondents from the other two communities. This could be attributed to the former finding on education: the findings reveal that Betit respondents were the most educated at 23% having higher education levels (secondary schooling and above). Although this finding is not related to migrants' experiences per se, it is more about the inherent characteristics of the wage workers, mainly those with higher education. However, this can also mean that the migration process is likely to reproduce or widen the gender gap in decision making with lack of awareness building on the importance of gender equality.

In the sending community of Ain Jemaa, titles of land distributed in 2006 were often allowed to be in one person's name, usually the elder brother. Key informants in the community reported that this has led to fights between male brothers over using the land. There was even an incident of a male individual stabbing his brother as he felt left out over the land profits. This could also explain women's marginalization from land ownership and control as being limited to women to a greater degree.

Conclusion

In response to climate change, governments in the MENA region have focused on tapping new water sources, from aquifers to seawater desalination (El-Sadek, 2010; Sowers *et al.*, 2011; Zeitoun *et al.*, 2012; Malm and Esmailian, 2013). Morocco is no exception; the GMP aims to both save water and develop new sources (Al-Monitor, 2014). However, while the Moroccan government's investment in agriculture – particularly in irrigation, as an adaptation strategy to cope with climate change – created a surge of employment in the agricultural wage sector for men and women, it is also important to pay attention to social aspects and impacts of adaptation strategies, such as tenure security for those who are employed. Our findings challenge the stereotype of left-behind women in the migration literature of the region by revealing that men and women are coping with drought vulnerability and subsequent unemployment, by engaging in rural–rural migration from areas where drought is prevalent to those where resources such as irrigation are more abundant, and labor demand greater. Women (most of whom were HHs) overcame their lack of wage labor opportunities in their original home towns, and prevalence of drought, by moving to areas more accepting of women's employment, primarily in areas with abundant low-paying jobs that men are reluctant to perform. As the findings of the current study show, migration-induced agricultural feminization comes with some benefits. It is important to leverage these benefits alongside others such as women's access to secure independent or joint land titles and better autonomy over their decision making regarding employment and resource use.

Conversely, the impacts of climate migration in sending communities presents a grim future for women. With lack of land titles and employment opportunities, women left behind in sending communities are limited in their ability to adopt new technologies and strengthen their resilience to climate change in agriculture. Extension programs are well advised to recognize that women can be farmers in their own right and to provide them with information they need, as well as consult with them in ongoing experimental work. This study also found that due to cultural norms, women in sending communities are far less likely than men to own landed property and more likely to own liquid assets, especially livestock, with limited long-term security (Najjar *et al.*, 2020). Strengthening property rights for women is an essential step towards building resilience in rural communities in light of women's increased roles in managerial agriculture.

Notes

[1] See Najjar *et al.* (2017) for a more detailed discussion of regression methods and findings.
[2] See Najjar *et al.* (2017) for a more detailed discussion of regression methods and findings.

References

Abdelali-Martini, M. and Dey de Pryck, J. (2015) Does the feminisation of agricultural labor empower women? Insights from female labor contractors and workers in northwest Syria. *Journal of International Development* 27, 898–916. doi.org/10.1002/jid.3007.

Abdelali-Martini, M. and Hamza, R. (2014) How do migration remittances affect rural livelihoods in drylands? *Journal of International Development* 26, 454–470. doi.org/10.1002/jid.2896.

Abdelali-Martini, M., Goldey, P., Jones, G. and Bailey, E. (2003) Towards a feminization of agricultural labor in northwest Syria. *The Journal of Peasant Studies* 30, 71–94.

Acosta, M., van Wessel, M., van Bommel, S., Ampaire, E.L., Twyman, J., Jassogne, L. and Feindt, P.H. (2019) What does it mean to make a 'joint' decision? Unpacking intra-household decision making in agriculture: implications for policy and practice. *The Journal of Development Studies* 56, 1210–1229.

Al-Monitor (2014) Green Morocco Plan focuses on sustainable agriculture. Available at [link requires free registration]: www.al-monitor.com/pulse/business/2014/10/httpalhayatcomarticles4906517----.html (accessed 1 February 2021).

Bernard, H.R. (2011) *Research Methods in Anthropology: Qualitative and Quantitative Approaches*. Rowman Altamira, Lanham, MD.

Bernard, T., Doss, C., Hidrobo, M., Hoel, J. and Kieran, C. (2020) Ask me why: patterns of intrahousehold decision-making. *World Development* 125:104671.

Bossenbroek, L. and Zwarteveen, M.Z. (2015) 'One doesn't sell one's parents': gendered experiences of shifting tenure regimes in the agricultural plain of the Sais of Morocco. In: Archambault, C.S. and Zoomers, A. (eds) *Global Trends in Land Tenure Reform*. Routledge Studies in Gender and Development, London and New York.

de Brauw, A., Li, Q., Liu, C., Rozelle, S. and Zhang, L. (2008) Feminization of agriculture in China? Myths surrounding women's participation in farming. *The China Quarterly* 194, 327–348. doi.org/10.1017/S0305741008000404.

de Haas, H. (2006) Migration, remittances and regional development in southern Morocco. *Geoforum* 37, 565–580.

de Haas, H. and Fokkema, T. (2010) Intra-household conflicts in migration decision making: return and pendulum migration in Morocco. *Population and Development Review* 36, 541–561. doi.org/10.1111/j.1728-4457.2010.00345.x.

de Haas, H. and van Rooij, A. (2010) Migration as emancipation? The impact of internal and international migration on the position of women left behind in rural Morocco. *Oxford Development Studies* 38, 43–62.

El-Sadek, A. (2010) Water desalination: an imperative measure for water security in Egypt. *Desalination* 250, 876–884.

Gartaula, H.N., Niehof, A. and Visser, L. (2010) Feminisation of agriculture as an effect of male out-migration: unexpected outcomes from Jhapa district, eastern Nepal. *The International Journal of Interdisciplinary Social Sciences: Annual Review* 5, 565–578.

GMP (2016) *Green Morocco Plan*. Available at: www.maroc.ma/en/content/green-morocoo-plan (accessed 3 February 2021).

Haddad, N. and Shideed, K. (2013) Mainstreaming adaptation to climate change into the development agenda. In: Sivakumar, M.V.K., Lal, R., Selvaraju, R. and Hamdan, R. (eds) *Climate Change and Food Security in West Asia and North Africa*. Springer, Amsterdam, The Netherlands, pp. 301–315.

IPCC (2014) *Fifth Assessment Report*. Intergovernmental Panel on Climate Change, Cambridge University Press, Cambridge, UK. Available at: www.ipcc.ch/report/ar5/ (accessed 3 February 2021).

Kandiyoti, D. (1988) Bargaining with patriarchy. *Gender & Society* 2, 274–290.

Malm, A. and Esmailian, S. (2013) Ways in and out of vulnerability to climate change: abandoning the Mubarak Project in the northern Nile Delta, Egypt. *Antipode* 45, 474–492.

Najjar, D. (2015) Women's contributions to climate change adaptation in Egypt's Mubarak Resettlement Scheme through cactus cultivation and adjusted irrigation. In: Buechler, S. and Hanson, A.M. (eds) *A Political Ecology of Women, Water and Global Environmental Change*. Routledge, London and New York, Chapter 8.

Najjar, D., Dhehibi, B., Aw-Hassan, A. and Bentaibi, A. (2017) Climate change, gender, decision-making power, and migration into the Saiss region of Morocco. *Working Paper No. 1102*. The Economic Research Forum (ERF), Giza, Egypt.

Najjar, D., Baruah, B., Aw-Hassan, A., Bentaibi, A. and Kassie, G.T. (2018) Women, work, and wage equity in agricultural labor in Saiss, Morocco. *Development in Practice* 28, 525–540.

Najjar, D., Baruah, B. and El Garhi, A. (2020) Gender and asset ownership in the old and new lands of Egypt. *Feminist Economics* 26, 119–143.

Pattnaik, I., Lahiri-Dutt, K., Lockie, S. and Pritchard, B. (2018) The feminization of agriculture or the feminization of agrarian distress? Tracking the trajectory of women in agriculture in India. *Journal of the Asia Pacific Economy* 23, 138–155. doi.org/10.1080/13547860.2017.1394569.

Pedrick, C. (2012) *Strategies for combating climate change in drylands agriculture: synthesis of dialogues and evidence presented at the International Conference on Food Security in Dry Lands, Doha, Qatar, November 2012*. The International Center for Agricultural Research in the Dry Areas (ICARDA), Aleppo, Syria and CGIAR Research Program on Climate Change, Agriculture and Food Security (CCAFS), Copenhagen, 36 pp.

Razavi, S. (2009) Engendering the political economy of agrarian change. *The Journal of Peasant Studies* 36, 197–226.

Schilling, J., Freier, K.P., Hertig, E. and Scheffran, J. (2012) Climate change, vulnerability and adaptation in North Africa with focus on Morocco. *Agriculture, Ecosystems & Environment* 156, 12–26.

Schmalzbauer, L. (2004) Searching for wages and mothering from afar: the case of Honduran transnational families. *Journal of Marriage and Family* 66, 1317–1331.

Sowers, J., Vengosh, A. and Weinthal, E. (2011) Climate change, water resources, and the politics of adaptation in the Middle East and North Africa. *Climatic Change* 104, 599–627.

Waterbury, J. (2013) The political economy of climate change in the Arab Region. *Arab Human Development Report Research Paper Series*. United Nations Development Programme, New York. Available at: www.arab-hdr.org/publications/other/ahdrps/AHDR%20ENG%20Waterbury%20v3.pdf (accessed 24 April 2014).

Yeoh, B.S.A. and Ramdas, K. (2014) Gender, migration, mobility and transnationalism. *Gender, Place & Culture*, 21(10), 1197–1213.

Zeitoun, M., Allan, T., Al Aulaqi, N., Jabarin, A. and Laamrani, H. (2012) Water demand management in Yemen and Jordan: addressing power and interests. *The Geographical Journal* 178, 54–66.

15 Bringing Women's Livelihood and Care Perspectives into Climate Decision Making

P.E. Perkins and B. Osman

Introduction: Intersectional, Gendered Impacts of Climate Change Affect Everyone's Lives and Livelihoods

It is well-understood that the impacts of climate change fall hardest on those who are already disadvantaged due to existing inequities linked to gender, race, ethnicity, dis/ability, sexuality, and other identity-based differences, which are also often related to geographic location (Jafry, 2019). Disproportional climate change impacts on women – half of humanity – stem not only from gender-based workplace discrimination and women's concentration in lower-paying health, care, education, and other service occupations, but also from their unpaid or underpaid home, care, community, and intergenerational skills transmission roles – all part of unjust patriarchal social and economic systems (MacGregor, 2017; Perkins, 2019). An intersectional and non-binary gender perspective expands this view to include discrimination of all kinds, and calls for all groups affected by climate change to be included in climate decision making, as climate change worsens pre-existing vulnerabilities. While this chapter's examples focus primarily on women, the analysis extends to gender more broadly. Evidence is mounting that the effectiveness of climate governance depends on the inclusion and wide participation of all, especially those who are most vulnerable. Climate impacts should also be understood broadly, especially from the viewpoint of people whose lives and livelihoods are increasingly being disrupted. In this chapter, we provide examples of such gendered impacts, what they imply for climate governance, and potential steps towards addressing them.

For example, the terrible impacts of three recent mine tailings dam collapses in Canada and Brazil (partly induced by climate change-related rainfall extremes) were very gendered. Women fed and cared for the injured, traumatized, and homeless, cleaned up the mud and ruined homes, and rebuilt communities and social institutions. Many women also had to seek paid work to supplement household income when men's jobs in the mines were suspended. Women's triple workdays (child and elder care, homemaking and food production, informal sector earnings) roughly doubled due to the unpaid work they carried out to help neighbors and relatives with community reconstruction, as well as their efforts in seeking alternative means of livelihood preservation (Maso, 2017; Shandro *et al.*, 2017; Marshall, 2018; Almeida *et al.*, 2019; McAllister, 2019; Marshall and Campolino, 2020). Obviously gendered pressures like these are unsustainable, but climate change is expanding the scale and incidence of such crises.

Female-headed households in South Africa and other countries are threatened by a 'triple burden' comprising gender discrimination in assets and access to productive resources; lone household headship requiring both domestic

 Gender, Climate Change and Livelihoods: Vulnerabilities and Adaptations (eds J. Eastin and K. Dupuy)
DOI: 10.1079/9781789247053.0015

duties and breadwinning; and relatively high numbers of dependents (Flatø *et al.*, 2017). This situation heightens the sensitivity of some female-headed households to environmental factors like variations in rainfall which reduce agricultural production. Thus, to address women's worsening climate vulnerability, 'strong, coordinated interventions by various ministries and government departments, specifically those in the social, economic, and environmental sectors' are needed (Flatø *et al.*, 2017, p. 61). Starting with the livelihoods of the marginalized reveals interconnections and important intervention pathways that might not be apparent from the top down.

In another example, drawing from how climate change affects people in the Hindu Kush, class, caste, and ethnicity-based socio-cultural institutions, along with religion, age, health, education, disability, language, and other features of identities, all interact with gender to produce differing climate vulnerabilities in different geographic locations, economic settings, and political environments. Since social contexts determine power relations, which act at all scales, from the household to formal institutions including the state and the market, intersectional and gendered climate vulnerabilities are produced by a wide variety of conditions and are embedded in 'everyday patterns of social interaction and organization and of access to resources' (Goodrich *et al.*, 2019, p. 12). In this context, livelihood perspectives offer indications of vulnerabilities which constrain women's options at ever-widening scales, and offer insights about effective policy approaches, as well as strategies available to women (Dankelman *et al.*, 2008).

This chapter explores livelihood and care implications of the climate crisis from a gendered viewpoint that includes the implications of this approach for climate decision making at multiple scales, from local to global. The focus is on grassroots political organizing, activism, and movements as well as women's community-based actions to (re)build social resilience in the face of climate chaos. Challenges and policy implications are discussed as governments struggle to meaningfully and equitably address climate change. We also highlight the transformational imperatives of care and livelihood priorities which cast into stark relief the unsustainability of the long-established gender inequities that serve as the foundation for economic systems everywhere.

In this chapter, 'women's livelihood perspectives' are referred to, which simply means starting with women's daily lives and basic survival strategies to understand how climate change is affecting them and what they can do about it. The authors recognize that there is no one 'livelihood perspective' because contexts, women, and livelihoods are dynamic and diverse. This framing is different from other uses of the term 'livelihoods' in development and gender literatures. These include the 'sustainable livelihood approach,' which 'evolved within the context of... intentional development... by which development practitioners were seeking to maximize the effectiveness of their interventions to help the disadvantaged' (Morse and McNamara, 2013, p. 17). The sustainable livelihood approach is discussed and critiqued in a substantial literature, as 'an analysis of people's current livelihood and what is needed for an 'enhancement'... useful in avoiding the inappropriate interventions critiqued by the post-developmentalists' (Morse and McNamara, 2013, p. 18; see also Elasha *et al.*, 2005). The framework's original neglect of power relations, and subsequent attempts to include how power affects livelihood enhancements, have led more recently to research focused on meta-analyses of livelihood studies (De Haan, 2012; see also Quandt, 2018). The recent turn to the 'everyday' in political science and political ecology acknowledges the importance of formerly unrecognized work, care, activism, and consumption, the invisibility of which is directly tied to patriarchy (Elias and Roberts, 2016; Saave *et al.*, 2020).

Another rich, growing literature uses various forms of livelihood approaches to assess climate change impacts in many specific contexts (e.g. see Hahn *et al.*, 2009; Below *et al.*, 2012; Shah *et al.*, 2013; Tanner *et al.*, 2015; Mersha and Van Laerhoven, 2016; Panthi *et al.*, 2016; Ravera *et al.*, 2016; Antwi-Agyeh *et al.*, 2017; Adzawla *et al.*, 2019; Alhassan *et al.*, 2019). This chapter's use of the phrase 'women's livelihood perspectives' is close to the 'subsistence perspective' advanced by Maria Mies and Veronika Bennholdt-Thomsen (see also Mies, 2014), which they define as:

> A perspective from below, from what is necessary... [including] everyday life and its politics, the strategies of women to keep life going... [based in their] control over means of subsistence... dignity... [understanding that] what is good for the village women should be

> good for the whole society… [and that this is] equally valid for the so-called developed countries and classes.
>
> (Mies and Bennholdt-Thomsen, 2000, pp. 3–4)

The global climate crisis emphasizes this sense of urgency and shared livelihoods. Research demonstrates how women's vital but unpaid work in the home and in communities to create, educate, socialize, feed, and care for the labor force so it can undertake economic activity, and which also sustains economies in many other ways (O'Hara 1995, 1997), involves worker-related carbon emissions just like other forms of production. Those emissions, which are much higher in developed countries because of the carbon intensity of lifestyles and work-saving household technologies, must be addressed as part of the energy transition to low-carbon energy sources (Horen Greenford *et al.*, 2020; see also Saave *et al.*, 2020). Ignoring women's work simply because it is unpaid both perpetuates gender injustice and hides important tools for addressing climate change. Global equity and development require that women have access to energy sources to reduce household drudgery such as for water provision, laundry, and clean cooking, and these technologies should be prioritized alongside others that contribute to economic livelihoods in factories, transportation, and infrastructure. However, as long as women's work is viewed as 'free,' this priority may be invisible and ignored. Moreover, home-based energy conservation practices (such as installing LED lights, turning off lights when not in use, and limiting use of heavy appliances to evenings and weekends) in effect shift the responsibility for maintaining a healthy atmosphere from industries (costed) on to households (uncosted) where in some cases this work may be done 'for free' by women, similar to how recycling enrolls households in work formerly done by industries (Thoyre, 2020; see also Johnson *et al.*, 2020). The resulting increases in the 'gender–climate gap' signal that equity-enhancing climate policies, and the knowledge and data (including input from women) required to develop them, are crucial to tap into the synergies between development imperatives, social equity, trust, and enhanced climate action (see Rojas and Roth, 2016). Such complexities and possibilities can only be revealed using livelihood and care perspectives.

The gendered realities of women's working lives generally give them special insights, grounded in their lived experience, about effective ways to prioritize and structure climate risk reduction initiatives. Since women are on the front lines of climate impacts, their perspectives are vital for efficient, equitable, and informed climate policy (Lieu *et al.*, 2020). Otherwise, grassroots action to protect livelihoods for all in the face of climate change inevitably takes precedence, sidelining or counteracting government policy. In any case, the actions women are taking in households, communities, jurisdictions at all scales, and global networks are greatly affecting carbon emissions and the energy transition, as part of the changes in societies, social norms, governance systems, and methods of providing care, food, shelter, clothing, skills transmission, and cultural cohesion as the world faces the climate crisis.

In the next section of this chapter, the authors discuss how women all over the world, in various ways, are putting their livelihood perspectives to use in climate change organizing, action, and (where possible) policy development as part of the energy transition. The third section provides a range of examples from around the world of women's climate agency and its areas of focus – community-based education and networking, citizen science for better data collection on the climate crisis, youth movements, and commoning – re/creating and organizing shared use, jointly owned resources, and initiatives for community resilience (Fournier, 2013; Federici, 2019). Some lessons are drawn on the characteristics, challenges, and climate policy implications of livelihood and care perspectives in the chapter's concluding section.

Women's Organizing and Participation Drive Climate Change Activism and Emissions Reductions

Probably because of their lived experiences and commitment to address the kinds of differential impacts noted above, women have long been leaders in environmental movements and activism (Mellor, 1997; Perkins, 2013). Women are more likely than men to be environmental and climate activists (Stein, 2004; Dankelman and Jansen, 2010; McCright and Xiao, 2014; Whyte, 2014; Wickramasinghe, 2015; Black, 2016; Vinyeta *et al.*, 2016; Pearse, 2017). Women, and marginalized people in general, place a stronger priority on addressing climate change than other groups, and are more likely to take personal and political climate action (Resurrección, 2013; Nagel, 2015, pp. 166–182; Glemarec *et al.*, 2016; Godfrey and Torres, 2016; Perkins, 2017b; Bryan *et al.*, 2018; Crawford, 2019). Environmental and climate justice activists often use organizing techniques developed by the feminist movement, such as consciousness raising, grounded theory, and contextual reasoning, which result from starting with women's lived experiences rather than political abstractions (Weiss, 2012). When climate change is seen as a scientific problem, to be solved using technical methods, this excludes and marginalizes women's contributions (Gaard, 2015), but intersectional organizing counters this technocratic approach to climate change through grassroots climate justice activism.

Engagement with environmental and climate justice activism also helps women confront and deepen their understanding of the wider gendered economic and political constraints they face, which can lead to activism for broader change towards more social justice (Weiss, 2012; see also Sangita, 2016; Perkins, 2018). The global climate justice movement is organized as a decentralized polycentric network of semi-autonomous, coordinated units, allowing it to simultaneously influence multiple sites of environmental governance from the local to the global (Tormos-Aponte and García-López, 2018).

Carbon emissions per capita are lower in countries where women have more political voice (Ergas and York, 2012; Mavisakalyan and Tarverdi, 2019). Men's lower rates of environmental concern and activism, on average, are often understood as being linked to their relative social position and sense of invulnerability (Goldsmith *et al.*, 2013, p. 161; Nagel, 2015, p. 168). Thus, increasing political equity for women is a climate-friendly step as well as an advance on moral, ethical, and human rights grounds.

Women's Collective Activism Prioritizes Climate Change from Livelihoods and Care Perspectives

Irene Dankelman, one of the earliest gender and climate researcher-activists, summarizes the reasons for women's engagement in climate activism as follows:

> Women play a key role in protecting, managing and recovering their household and assets during a disaster. They have been strong advocates for preparedness measures at the community level because they understand what disaster means to the day-to-day, realities of life.
> (Dankelman *et al.*, 2008, p. 61)

Dankelman further notes women's capacities in strengthening preparedness and adaptation to the changing nature of disasters. Women have not only the knowledge for developing innovative strategies to address climate change impacts but are also resourceful community mobilizers in disaster response. Most women demonstrate diverse adaptation and coping strategies and mechanisms, including moving to safer places, saving their assets, adopting strict dietary adaptations, energy saving, adapting agricultural practices, earning income from alternative sources, seeking alternative healthcare, organization, and collective action. Indeed, women have shown and continue to show a clear sense of the capacities needed to adapt to climate change.

The authors have gathered some examples of how livelihood and care perspectives are expressed collectively through women's activism to address climate change. They are organized into sections on community-based education and networking; citizen science and data collection; youth movements; and commoning. There are many more examples and categories discussed elsewhere in the literature which could have been cited; this discussion is organized around four categories which demonstrate priorities for action as well as women's political agency in climate decision making.

Community-based education and networking

Women tend to rely on 'bonding' forms of social capital, including informal connections to family, relatives, and friends, rather than the 'bridging' or 'linking' of social capital that men are more likely to rely on (Perez *et al.*, 2015, p. 101). Throughout Africa and in many parts of Asia and Latin America, women's collaborative and reciprocal sharing of work, food, fuel, childcare, and community celebrations is the norm. Collective sharing and assistance through micro-lending credit pools, schools, seed-sharing, labor pools, insurance, and international links via diaspora communities aid resilience for individuals and communities in preparing for, adapting to, and recovering from climate-induced crises (Perez *et al.*, 2015).

In the case of women's community-building roles and social trust, data from the European Bank for Reconstruction and Development's 'Life in Transition' survey, which includes detailed interviews about people's life circumstances, show that women are more likely than men to take personal action to help fight climate change, especially when they are married, and thus more likely to be responsible for household management, child and elder care, and purchasing decisions. Social dilemmas and social trust play a critical role for women's personal engagement in pro-environmental behaviors irrespective of the burden. Strengthening social trust – resulting in strong community cohesion – was found to be as important as environmental policies in shaping women's environmental behaviors, even when these were time-consuming (Gür, 2020).

In the Sahelian country of Mali, firewood and charcoal production, which are an important source of income for many women, cause environmental degradation by intensifying deforestation, soil erosion, and exposure to flooding. Sinsibere Cooperative, a women's grassroots organization in the villages of Bougoula, Sanankoroba, and Dialakoroba, works to counteract these effects and bring a win–win benefit for women's livelihoods and the environment. The cooperative educates members on tree planting and provides training and microfinance support for developing alternative sources of incomes from shea butter production, soap making, and vegetable gardening. The program has benefited over 600 women and about 80% have stopped or significantly reduced firewood extraction (UN/ISDR, 2008; Wenden, 2011; Sinsibere, 2013; Alam *et al.*, 2015; Mali Folkecenter, 2015).

Another example of women's community-based organizing with positive climate impacts is the Alianza Internacional de Reforestación (AIRES), a non-governmental organization based in Iztapa, Guatemala, founded out of concern with deforestation. For more than 20 years, AIRES has relied on the support of Indigenous Maya women and forestry professionals to implement afforestation programs to fight erosion and other environmental hazards. Through the partnership, the Indigenous women have planted millions of trees on previously deforested mountain slopes to prevent deadly mudslides. AIRES provides these women, who are mainly small-scale farmers, with the seeds and tools to implement and maintain the reforestation activities.

Although women in Mali and those from Maya communities in Guatemala have key roles to cater for their families' needs, their decisions to plant trees and reduce firewood cutting illustrate their concerns to protect the ecological resources they depend on for their livelihoods. These actions also indicate the many contributions community members can make to sequester carbon and mitigate the threats of climate change (Hallum-Montes, 2009; UNFCCC, 2014; Alam *et al.*, 2015). Overcoming climate change problems related to livelihoods also involves championing environmentally sustainable sources of income, and upgrading and ensuring that energy for household cooking and lighting is efficient. In southern Kenya, women are taking leading roles in reducing the use of kerosene lanterns through an initiative launched by the Maasai Wilderness Conservation Trust (MWCT) and the organization New Course. Women's groups receive solar lanterns and put half the money saved in kerosene costs into funding for their own collective enterprises (MWCT, 2015; Multer, 2016). Another network of women across Tanzania, Uganda, and Nigeria known as Solar Sister is bringing clean energy, including environmentally efficient cooking stoves and lights, to off-grid communities. The network currently consists of about 4565 women entrepreneurs reaching out to about 2 million people (Solar Sister, 2020). Solar Sister recruits, trains, and supports new entrepreneurs in marketing durable, affordable solar-powered household items. Entrepreneurs earn income selling environmentally sustainable products at affordable prices directly to community members. These two women-led solar entrepreneurship initiatives enable poor families to gain economic and financial stability, in addition to the broader benefits to the environment. Households' use of renewable and sustainable sources of energy contributes to reducing carbon emissions (Alam *et al.*, 2015; Bechtel, 2015; Heuër, 2017). None the less, the reach of such initiatives is still relatively low, and thus partnership with environmental financing agencies is necessary to scale up these win–win climate mitigation and adaptation approaches.

Citizen science and data collection

Governments often do not collect the kind of data that is most useful to document who is being harmed by climate crises, how they are acting in response, and who they rely on for assistance (and therefore how a climate resilience strategy should be designed). In particular, the benefits of the kinds of multitasking that women are often experts in, for providing basic community necessities of food, shelter, water/sanitation, care, and education with low levels of fuel/energy and material energy inputs, often elude government data-gathering mechanisms which may be stuck in sectoral silos, blind to the value of skills transmission, care, and/or social reciprocity, and unwilling to recognize unpaid work.

During the COVID-19 crisis, Dr Nancy Folbre of the University of Massachusetts, an expert on unpaid economic contributions and their limits, invited international participants to webinars organized by the International Association for Feminist Economics (IAFFE) to gather 'citizen science' time-use data on how childcare and other household work was being done at home during the lockdown, since government agencies were not collecting this important data (IAFFE, 2020). Without time-use data on household activities, care, food processing, and overlapping work responsibilities, it is impossible to plan or even understand the constraints on emissions reductions involving such things as materials recycling, composting, reduced food waste, efficient low-polluting cooking stoves, solar water heating, rainwater harvesting, energy storage in water tanks, solar lighting and refrigeration, and other techniques that require understanding and work by users. By collecting and documenting the various data on household activities in times of the pandemic-induced lockdown, Dr Folbre is harnessing the opportunity to provide governments with evidence for policy making and planning.

When women play a leadership role in early warning systems and reconstruction, this aids community disaster outcomes. Women tend to share information related to community well-being, choose less polluting energy sources, and adapt more easily to environmental changes when their family's survival is at stake. During hurricane Mitch in 1998, for example, women trained in early warning disaster reduction made a big difference in La Masica, a village in Honduras that, unlike nearby communities, reported no deaths during the hurricane. Prior to the disaster, La Masica received gender-sensitive disaster preparedness education on early warning systems and hazard management through a program implemented by the Central America Disaster Preparedness Agency. The training focused on gender-disaggregated vulnerability analysis and capacity building activities, and compelled women to take active roles over the abandoned and previously men-held task of continuously monitoring the early warning system. The women's roles in disseminating early warning information facilitated the municipality's ability to evacuate the area promptly before hurricane Mitch struck. Integrating gender perspectives in the design and implementation of policies and programs also helps curtail the gender-differentiated impacts of environmental degradation – shortage of water, deforestation, desertification – which are exacerbated by climate change (ILO, 2008, p. 3).

South African water activist Ferrial Adam, working with the Johannesburg-based Co-Operative and Policy Alternative Centre (COPAC), trains local women's organizations to use 'people's science' to generate new scientific data to help reclaim water commons and protect community livelihoods in times of climate change. The Anti-Privatisation Forum (APF), also in Johannesburg, challenged private firms which had been put in charge of water management, calling for water to be recognized as a public good. The APF used various approaches including mobilization, education, and mass struggles to reclaim human and constitutional rights to water. One of their grassroots campaigns, Operation Vulamanzi ('water for all'), saw communities take control over systems that had been forced on them such as trickler systems, rerouted water piping, and prepaid meters, to gain access to water supplies. The campaign changed water governance in Johannesburg: it forced the African National Congress (ANC) government to implement a partial free water policy in late 2002, and contributed to both the failure and renegotiation of many South African water privatization projects (Adam, 2017).

Women may be less familiar than men with scientific terminology and technical/policy approaches although their contributions to environmental activism, as presented in the case studies, are invaluable. One possible way to strengthen women's active role, inclusion, and continuous recognition in policy processes is to start with discussions on personal experiences relating to health, environment, climate, and collective responses in livelihoods-oriented community-based data collection and organizing. Although there is a likely peril of self-exclusion, especially for busy women in developing countries, continuous support from government to strengthen women's agency and ownership of the policy discussions is an important way to overcome these potential challenges (Gasior Altman *et al.*, 2008; see also Perkins and Walker, 2015).

Community-based data collection via 'citizen science' allows women to specify and share knowledge that is fundamental for inclusive climate action, thus integrating their lived experience in how climate change is addressed, beginning at the household and local levels.

Youth movements

Youth climate movements founded and led by young women include School Strike for the Climate/Global Climate Strike (2020), Fridays For Future (2020), Mother Earth Project (2020) and Parachutes for the Planet, and many others worldwide (Kaplan, 2019; Nevett, 2020; Parker, 2020; UN Women, 2020). As the motivation for their activism, young women cite home and community concerns related to consumption and transportation choices, environmental damage in local communities, food security and health – in other words, livelihood issues.

Fridays For Future, for example, a Swedish-based movement, was born out of strikes by Greta Thunberg and other young people who took their frustrations to the front of the Swedish parliament every school day for 3 weeks in August 2018. The young activists continue to demand political actions to speed the transition away from fossil fuel-based economy and lifestyles; it has become a global movement of school students, particularly from the minority world countries, including the UK and Australia, who take time off from class on Fridays to participate in these demonstrations (Fridays For Future, 2020). While the activities of these young people have attracted the attention and support of many parents, teachers, and adult groups, conservative politicians have described the strikes as truancy and punished students for joining such calls for action. Nevertheless, these young people, like many other strikers, bring solidarity in the fight against climate change. The Global Climate strikers who captivated the world in September 2019 are still connecting, organizing, and reaching out to new strikers online to increase awareness of the need for action on climate change. Since climate discussions have long been the preserve of adults and bureaucrats, the activities of young activists represent a paradigm shift.

While youth activism, especially by girls, is not new and is complicated and contested, global networks of youth activists show 'resilience... in rising against patriarchy and intersecting oppressions with fervor and urgency, particularly when exercised communally and collaboratively' (Vanner and Dugal, 2020, p. xii; see also Budgeon, 2001). Climate activists who are young women have become widely recognizable; they have become celebrated cultural figures 'through a potent combination of hopefulness, harmlessness, and heroism' (Taft, 2020, p. 13). As Greta Thunberg was quoted as saying, somewhat sarcastically, 'Oh you children, you young people are the hope. You will save the world.' She added, 'I think it would be helpful if you could help us just a little bit' (Sengupta, 2019; quoted in Taft, 2020, p. 13). Young women's climate activism is global, and includes movements led by young women of color in Uganda, Brazil, New Zealand, Fiji, Samoa, the Philippines, Kenya, Nigeria, and India as well as Canada, Germany, and the USA (Global Citizen, 2019; Amnesty International UK, 2020).

Young Indigenous women are at the forefront of many climate and environmental justice movements, calling attention to the essentiality of water, land, and care for human survival and the importance of Indigenous wisdom regarding these relationships (Whyte, 2014; Konsmo and Pacheco, 2016; Vinyeta *et al.*, 2016). Indigenous women are respected and effective leaders of networked environmental and climate struggles worldwide (Gorecki, 2014; Thomas-Muller, 2014; Manuel, 2015, p. 211; Nixon, 2015; Giacomini, 2016; Acha, 2017; Whyte, 2018).

Young women's leadership highlights intergenerational justice as a complement to intragenerational justice, the traditional focus of climate justice movements. This is part of 'a growing movement of community-scale demands for accountability and action, particularly at the local level' (Cretney and Nissen, 2019). When male leaders and voters do not care or act to improve environmental protection, the resulting delays in developing and implementing effective climate change policies increase the intensity and impacts of climate change, which fall more heavily on women, in part because they tend to live longer – an intergenerational injustice feedback loop. This cycle of youth and age, care and interpersonal empathy, is also central to women's livelihood struggles worldwide.

Commoning

'Commoning' means building and preserving the means of sharing livelihood necessities (e.g. land, food, water, care, skills, work) as a communal approach to protecting collective social values and material and ecological resilience (Mies and Bennholdt-Thomsen, 2001). For women, commoning can mean the difference between life and death in times of crisis: 'Women have less access than men to common property resources' (Perez *et al.*, 2015, p. 95).

Many kinds of commons are being (re)developed and protected worldwide, through women's partnerships and global solidarity, to help address the challenges of climate change at the local/urban level (Röhr *et al.*, 2008, 2010; Ostrom, 2014; Alber *et al.*, 2017; Perkins, 2017a). These initiatives and models include equitable local economy institutions, cooperatives and land trusts, community gardens and food programs, childcare and elder care cooperatives, support for victims of gender-based violence, water harvesting schemes, community shelters, agroforestry projects, and many other collective livelihood and care initiatives, appropriately adapted for local socio-ecological conditions (Gibson-Graham, 2006; Kaufman, 2012; Klein, 2014). All bring people together to build community resilience in the face of climate change while developing the skills and relationships necessary for equitable and sustainable commons governance.

In Kenya, where firewood is used to meet three-quarters of household energy needs, the Greenbelt Movement started by Wangari Maathai in 1977, under the auspices of the National Council of Women of Kenya, has organized women to plant more than 20 million trees throughout Africa. In the process, they have strengthened their claims to the land, learned and used business strategies and traditional ecological knowledge, built food sovereignty, and expressed their political agency and resolve to heal environmental damage collectively (Brownhill, 2007).

La Via Campesina and It Takes Roots climate activists, building on years of experience in women's feminist struggles, understand their movement as part of 'the transition from life-denying capitalism to a life-affirming, post-capitalist commons... defending life by strengthening commons systems including, centrally, Indigenous people's long-established political economies and cultures' (Giacomini, 2020, pp. 195–196). Since commons have always allowed marginalized people to survive through mutual aid, care, and food provision (Shenaz Hossain, 2018; Federici, 2019), the climate crisis is activating long-held social traditions and growing global commons networks (Leroy, 2016) which also help to spread information and foster commons development at community and local scales.

Conclusion: Women's Bottom-Up Climate Action Draws Strength from a Focus on Livelihoods and Care

This chapter began by advancing the idea that adopting livelihood and care perspectives could be a way to 'bring women in' to decision making processes that affect climate change mitigation and adaptation. However, the examples cited above show that, in fact, livelihood perspectives are already central to climate decision making.

Seen from the vantage point of everyday life (the kitchen table), decisions about food, clothing, shelter, transport to work or school, energy for cooking and perhaps space heating, water provision, and care are the crux of basic human needs which access to energy makes manageable – shaped by the infrastructures each society has built. From this vantage point, it is clear that livelihood necessities are fundamental, and of a much higher priority than alternate, wasteful uses of energy such as air travel, luxury consumption, or rapid turnover resource use.

Around the kitchen table, discussions about climate change, its causes and impacts, have a rationale that is incontrovertible. If this kitchen-table rationale should come into conflict with a very different policy-table view – for example, that rights to consume energy and emit carbon should be allocated according to wealth, rather than for livelihood necessities – this would probably not bode well for the policy-table approach. There are, after all, many more kitchen tables than policy tables.

Attempts to bring gendered livelihood and care perspectives into formal, top-down policy processes have a long and fraught history. In patriarchal societies and economies which depend on women's exploitation, this is no surprise. The COVID-19 pandemic – showing the grim toll society pays for undercutting and underpaying for care, and the unjust burden these failures place on women and marginalized communities – has heightened awareness of the threat that exploitation of women poses for everyone (McLaren *et al.*, 2020; Priola and Pecis, 2020; Wenham and Morgan, 2020). Government expenditures to restart economies reeling from the pandemic could address both climate mitigation and adaptation/community resilience by focusing on care, social services, renewable distributed energy access, and provision of energy efficient housing, transport, and food security for all (Di Chiro, 2019; Loske, 2020; Van den Berg *et al.*, 2020).

Requiring that women be fairly represented even just as participants in government-initiated climate interventions, through gender quotas, improves not only the interventions' equity outcomes but also their effectiveness more broadly (Cook *et al.*, 2020).

The International Labor Organization has stated that for women to be able to assume a fair share of the jobs and responsibilities connected with global climate change, the following elements must be in place for them (Bäthge, 2010, p. 7; see also ILO, 2016):

- access to education, training, and upgrading;
- access to and control over productive resources including access to land and ownership rights;
- access to markets (land, labor, financial, and product markets);
- access to services;
- benefits from the use of public funds, particularly for infrastructure, and access to public goods;
- means of enforcing claims for unpaid/reproductive work and redistribution/remuneration for such work; and
- the possibility of generating income from the use of their own labor.

Can women's livelihood and care perspectives be included in climate policy decision making itself, however? The 'add women and stir' approach, of getting some women to sit at the policy table, is notably unpromising. Women's token participation, in fact, may solidify power in the hands of those who hold it; communication and governance processes must themselves be critically examined and changed if they are to integrate bottom-up, community-driven priorities (Roncoli *et al.*, 2010; Kythreotis *et al.*, 2019).

If governments truly aim to include women's livelihood and care perspectives more fully in climate policy decision making, there are three important first steps:

1. *Obtain good data* on women's economic contributions and climate-related work. This means recognizing that much of this work is unpaid, informal, and/or carried out in synergistic, overlapping ways rather than easily quantifiable 'jobs.' It also requires inviting and paying women to help develop ways of measuring and valuing this work, and believing and disseminating the resulting statistics in policy circles rather than burying them.

2. *Use that data* for policy development to facilitate women's synergistic contributions to care, livelihoods, and climate mitigation/adaptation, again by consulting with women who are experts in these areas and listening to their advice, in equitable contexts that ensure power sharing, support, security, and respect for their contributions.

3. *Open the doors* for women's participation at all levels of governance, from basic voting rights to grassroots, neighborhood, and community decision making to regional, national, and global positions of leadership. This requires equitable education access, mentorship, and structural changes in work regimes that make it possible for women to flourish.

Fortunately, as described in the examples above, initiatives grounded in women's livelihoods perspectives are already under way everywhere, effectively addressing the climate crisis in multiple appropriate and creative ways.

References

Acha, M.A.R. (2017) We have to wake up, humankind! Women's struggles for survival and climate and environmental justice. *Development*, 60, 32–39.

Adam, F. (2017) Building people's power for water sovereignty: an activist guide. COPAC, Johannesburg. Available at: www.safsc.org.za/wp-content/uploads/2017/11/Water-Guide-Final-Web_colour.pdf (accessed 3 February 2021).

Adzawla, W., Azumah, S.B., Anani, P.Y. and Donkoh, S.A. (2019) Gender perspectives of climate change adaptation in two selected districts of Ghana. *Heliyon* 5(11), e02854.

Alam, M., Bhatia, R., and Mawby, B. (2015) Women and climate change: impact and agency in human rights, security, and economic development. Georgetown Institute for Women, Peace and Security. Available at: www.giwps.georgetown.edu/resource/women-and-climate-change/ (accessed 3 February 2021).

Alber, G., Cahoon, K. and Röhr, U. (2017) Gender and urban climate change policy: tackling cross-cutting issues towards equitable, sustainable cities. In: Buckingham, S. and Le Masson, V. (eds) *Understanding Climate Change Through Gender Relations*. Routledge, Oxon and New York, pp. 64–86.

Alhassan, S.I., Kuwornu, J.K.M., and Osei-Asare, Y.B. (2019) Gender dimension of vulnerability to climate change and variability: empirical evidence of smallholder farming households in Ghana. *International Journal of Climate Change Strategies and Management* 11(2), 195–214.

Almeida, I.M., Filho, J.M.J., and Vilela, R.A.G. (2019) Reasons for investigating the organizational dynamics of the Vale tailings dam disaster in Brumadinho, Minas Gerais State, Brazil. *Cadernos de Saúde Pública* 35(4), 2 May. Available at: www.scielo.br/scielo.php?pid=S0102-311X2019000400301&script=sci_arttext&tlng=en (accessed 3 February 2021).

Amnesty International UK (2020) Five young women of color fighting climate change worldwide. Available at: www.amnesty.org.uk/climate/five-women-colour-fighting-climate-change-worldwide (accessed 3 February 2021).

Antwi-Agyeh, P., Quinn, C.H., Adiku, S.G.K., Codjoe, S.N.A., Dougill, A.J., Lamboll, R., and Dovie, D.B.K. (2017) Perceived stressors of climate vulnerability across scales in the Savannah zone of Ghana: a participatory approach. *Regional Environmental Change* 17, 213–227.

Bäthge, S. (2010) Climate change and gender: economic empowerment of women through climate mitigation and adaptation? OECD Governance and Democracy Division/GTZ, Eschborn, Germany.

Bechtel, J. (2015) What every environmentalist needs to know about time security. *The Huffington Post*, Aug. 7, 2015. Available at: www.huffingtonpost.com/jamie-bechtel/what-every-environmentalist-needs-to-know-about-time-security_b_7951920.html (accessed 3 February 2021).

Below, T.B., Mutabazi, K.D., Kirschke, D., Franke, C., Sieber, S., Siebert, R., and Tscherning, K. (2012) Can farmers' adaptation to climate change be explained by socio-economic household-level variables? *Global Environmental Change* 22(1), 223–235.

Black, T., (2016) Race, gender, and climate injustice: dimensions of social and environmental inequality. In: Godfrey, P. and Torres, D. (eds) *Systemic Crises of Global Climate Change: Intersections of Race, Class and Gender*. Routledge, New York, pp. 172–184.

Brownhill, L. (2007) Gendered struggles for the commons: food sovereignty, tree-planting and climate change. *Women and Environments International Magazine*, special issue on Women and Global Climate Change, No. 74/75, 34–37.

Bryan, E., Ringler, C., Okoba, B.O. and Roncoli, C. (2018) Adapting agriculture to climate change in Kenya: household strategies and determinants. *Journal of Environmental Management* 114C, 26–35.

Budgeon, S. (2001) Emergent feminist(?) identities: young women and the practice of micropolitics. *The European Journal of Women's Studies* 8(1), 7–28.

Cook, N., Grillos, T. and Andersson, K.P. (2020) Gender quotas increase the equality and effectiveness of climate policy interventions. *Nature Climate Change* 9, 330–334.

Crawford, N.C. (2019) Pentagon fuel use, climate change, and the costs of war. Watson Institute of International and Public Affairs, Brown University. Available at: www.watson.brown.edu/costsofwar/files/cow/imce/papers/Pentagon%20Fuel%20Use%2C%20Climate%20Change%20and%20the%20Costs%20of%20War%20Revised%20November%202019%20Crawford.pdf (accessed 3 February 2021).

Cretney, R., and Nissen, S. (2019) Climate politics ten years from Copenhagen: activism, emergencies, and possibilities. *Women Talking Politics*: A research magazine of the New Zealand Political Studies Association November, 15–19.

Dankelman, I., and Jansen, W.H.M. (2010) Gender, environment and climate change: understanding the linkages. In: Dankelman, I. (ed.) *Gender and Climate Change: An Introduction*. Routledge, London, pp. 21–54.

Dankelman, I., Alam, K., Ahmed, W. B., Geyey, Y.D., Fatema, N. and Mensah-Kutin, R. (2008) Gender, Climate Change and Human Security: Lessons from Bangladesh, Ghana and Senegal. Women's Environment and Development Organization (WEDO) with ABANTU for Development in Ghana, ActionAid Bangladesh and ENDA in Senegal. Available at: www.repository.ubn.ru.nl/bitstream/handle/2066/72456/72456.pdf (accessed 3 February 2021).

De Haan, L. (2012) The livelihood approach: a critical exploration. *Erdkunde* 66(4), 345–357.

Di Chiro, G. (2019) Care not growth: imagining a subsistence economy for all. *British Journal of Politics and International Relations* 21(2), 303–311.

Elasha, B.O., Elhassan, N.G., Ahmed, H., and Zakieldin, S. (2005) Sustainable livelihood approach for assessing community resilience to climate change: case studies from Sudan. *AIACC Working Paper No. 17*. Available at: www.researchgate.net/publication/228591150_Sustainable_Livelihood_Approach_for_Assessing_Community_Resilience_to_Climate_Change_Case_Studies_from_Sudan (accessed 2 July 2020).

Elias, J., and Roberts, A. (2016) Feminist global political economies of the everyday: from bananas to bingo. *Globalizations* 13(6), 787–800.

Ergas, C., and York, R. (2012) Women's status and carbon dioxide emissions: a quantitative cross-national analysis. *Social Science Research* 41(4), 965–976.

Federici, S. (2019) *Re-enchanting the World: Feminism and the Politics of the Commons*. PM Press, Oakland, CA.

Flatø, M., Muttarak, R. and Pelser, A. (2017) Women, weather, and woes: the triangular dynamics of female-headed households, economic vulnerability, and climate variability in South Africa. *World Development* 90, February, 41–62

Fournier, V. (2013) Commoning: on the social organisation of the commons. *M@n@gement* 4(16), 433–453.

Fridays For Future (2020) Available at: www.fridaysforfuture.org/ (accessed 3 February 2021).

Gaard, G. (2015) Ecofeminism and climate change. *Women's Studies International Forum* 49, 20–33.

Gasior-Altman, R., Morello-Frosch, R., Green Brody, J., Rudel, R., Brown, P. and Averick, M. (2008) Pollution comes home and gets personal: women's experience of household chemical exposure. *Journal of Health and Social Behaviour* 49(4), 417–435.

Giacomini, T. (2016) Ecofeminism and system change: women on the frontlines of struggle against fossil capitalism and for the solar commons. *Canadian Woman Studies* 31(1–2), 95–100.

Giacomini, T. (2020) An ecofeminist perspective on the just transition from capitalism to commons. In: Tokar, B. and Gilbertson, T. (eds) *Climate Justice and Community Renewal: Resistance and Grassroots Solutions*. Routledge, London and New York, pp. 194–205.

Gibson-Graham, J.K. (2006) *A Postcapitalist Politics*. University of Minnesota Press, Minneapolis.

Glemarec, Y., Bayat-Renoux, F. and Waissbein, O. (2016) Removing barriers to women entrepreneurs' engagement in decentralized sustainable energy solutions for the poor. *AIMS Energy* 4(1), pp. 136–172.

Global Citizen (2019) 12 female climate activists who are saving the planet. Available at: www.globalcitizen.org/en/content/female-activists-saving-planet/ (accessed 3 February 2021).

Global Climate Strike (2020) Available at: www.globalclimatestrike.net/ (accessed 3 February 2021).

Godfrey, P. and Torres, D. (eds) (2016) *Systemic Crises of Global Climate Change: Intersections of Race, Class and Gender*. Routledge, New York.

Goldsmith, R.E., Feygina, I. and Jost, J.T. (2013) The gender gap in environmental attitudes: a system justification perspective. In: Alston, M. and Whittenbury, K. (eds) *Research, Action, and Policy: Addressing the Gendered Impacts of Climate Change*. Springer, Dordrecht, The Netherlands, pp. 159–170.

Gorecki, J. (2014) No climate justice without gender justice: women at the forefront of the People's Climate March. Available at: https://thefeministwire.com/2014/09/climate-justice-without-gender-justice-women- (accessed 9 June 2021).
Goodrich, C.G., Udas, P.B. and Larrington-Spencer, H. (2019) Conceptualizing gendered vulnerability to climate change in the Hindu Kush Himalaya: contextual conditions and drivers of change, *Environmental Development* 31, September, 9–18.
Gür, N. (2020) Does social trust promote behaviour aimed at mitigating climate change? *Economic Affairs* 40(1), 36–49.
Hahn, M.B., Riederer, A.M. and Foster, S.O. (2009) The Livelihood Vulnerability Index: a pragmatic approach to assessing risks from climate variability and change – a case study in Mozambique. *Global Environmental Change* 19(1), pp. 74–88.
Hallum-Montes, R. (2009) Agricultural development, environmental degradation, and women's work in highland Guatemala. *The Latin Americanist*, 53(2) pp. 5–28.
Heuër, A. (2017) Women-to-women entrepreneurial energy networks: a pathway to green energy uptake at the base of pyramid. *Sustainable Energy Technologies and Assessments*, 22, 116–123.
Horen Greenford, D., Crownshaw, T., Lesk, C., Stadler, K. and Matthews, H.D. (2020) Shifting economic activity to services has limited potential to reduce global environmental impacts due to the household consumption of labor. *Environmental Research Letters*. Available at: www.academia.edu/42973203/Collaborative_Feminist_Degrowth_Pandemic_as_an_Opening_for_a_Care-Full_Radical_Transformation (accessed 3 February 2021).
IAFFE (2020) Feminist economics perspectives on COVID-19, recorded webinar, April 15. International Association for Feminist Economics. Available at: www.mailchi.mp/28a34cf496b4/iaffepresents-covid-1901 (accessed 3 February 2021).
ILO (2008) Green jobs: improving the climate for gender equality too! International Labor Organization. Available at: www.ilo.org/wcmsp5/groups/public/@dgreports/@gender/documents/publication/wcms_101505.pdf (accessed 3 February 2021).
ILO (2016) Gender, labour and a just transition towards environmentally sustainable economies and societies for all. International Labour Organization, Geneva, Switzerland.
Jafry, T. (ed.) (2019) *Routledge Handbook of Climate Justice*. Routledge, London and New York.
Johnson, O.W., Han, J.Y., Knight, A.-L., Mortensen, S., Aung, M.T., Boyland, M. and Resurrección, B.P. (2020) Assessing the gender and social equity dimensions of energy transitions. Stockholm Environment Institute, Stockholm.
Kaplan, S. (2019) Teen girls are leading the climate strikes and helping change the face of environmentalism. *The Washington Post*, 24 September.
Kaufman, C. (2012) *Getting past capitalism: history, vision, hope*. Lexington Book, Lanham, MD.
Klein, N. (2014) *This Changes Everything: Capitalism vs. the Climate*. Simone and Schuster, New York.
Konsmo, E.M. and Pacheco, K. (2016) Violence on the land, violence on our bodies. Native Youth Sexual Health Network/Women's Earth Alliance. Available at: www.nativeyouthsexualhealth.com/weanyshn-partnership.pdf (accessed 3 February 2021).
Kythreotis, A.P., Mantka-Pringle, C., Mercer, T.G., Whitmarch, L.E., Corner, A., Paavola, J., Chambers, C., Miller, B.A. and Castree, N. (2019) Citizen social science for more integrative and effective climate action: a science-policy perspective. *Frontiers in Environmental Science*, 5 February. doi.org/10.3389/fenvs.2019.00010.
Leroy, J.P. (2016) Markets or the commons? The role of Indigenous peoples, traditional communities and sectors of the peasantry in the environmental crisis. In: Issberner, L.-R. and Léna, P. (eds) *Brazil in the Anthropocene: conflicts between predatory development and environmental policies*. Routledge, London and New York. pp. 103–122.
Lieu, J., Sorman, A.H., Johnson, O.W., Virla, L.D. and Resurrección, B.P. (2020) Three sides to every story: Gender perspectives in energy transition pathways in Canada, Kenya and Spain. *Energy Research and Social Science* 68: 101550.
Loske, R. (2020) Re-embedding the economy in nature and society. *CUSANUS Working Papers No. 62*. Available at: www.econstor.eu/bitstream/10419/222291/1/1713956764.pdf (accessed 3 February 2021).
MacGregor, S. (2017) *Routledge Handbook of Gender and Environment*. Routledge, london and New York.
Mali Folkecenter (2015) The gender, energy and environment department. Mali Folkecenter, Bamako. Available at: www.mail.malifolkecenter.org/ (accessed 1 February 2021).

Manuel, Arthur (2015) *Unsettling Canada: A National Wake-Up Call*. Between the Lines, Toronto.

Marshall, J. (2018) Tailings dam spills at Mount Polley and Mariana: chronicles of disasters foretold. Corporate Mapping Project/Canadian Centre for Policy Alternatives, Vancouver. Available at: www.policyalternatives.ca/tailings-disasters (accessed 3 February 2021).

Marshall, J. and Campolino, D. (2020) Mine Tailings Dam Collapses in Canada and Brazil: who cares? Kairos webinar recording, 29 May, Toronto. Available at: www.kairoscanada.org/may-29-mining-tailings-dam-collapses-in-canada-and-brazil-who-cares (accessed 3 February 2021).

Maso, T. (2017) Women affected by the dam failure fighting against the power of big business. CIDSE Together for Global Justice, 15 December. Available at: www.cidse.org/2017/12/15/women-affected-by-the-dam-failure-fighting-against-the-power-of-big-business/ (accessed 3 February 2021).

Mavisakalyan, A. and Tarverdi, Y. (2019) Gender and climate change: do female parliamentarians make a difference? *European Journal of Political Economy* 56:C, pp. 151–164.

McAllister, S. (2019) Speaking with authority: gender and Indigenous politics in the Mount Polley Mine Disaster. Doctoral thesis, University of Victoria, Canada. Available at: www.dspace.library.uvic.ca/handle/1828/11133 (accessed 3 February 2021).

McCright, A. and Xiao, C. (2014) Gender and environmental concern: insights from recent work and for future research. *Society and Natural Resources* 27(10), 1109–1113.

McLaren, H.J., Wong, K.R., Nguyen, K.N. and Mahamadachchi, K.M. D. (2020) Covid-19 and Women's Triple Burden: Vignettes from Sri Lanka, Malaysia, Vietnam and Australia. *Social Sciences* 9(5), 87. doi.org/10.3390/socsci9050087.

Mellor, M. (1997) *Feminism and Ecology*. Wiley, London.

Mersha, A.A. and Van Laerhoven, F. (2016) A gender approach to understanding the differentiated impact of barriers to adaptation: responses to climate change in rural Ethiopia. *Regional Environmental Change* 16, pp. 1701–1713.

Mies, M. (2014) Housewifisation – globalization – subsistence-perspective. In: van der Linden M. and Roth, K.H. (eds) Beyond Marx: *Theorising the Global Labor Relations of the Twenty-First Century* Brill, Leiden and Boston. pp. 209–237.

Mies, M. and Bennholdt-Thomsen, V. (2000) *The Subsistence Perspective: Beyond the Globalised Economy*. Zed Books/Spinifex, London, New York and Australia.

Mies, M. and Bennholdt-Thomsen, V. (2001) Defending, reclaiming and reinventing the commons. *Canadian Journal of Development Studies* 22(4), 997–1023.

Morse, S. and McNamara, N. (2013) The theory behind the sustainable livelihood approach. In Morse, S. and McNamara, N., *Sustainable Livelihood Approach: a critique of theory and practice*. Springer, Dordrecht, The Netherlands, chapter 2, pp. 15–60.

Mother Earth Project (2020) Mother earth project: activating sustainability worldwide. Available at: www.motherearthproject.org/ (accessed 3 February 2021).

Multer, J. (2016) A little solar light sparks BIG changes for women in Kenya, July 18. Available at: www.mpowerd.com/blogs/bright-ideas/new-course-fgc (accessed 3 February 2021).

MWCT (2015) Conservation program update: luci lights initiate change. Maasai Wilderness Conservation Trust. Available at: www.maasaiwilderness.blogspot.com/2015/06/conservation-program-update-luci-lights.html (accessed 3 February 2021).

Nagel, J. (2015) *Gender and Climate Change: Impacts, Science, and Policy*. Taylor and Francis, London.

Nevett, J. (2020) The Greta effect? Meet the schoolgirl climate warriors. BBC News, 7 July. Available at: www.dhushara.com/Biocrisis/19/5/climate%20warrriors.Reduce%20to%20300%20dpi%20average%20quality%20-%20STANDARD%20COMPRESSION.pdf (accessed 3 February 2021).

Nixon, L. (2015) Ecofeminist appropriations of indigenous feminisms and environmental violence. Available at: thefeministwire.com (accessed 7 August 2020).

O'Hara, S. (1995) From production to sustainability: considering the whole household. *Journal of Consumer Policy* 18(2), 111–134. DOI: 10.1007/BF01016508

O'Hara, S. (1997) Toward a sustaining production theory. *Ecological Economics* 20(2), 141–154. DOI: 10.1016/S0921-8009(96)00024-9

Ostrom, E. (2014) A polycentric approach for coping with climate change. *Annals of Economics and Finance* 15(1), 97–134.

Panthi, J., Aryal, S., Dahal, P., Bhandari, P., Krakauer, N.Y. and Pandey, V.P. (2016) Livelihood vulnerability approach to assessing climate change impacts on mixed agro-livestock smallholders around the Gandaki River Basin in Nepal. *Regional Environmental Change* 16, 1121–1132.

Parker, L. (2020) Greta Thunberg wasn't the first to demand climate action. Meet more young activists. *National Geographic*, 25 March.

Pearse, R. (2017) Gender and climate change. *WIREs Climate Change* 8(2), doi-org.ezproxy.library.yorku.ca/10.1002/wcc.451.

Perez, C., Jones, E.M., Kristjanson, P., Cramer, L., Thornton, P.K., Förch, W. and Barahona, D. (2015) How resilient are farming households and communities to a changing climate in Africa? A gender-based perspective. *Global Environmental Change* 34, 95–107.

Perkins, P.E. (2013) Environmental activism and gender. In: Figart, D.M. and Warnecke, T.L. (eds) *Handbook of Research on Gender and Economic Life*. Edward Elgar, Cheltenham and Northampton, UK, pp. 504–521.

Perkins, P.E. (2017a) Gender justice and climate justice, building women's economic and political agency through global partnerships. In: Buckingham, S. and Le Masson, V. (eds) *Understanding Climate Change Through Gender Relations*. Routledge, Oxon and New York, pp. 45–63.

Perkins, P.E. (2017b) Canadian Indigenous female leadership and political agency on climate change. In: Cohen, M.G. (ed.) *Climate Change and Gender in Rich Countries: Work, Public Policy and Action*. Routledge, London and New York, pp. 283–296.

Perkins, P.E. (2018) Gender, Climate Justice, Indigenous Leadership and Political Agency in Canada. In: Cohen, M.G. (ed.) *Climate Change and Gender in Rich Countries: Work, Public Policy and Action*. Routledge, London and New York, pp. 282–296.

Perkins, P.E. (2019) Climate justice, gender, and intersectionality. In: Jafry, T. (ed.) *Routledge Handbook of Climate Justice*. Routledge, London and New York, pp. 349–358.

Perkins, P.E. and P.F. Walker (2015) International partnerships of women for sustainable watershed governance in times of climate change. In: Buechler, S. and Hanson, A.-M. (eds) *A Political Ecology of Women, Water and Global Environmental Change*. Routledge, London and New York, pp. 118–140.

Priola, V. and Pecis, L. (2020) Missing voices: the absence of women from Italy's Covid-19 pandemic response. *Gender in Management*. doi-org.ezproxy.library.yorku.ca/10.1108/GM-07-2020-0218.

Quandt, A. (2018) Measuring livelihood resilience: The Household Livelihood Resilience Approach (HLRA). *World Development* 107, July, 253–263.

Ravera, F., Martín-López, B., Pascual, U. and Drucker, A. (2016) The diversity of gendered adaptation strategies to climate change of Indian farmers: a feminist intersectional approach. *Ambio* 45, 335–351.

Resurrección, B. (2013) Persistent women and environment linkages in climate change and sustainable development agendas. *Women's Studies International Forum* 40, 33–43.

Röhr, U., Sitzner, M., Stiefel, E. and Winterfeld, U. (2008) Gender justice as the basis for sustainable climate policies. NGO Forum on Environment and Development, Bonn, Germany.

Röhr, U., Hemmati, M. and Lambrou, Y. (2010) Towards gender equality in climate change policy: challenges and perspectives for the future. In: Enarson, E. and Chakrabarti, P.G.D. (eds) *Women, gender and disaster*. Sage Publications, New Delhi, pp. 289–304.

Rojas, A., Siles, J. and Roth, M. (2016) Promote gender equality to realize the benefits of low emission development. Low Emission Development Strategies Global Partnership/IUCN. Available at: www.ledsgp.org/wp-content/uploads/2016/06/CDKN_LEDS_CB_Gender_final_web-res-1.pdf (accessed 3 February 2021).

Roncoli, C., Orlove, B.S., Kabugo, M.R. and Waiswa, M.W. (2010) Cultural styles of participation in farmers' discussions of seasonal climate forecasts in Uganda. *Agriculture and Human Values* 28, 123–138.

Saave, A., Dengler, C., Just, D., McDonald, E.R., Kouromichaki, E, Dannanberg, J., Temper, L., Hansen, L., Barbieri, L., Avlona, N.-R., Rutt, R., Sanniti, S., Barca, S., Paulson, S. (2020) Collaborative Feminist Degrowth: Pandemic as an Opening for a Care-Full Radical Transformation. https://www.degrowth.info/en/feminisms-and-degrowth-alliance-fada/collective-research-notebook/ (accessed 5 July 2020).

Sangita, K. (2016) Transnational feminism and women's activism: Building resilience to climate change impact through women's empowerment in climate smart agriculture. *Asian Journal of Women's Studies* 22(4), pp. 497–506.

Sengupta, S. (2019) Becoming Greta: 'Invisible Girl' to global climate activist, with bumps along the way. *New York Times*, 18 February.

Shah, K.U., Dulal, H.B., Johnson, C. and Baptiste, A. (2013) Understanding livelihood vulnerability to climate change: applying the livelihood vulnerability index in Trinidad and Tobago. *Geoforum* 47, 125–137.

Shandro, J., Jokinen, L., Stockwell, A. and Mazzei, F. (2017) Risks and impacts to First Nation Health and the Mount Polley Mine Tailings Dam Failure. *International Journal of Indigenous Health* 12(2), pp. 84–102.

Shenaz Hossain, C. (2018) *The Black Social Economy in the Americas: Exploring Diverse Community-Based Markets*. Springer, Dodrecht, The Netherlands.

Sinsibere (2013) Sinsibere – Combating desertification with women's sustainable livelihoods. Ministry for Foreign Affairs of Finland. Available at: www.um.fi/latest-funding-decisions/-/asset_publisher/SYmYYmTYh0sD/ahaKytInterventionType/id/1159610 (accessed 1 February 2021).

Solar Sister (2020) Our Impacts – Solar Sister entrepreneurs create a ripple effect. Available at: www.solarsister.org/what-we-do/our-impact/ (accessed 3 February 2021).

Stein, R. (2004) *New perspectives on environmental justice: Gender, sexuality, and activism*. Rutgers University Press, New Brunswick, NY.

Taft, J.K. (2020) Hopeful, harmless, and heroic: figuring the girl activist as global savior. *Girlhood Studies* 13(2), 1–17.

Tanner, T., Lewis, D., Wrathall, D., Bronen, R., Cradock-Henry, N., Huq, S., Lawless, C., Nawrotzki, R., Prasad, V., Rahman, M.A., Alaniz, R., King, K., McNamara, K., Nadiruzzaman, M., Henly-Shepard, S. and Thomalla, F. (2015) Livelihood resilience in the face of climate change. *Nature Climate Change* 5, 23–25.

Thomas-Muller, C. (2014) The rise of the native rights-based strategic framework. In: D'Arcy, S., Black, T., Weis, T. and Russell, J.K. (eds) *A Line in the Tar Sands: Struggles for Environmental Justice*. Between the Lines, Toronto, pp. 240–252.

Thoyre, A. (2020) Home climate change mitigation practices as gendered labor. *Women's Studies International Forum* 78:102314. doi.org/10.1016/j.wsif.2019.102314.

Tormos-Aponte, F. and García-López, G.A. (2018) Polycentric struggles: the experience of the global climate justice movement. *Environmental Policy and Governance* 28(4).

UN Women (2020) Advice from young activists: How COVID-19 is changing climate activism for young women. 21 April. Available at: www.unwomen.org/en/news/stories/2020/4/compilation-covid-19-and-climate-activism (accessed 3 February 2021).

UNFCCC (2014) Women Farmers in Itzapa, and AIRES: Engaged in Agro-Forestry for the Earth I Guatemala. United Nations Framework Convention on Climate Change. Available at: www.unfccc.int/secretariat/momentum_for_change/ items/7843.php (accessed 1 February 2021).

UN/ISDR (2008) Gender Perspectives: Integrating Disaster Risk Reduction into Climate Change Adaptation. United Nations Office for Disaster Risk Reduction. Available at: www.sdg.iisd.org/news/unisdr-publication-gender-perspectives-integrating-disaster-risk-reduction-into-climate-change-adaptation-good-practices-and-lessons-learned/ (accessed 3 February 2021).

Van den Berg, R., Sorkin, L.N., Molenaar, A. and Tuts, R. (2020) Building climate-resilient and equitable cities during COVID-19. World Resources Institute Commentary. Available at: www.wri.org/news/building-climate-resilient-and-equitable-cities-during-covid-19#looking-ahead-post-covid (accessed 3 February 2021).

Vanner, C. and Dugal, A. (2020) Personal, powerful, political: Activist networks by, for, and with girls and young women. *Girlhood Studies* 13(2), i–xv.

Vinyeta, K., Whyte, K.P. and Lynn, K. (2016) Climate change through an intersectional lens: gendered vulnerability and resilience in Indigenous communities in the United States. General Technical Report PNWGTR-923, United States Department of Agriculture, U.S. Forest Service, Pacific Northwest Research Station. Available at: www.researchgate.net/publication/291523872_Climate_change_through_an_intersectional_lens_Gendered_vulnerability_and_resilience_in_indigenous_communities_in_the_United_States (accessed 3 February 2021).

Weiss, C. (2012) Women and environmental justice: a literature review. Women's Health in the North (WHIN), Australia. Available at: www.semanticscholar.org/paper/Women-and-Environmental-Justice-%3A-a-literature-Weiss/09ef8eee40e4b6f393ef47d0951e838f423717fa (accessed 1 February 2021).

Wenden, A.L. (2011) Women and climate change: Vulnerabilities and challenges. In: Weissbecker, I. (ed.) *Climate Change and Human Well-Being: Global Challenges and Opportunities*. Springer, New York, pp. 119–133.

Wenham, C., Smith, J., and Morgan, R. (2020) COVID-19: the gendered impacts of the outbreak. *The Lancet* 395(10227), 846–848.

Whyte, K.P. (2014) Indigenous women, climate change impacts, and collective action. *Hypatia* 29(3), 599–616. doi-org.ezproxy.library.yorku.ca/10.1111/hypa.12089.

Whyte, K.P. (2018) Settler colonialism, ecology, and environmental injustice. *Environment and Society* 9, 125–144.

Wickramasinghe, A. (2015) Energy for rural women: beyond energy access. In: Guruswamy, L. (ed.) *International Energy and Poverty: the emerging contours*. Routledge, London and New York, pp. 231–244.

16 Gender Mainstreaming in Climate Change Adaptation Strategies in Bangladesh and Nepal

S. Shehwar

Introduction

There is a large body of evidence suggesting that women are more vulnerable to the detrimental effects of climate change than men (Elford, 2013; Olson, 2014; Shabib and Khan, 2014; Sugden *et al.*, 2014; Sustainable Development Institute, 2017). Women constitute the majority of the world's poor and are more dependent for their livelihood on natural resources that are threatened by climate change. Moreover, women face social, economic, and political barriers, as well as unequal access to resources and decision making processes that limit their coping capacity. However, there has been growing awareness that women should not be seen just as victims of climate change but also be considered powerful agents of change, and any responses to the effects of climate change should be gender-sensitive.[1]

This chapter discusses the state of gender mainstreaming in climate action activities and policies in two South Asian neighbors, Bangladesh and Nepal, based on a review of key climate change policy documents. UN Women defines gender mainstreaming as '... a strategy for promoting gender equality... [which]... involves ensuring that gender perspectives and attention to the goal of gender equality are central to all activities' (UN Women, 2001). Gender mainstreaming is an important component of countries' national adaptation plans (NAP) and actions because it ensures that both women's and men's concerns are integral to the design, implementation, monitoring, and evaluation of the plan. The result can increase the effectiveness of NAPs and reduce gender inequalities. By partaking in gender mainstreaming, countries can also reveal gaps in existing strategies and frameworks and help foster organizational environments that are more gender-equal. Gender mainstreaming is especially important to addressing challenges for impoverished women in developing countries. According to a Food and Agriculture Organization (FAO) report, women comprise just over 40% of the agricultural labor force in the developing world. In Bangladesh this share exceeds 50%, while in Nepal female labor accounts for about 65% of all agricultural activities (Raney *et al.*, 2011).

The chapter addresses the following three questions:

1. How do Bangladesh and Nepal mitigate the detrimental effects of climate change for rural women?
2. How do climate policies and programs in Bangladesh and Nepal respond to the different needs and concerns of these women within their national adaptation strategies?
3. What are the opportunities and challenges of mainstreaming gender in climate action policies and programs in Bangladesh and Nepal?

The following sections of this chapter present the current state of climate change adaptation

 Gender, Climate Change and Livelihoods: Vulnerabilities and Adaptations (eds J. Eastin and K. Dupuy)
DOI: 10.1079/9781789247053.0016

policies and programs in the case countries. The achievements, opportunities, and challenges associated with gender mainstreaming are then discussed, and the final section concludes. A key argument of this chapter is that climate change is not gender-neutral, and it has become increasingly necessary for Bangladesh and Nepal to learn from one another in order to build gender-sensitive strategies that are cognizant of the needs of rural women.

Climate Change and Gender in Bangladesh and Nepal

Bangladesh and Nepal are among the poorest and most climate-vulnerable countries in the world. Indeed, Nepal ranked fourth and Bangladesh ranked ninth among the ten countries most affected by extreme weather events between 1998 and 2017, according to the 2019 Global Climate Rate Index compiled by the German environmental think tank, GermanWatch (Eckstein *et al.*, 2019). In Bangladesh, the majority of the densely populated urban centers are susceptible to increased flooding related to storm surges and sea level rise (USAID, 2018). The country is also exposed to tropical cyclones, which displace large populations of Bengalis in coastal areas. In Nepal, floods and landslides due to heavy rains are the most common hazards, while droughts are becoming increasingly frequent (USAID, 2017). In general, both countries are affected by and are especially vulnerable to changes in precipitation that can adversely affect agriculture. Both countries are also adversely affected by the Asian monsoon, as cyclical flooding destroys infrastructure and leads to regular losses of lives and livelihoods. Furthermore, agriculture is the main source of income in both Bangladesh and Nepal, and as a sector faces dire risks associated with climate change. According to the United Nations (UN), the gross domestic product (GDP) per capita of Bangladesh in 2019 was about US$752.60, with agriculture accounting for 36.4% of the country's total employment and providing for 14.2% of the gross value added (GVA) of the economy (UN, undated_a). In the same year, Nepal's GDP per capita was US$724.5, with agriculture accounting for 29.4% of the country's GVA and providing employment to almost 56% of the country's total employed population (UN, undated_b).

Climate change policies and gender mainstreaming in Bangladesh

Bangladesh launched its National Adaptation Program of Action (NAPA) in 2005 as per the requirements of the United Nations Framework Convention on Climate Change (UNFCCC, 2014). The NAPA identified floods, salinity intrusions, and droughts that drastically affect crop productivity every year as the most damaging climate change impacts. Waterlogging and drainage congestion, early and untimely floods, localized inundation, and flash flooding were identified as key climate-related issues. Among the 15 strategies identified in the NAPA, at least six were directed at the adaptation of agriculture and fisheries to floods and salinity intrusions. Although the NAPA pointed to gender-differentiated impacts of climate change, these were not specifically identified (Bangladesh Ministry of Environment and Forests, 2005, p. 34).

Bangladesh released its Climate Change Strategy and Action Plan (BCCSAP) in 2009 with a comprehensive strategy for addressing the climate change challenges in six thematic areas that included food security, social protection, and health. BCCSAP outlined a 10-year (2009–2018) climate change action plan with 44 programs. Some of these programs concerned the adaptation of agriculture, fisheries, and livestock sectors to meet changing conditions, especially an increased likelihood of flooding, which was identified as a special climate change hazard (Bangladesh Ministry of Environment and Forests, 2009, p. 52). None of the programs, however, recognized gendered impacts of climate change or identified women as agents of change. In fact, women were only identified among vulnerable groups in one program out of the 44 total: 'Livelihood protection of vulnerable socio-economic groups.' This program included a comprehensive study of the impact of climate change on women and gender relations (Bangladesh Ministry of Environment and Forests, 2009, p. 61). Although the BCCSAP also included an exclusive program that recognized women's major roles in rural communities and highlighted the significance of women's special roles in the management of natural disasters, it did not identify what these roles were or how they should be executed (Bangladesh Ministry of Environment and Forests, 2009, p. 52).

By excluding women, the plan falls short on its foundational pillars of capacity building and institutional strengthening, as it does not fully prioritize the needs of poor and vulnerable women. Some of the other programs would also benefit from an added dimension focusing on gender. For instance, the 'Adaptation in health sector' program, which is also included in the plan, has as objectives the research and monitoring of climate change impacts on disease pattern and the social and economic costs of developing adaptative measures to disease (Bangladesh Ministry of Environment and Forests, 2009, p. 5). However, climate change impacts on health – including increased exposure to heat, poor air quality, extreme weather events, and decreased food security – affect women and men differently, depending on geographic factors and social determinants of health. Therefore, without including a gender dimension to this program, adaptative strategy development may have a diminished impact on understanding disease patterns and on protecting the health of vulnerable women.

Bangladesh integrates its economic development and poverty reduction goals with its environment and climate change actions in the country's 5-year plans. In its sixth 5-year plan (2011–2015; 6FYP), the government undertook several strategies and policy initiatives to improve the economic, political, and social inclusion and empowerment of women, taking the country's Constitution as the basis and in conformity with the government's commitments to various international conventions for gender equality. The 6FYP also undertook to reform laws, rules and regulations, institutional mechanisms, policies, projects, and programs to make them gender-sensitive (Bangladesh Planning Commission, 2011, p. 441).

In the current seventh 5-year plan (2016–2020; 7FYP), Bangladesh deepens its recognition of the gendered impacts of climate change by recognizing that patriarchal norms and practices put women and girls in a vulnerable state, which intensifies during climate-induced disasters (Bangladesh Planning Commission, 2011, p. 472). For instance, the plan states that following natural disasters, women more than men are subject to excruciating circumstances. According to a recent Oxfam–Monash Partnership project, declining yields from agriculture and fisheries have driven a significant number of men to out-migrate for employment, leaving women to cope with and adapt to the changes in their communities (Alston *et al.*, 2014). This has affected family relationships with increased domestic violence against women, on top of the women's disproportionately higher suffering from shortages of food, drinking water, and fuel. With increasing financial hardships, some families are pulling their daughters out of school in favor of early marriage. These findings are consistent with those of Alston *et al.* (2014), who established a link between children and forced marriage, dowry, and climate changes in Bangladesh.

With the recognition of such gendered impacts, the 7FYP goes into a more proactive mode than previous plans to integrate gender concerns in the designs of adaptation and mitigation projects. The 7FYP stresses that while national capacity and expertise to address environment, climate change, and disaster management concerns have increased, there is still the challenge of effectively implementing gender mainstreaming policies. The absence of gender mainstreaming in Bangladesh is problematic because vulnerability to climate change continues to have concrete gender dimensions. Women face greater difficulties than men before and after natural disasters, demonstrating gendered differences in adaptive capacity. Women also face increased domestic violence as a result of the impacts of climate change in their communities and have long had limited control over capital. The implication of not focusing on gender issues as part of climate change adaptation perpetuates harmful patriarchal societal norms that constrain women's participation in community decision making on climate change adaptation and leaves women in especially vulnerable conditions during natural disaster events.

Climate change policies and gender mainstreaming in Nepal

Nepal completed its NAPA in 2010, identifying agriculture, freshwater supply, disaster risk management, forestry, human health, and urban areas as key sectors of focus. It gave highest priority to agriculture and food security as the main vulnerable sectors affected by climate change, due to changes in precipitation, rising temperatures, flooding, and erratic monsoon rainfall (Government of Nepal, 2010, p. 48).

The NAPA emphasized stakeholder involvement in the design and implementation of priority adaptation activities and highlighted the importance of local ownership and capacity building; social and cultural acceptance, and inclusiveness of gender; and Indigenous and Dalit communities (Government of Nepal, 2010, p. 30). The NAPA also encouraged the Government of Nepal to commission a Local Adaptation Program of Action (LAPA) – a community-centric planning approach. This encouragement helped start a wider discussion about the government's role in helping local communities adapt to climate change (Government of Nepal, 2010, p. 36; Lamsal *et al.*, 2014, p. 1).

The NAPA proposed an implementation framework with 80% of available climate funds to be used for field level climate change activities (Government of Nepal, 2010, p. 37). The NAPA also conducted a gender sensitivity analysis of the impacts of climate change and examined gender differentiated climate change effects (Government of Nepal, 2010, pp. 30, 77, Annex 5). The gender sensitivity analysis recognized women as custodians of local knowledge, agricultural skills, and agricultural practices (e.g. seed preservation). It was noted that climate change disasters (e.g. floods) often lead to male outmigration imposing additional workloads on women. Women left behind experience increased vulnerability to climate change impacts due to their limited access to credit, market, land, and agricultural extension services.

Nepal released its National Climate Change Policy in 2011, which recognized the importance of local communities in developing and implementing local adaptation plans and formalized the provision of climate action funds for implementation of climate change activities through LAPA (Government of Nepal, 2010, p. 4). The policy also sought to ensure the participation of poor people, Dalits, marginalized Indigenous communities, and women in the implementation of climate adaptation and climate change related programs. Since this policy has been released, the Government of Nepal has initiated several activities to implement mitigation and adaptation measures to combat climate change problems. One such measure is the Reducing Emissions from Deforestation and Degradation (REDD) of forests initiative. REDD includes various programs and activities that mobilize local institutions. The REDD Preparedness Plan (RPP) involves consultations with multiple stakeholders at the community level and is informed by components such as 'ensuring a participatory and inclusive process; mainstreaming gender and equity concerns at all levels; multi-stakeholder collaboration; and integration with safeguard measures, (Ojha, 2011, p. 8).

Achievements, Challenges and Opportunities

Bangladesh and Nepal have developed climate change policies aimed at helping adaptation of the two countries' most vulnerable populations that depend on agriculture for their main livelihoods. Both countries have constitutional and legal policies and frameworks, and both are signatories to international conventions to ensure gender equality in livelihoods. The two countries' key climate change policies generally consider the gendered impacts of climate change, although, as this chapter suggests, there remain numerous opportunities in the plans of both to better ensure that gender considerations lead to meaningful results in the future.

While policies, strategies, and frameworks have been put in place, climate change policies in both countries, but especially in Nepal, consider women as vulnerable victims of climate change rather than active agents and partners of adaptation and mitigation strategies. This view is a limitation as women should be incorporated into decision making roles for climate adaptation to enable gender equality. The reality in many countries is that women are underrepresented in decision making in areas relevant to climate change adaptation. This lack of representation limits women's interests from being understood and acted upon in a consistent and accurate way. Mainlay and Tan (2012) interviewed several experts, women's rights advocates, non-governmental organizations (NGO), and gender consultants, noting: 'Nepal's NAPA acknowledges gender-specific vulnerabilities and women are recognized as a vulnerable group, but it does not provide any targets for women's involvement or capacity building, nor does it contain any gender-specific projects.' Nepal's 2011 Climate Change Policy translated the frameworks and priorities identified in the NAPA into a policy, but without any indication of how to modify these situations (Nepal Ministry of Environment, 2011).

The NAPA of Bangladesh also considered women as victims of climate change, rather than as agents of change. The BCCSAP, however, included one exclusive program, 'Strengthening gender consideration in climate change management,' that was aimed at building the capacity of gender focal points in all ministries and agencies to incorporate gender issues in their climate response activities. A key achievement in Bangladesh is that the country has integrated its climate change actions and activities into its national 5-year plans. This integration allows government planners to track the achievements made and to revise actions in future plans accordingly. Thus, the 6FYP of Bangladesh undertook to implement the gender consideration program. The sixth plan included a strategy for enhancing the livelihoods of poor, rural women by focusing on ensuring improved access to irrigated water, fertilizer, electricity, rural roads, and institutional finance as part of an overall objective to reduce income inequality owing to unequal distribution of physical assets and access to means of production (Bangladesh Ministry of Planning, 2013, p. 6). A pilot review of this program in 2014 found that it achieved major targets related to economic growth, employment, and poverty reduction. However, while poverty declined at an 'appreciable rate,' more focus was needed on the implementation of gender-related laws to eliminate gender-based violence and socio-economic discrimination, including eliminating wage discrimination against female workers and the encouragement of women to increase labor force participation (Bangladesh Planning Commission, 2015, p. 19).

The current 7FYP has further advanced gender implementation programming by integrating gender sensitivities in project designs. In addition, the 7FYP has undertaken projects to promote women's participation at the implementation, design, and monitoring levels, and through decision making processes. In the country's Social Afforestation Programs, for example, 305 of the 109,000 rural beneficiaries are women. Among 38 Integrated Protected Areas Co-management projects, 50% of the beneficiaries are poor women who are involved in participatory biodiversity preservation activities. In addition, 30% of women's representation has been ensured in local project planning committees (Office of the United Nations High Commissioner for Human Rights, undated). Most of the chapters in the seventh plan include a focus on gender issues. The governance and human rights 5-year plan presented in the 7FYP also presents a list of essential policy items to be undertaken over a 5-year period (2015–2020), which can be described as high priority, essential, primary, and long term. These items involve consulting with multiple stakeholders, including public officials, civil society personalities, academics, professionals, and international development partners. Nepal can learn from Bangladesh to integrate climate change actions into its 5-year plans and include gender sensitivities into all project designs, as this has led to improved integration of women's needs in climate adaptation planning.

Nepal's climate change action agenda appears to be a good blend of bottom-up and top-down approaches, a prerequisite for ensuring not only sufficient ownership and engagement, but also the flow of resources from the center to the communities who are at the front lines of climate change. The alignment of LAPA and NAPA in the country's climate change policy articulates a vertical linkage between the national scale, top-down assessments of current and future climate risks, with bottom-up planning of adaptation needs, options, and priorities assessed by community members and informed by local knowledge and geographical specificity. This approach aligned with the country's Local Self-Governance Act (LSGA) enacted in 1999, which provides for women's participation in local decision making by law, gives a great opportunity for mainstreaming gender issues into activities to address climate change. This is where Bangladesh can learn from Nepal's experience. In Nepal, the LSGA legislates women's participation in local governments, including their representation on Village Development Committees (VDC) and councils. The LSGA requires VDCs to give priorities to plans and programs that benefit women and develop income generating activities for women (Mainlay and Tan, 2012, p. 8). Additionally, district and village committees are charged with advancing women's causes and the 'elimination of social ill practices against women and girls' (Mainlay and Tan, 2012, p. 6).

Bangladesh claims that it has pioneered community-based approaches to reducing vulnerability to climate change (Bangladesh Ministry of Environment and Forests, 2009, p. 41).[2] However, micro-level individual initiatives in Bangladesh fall short of Nepal's macro-level national and local adaptation planning linkages established by the NAPA–LAPA approach, with 80% of the climate action budgets allocated to LAPA. In its 7FYP, Bangladesh appears to be contemplating adoption of a community-based planning and adaptation approach along Nepal's NAPA–LAPA line (Bangladesh Planning Commission, 2015, p. 476). This would be in the country's best interest, with Nepal's LSGA emerging as a best practice for mainstreaming and for codifying women's issues into law.

The socio-political structures of Bangladesh and Nepal pose challenges to the implementation of effective climate change adaptation programs. Bangladesh has a political system that is democratic but marked by significant political intolerance and instabilities. Nepal's political condition is no better: politics of the country, that until 2008 was a constitutional monarchy, has led Nepal to function as a republic within a multiparty system. None the less, considerable political turmoil and frequent changes in government hamper the country's economic development.

Significant political instabilities in both Bangladesh and Nepal constrain the development of sound, accountable, and transparent democratic institutions and governance systems that can support sustained economic growth and enable implementation of effective climate action plans. Weak, non-transparent, and unaccountable institutions can contribute to corrupt governance systems. The Corruption Perceptions Index for 2017 compiled by Transparency International (2018) puts Nepal at 122 and Bangladesh at 143 among 180 countries ranked from the lowest to the highest level of corruption perception. Corruption limits the political capacity of these countries to work towards sustainable changes. Bangladesh's 7FYP recognizes that weak, inefficient, and unaccountable government institutions, especially at local levels, are serious hindrances towards effective climate change adaptation and mitigation efforts. Such institutional weaknesses are critical barriers towards implementation of decentralized and community-based approaches to climate change adaptations strategies (see Bangladesh Planning Commission, 2015, p. 473). In the Nepal context, Bahadur *et al.* (2017, p. 7) noted that many municipalities and VDCs lack technical capacity to plan and implement climate change adaptation actions.

However, it is important to note that these observations are not meant to imply that the countries' plans are without merit or potential, but rather that their socio-political structures cause constraints that must also be sufficiently addressed. One of the ways that the challenge of weak and non-transparent institutions and governance are addressed is with the growing presence and influence of numerous NGOs in the implementation of climate change adaptation programs, both in Bangladesh and Nepal. NGOs are primarily engaged by foreign donor agencies and they are recognized to play important roles in broad development efforts of the two countries. The main challenge associated with the role and presence of NGOs is that with weak and non-transparent government institutions, coordination, monitoring, and evaluation of NGO activities can be extremely cumbersome. Moreover, it involves a trade-off against government capacity building and ownership.

In both Bangladesh and Nepal, NGOs play a significant role in developing and implementing climate adaptation. In Bangladesh, NGOs are generally credited for effective implementation of many gender-responsive climate adaptation programs and for imparting important lessons in the country's climate change adaptation. As Ahmad (2012, p. 28) notes, the country gained significant learning for its climate change adaptation agenda from three pilot programs supported by donors and implemented by NGOs: the Comprehensive Disaster Management Program (CDMP) Phase I and II; Reducing Vulnerability through Climate Change (RVCC); and the Char Livelihoods Program (CLP). These pilots aimed to reduce community vulnerability to natural hazards. Furthermore, in its 7FYP, Bangladesh deepened its partnerships with NGOs to create wider awareness on environmental issues, stating that the government would facilitate effective participation of NGOs in environmental decision making and community mobilization (Bangladesh Planning Commission, 2015, p. 499).

Bangladesh's civil society sector has been a key factor in mainstreaming women's issues through programs that focus on enhancing women's social and economic status. Bangladesh is reported to have a long tradition of women's organizations, including Mahila Parishad, which claims to be the largest women's organization in Bangladesh with a membership of 130,000 individuals. Bangladesh has around 23,000 registered NGOs, some of which began as organizations for relief distribution in newly independent Bangladesh, before shifting attention towards gender-based violence and women's political participation (UN Refugee Agency, 2007, p. 4). With NGO assistance, Bangladesh has made commendable progress in areas such as women's economic empowerment, partially credited to the success of microcredit schemes, as well as a surge in women's labor force participation, most notably in the garments industry. Some notable organizations that have advanced women's causes in Bangladesh include Naripokko, Steps Toward Development, and the Grameen Bank.

None the less, the Government of Bangladesh has been criticized for poor implementation of its NAPA and many civil society actors have been vocal in protesting the 'low priority given to women's issues at the national level' and in the frustration at not being consulted regarding changes in policy formulations (Sørvald, 2005, p. 9). This outcome reflects a need for more coordination between the country's political and civil society, but also reveals a tension between the growth of women's rights organizations and a male-dominated political system. Women's participation in politics remains largely marginal despite state interventions to enhance it.

In Nepal, NGOs play significant roles in the implementation of the climate adaptation agenda. The country's Strategic Program for Climate Resilience (SPCR) is supported by climate investment funds managed by multilateral development banks (MDB), who engage NGOs to implement the program. This support gives the MDBs a fast turnaround for a results-driven approach (Ayers *et al.*, 2011). In addition, gender is being mainstreamed in various ministries, district- and local-level government departments, and public sector projects. For example, the Armed Police Force contains around 2000 women; more than 200 police posts with gender-friendly facilities have been constructed since the beginning of the peace process (Baniya *et al.*, 2017, p. 11).

In comparison to Bangladesh, Nepal's civil society actors have not achieved a similar foothold towards advancing women's issues. This failure can be attributed in part to civil strife and political instability, though the post-war era in Nepal has indicated some gains for women in terms of increased political participation and representation. Nepal's return to multiparty democracy in 1990 has introduced a requirement for equal pay for women and men, and has led to the establishment of civil society groups to speak for women's rights (Baniya *et al.*, 2017, p. 8). Despite these gains, as in Bangladesh, Nepal's male-dominated political system monopolizes policy negotiations; women's issues need to be better integrated. In addition, civil society organizations are largely dominated by 'high caste' women, and so they do not bring due attention to the needs and priorities of women from marginalized communities.

Dalits and other marginalized women's groups are working to gain traction in Nepal's civil society landscape in spite of intersectional barriers. Some notable organizations in Nepal include the Rural Women's Network (RUWON) Nepal, Women Act, and the Feminist Dalit Organization (FEDO). These organizations work to help grassroots women, mainly through training initiatives and advocacy work. Generally, NGOs in Nepal are not as well-established as those in Bangladesh, but on a smaller scale, they have been able to create improvements to rural women's livelihoods. For instance, RUWON is fostering a microfinance program in Nepal for the communities of the rural district of Sindhuli (RUWON Nepal, undated).

Concluding Remarks

Bangladesh and Nepal have put in place policies, frameworks, and programs to mitigate the detrimental effects of climate change for rural women, who are especially dependent on agriculture.

The two countries' policies and programs have made attempts to address gendered impacts of climate change. The discussion in this chapter indicates that the two countries have made some progress in mainstreaming gender in their climate policies and programs. It is evident that the two countries can learn from each other's policy tools and strategies for effective gender mainstreaming.

Although Bangladesh appears to have made better progress in implementing programs with gender mainstreaming, Nepal's policy tools, legal framework, NAPA–LAPA structure, in conjunction with the country's LSGA, renders greater opportunity for mainstreaming gender issues into activities to address climate change. Success, however, requires targeted programs which enable women's participation in local, regional, and national political institutions. While underdeveloped and unaccountable institutions and governance structures pose significant barriers, Bangladesh should in the long run consider a structure more similar to that of Nepal.

Notes

[1] See the moderator's summary of the expert panel on 'Gender Perspectives on Climate Change' at the 52nd session of the UN Commission on the Status of Women, 25 February to 7 March 2008. The summary is available at: www.un.org/womenwatch/daw/csw/csw52/panels/summaries/Summary%20Emerging%20 Climate%20Change%206%20MAR.pdf (accessed 10 February 2020).

[2] As an example, the BCCSAP refers to a community-drawn plan, with the support of a local NGO, funded by the UK through its Comprehensive Disaster Management Program, to save a village from recurrent floods.

References

Ahmad, N. (2012) Gender and Climate Change in Bangladesh: The Role of Institutions in Reducing Gender Gaps in Adaptation Program. World Bank, Washington, DC.

Alston, M., Whittenbury, K., Haynes, A. and Godden, N. (2014) Are climate challenges reinforcing child and forced marriage and dowry as adaptation strategies in the context of Bangladesh? *Women's Studies International Forum* 47, 137–144.

Ayers, J., Kaur, N. and Anderson, S. (2011) Negotiating climate resilience in Nepal. *IDS Bulletin* 42, 70–79.

Bahadur, T., Air, A., Uprety, B.K. and Midha, N. (2017) Nepal's approach to climate change adaptation with Local Adaptation Plans for Action (LAPAs): a water resource perspective. Global Water Partnership, Stockholm.

Bangladesh Ministry of Environment and Forests (2005) National Adaptation Programme of Action (NAPA). Government of the People's Republic of Bangladesh, Dhaka.

Bangladesh Ministry of Environment and Forests (2009) Bangladesh Climate Change Strategy and Action Plan. Government of the People's Republic of Bangladesh, Dhaka.

Bangladesh Ministry of Planning (2013) Sixth Five Year Plan: FY2011–FY2015. International Monetary Fund, Washington, DC.

Bangladesh Planning Commission (2011) Sixth Five Year Plan FY2011–FY2015, Part 2. Government of the People's Republic of Bangladesh, Dhaka.

Bangladesh Planning Commission (2015) Seventh Five Year Plan FY2016–FY2020. Government of the People's Republic of Bangladesh, Dhaka.

Baniya, J., Kharel, S., Thapa, D. and Ramsbotham, A. (2017) Gender and Nepal's Transition from War. Conciliation Resources, London.

Eckstein, D., Hutfils, M.-L. and Winges, M. (2019) Global Climate Risk Index 2019: who suffers from extreme weather events? Weather-related loss events in 2017 and 1998–2017. GermanWatch, Bonn, Germany.

Elford, L. (2013) Mid-Partnership Report 2010-2013. Oxfam-Monash Partnership, Clayton, VIC, Australia.

Government of Nepal (2010) National Adaptation Programme of Action (NAPA). Government of Nepal, Kathmandu.

Lamsal, K., Chaudhary, P. and Bhandari, K.H. (2014) Climate Change Policy in Nepal: Challenges, Opportunities, and Imperative. Local Initiatives for Biodiversity, Research and Development (LI-BIRD), Pokhara, Nepal.

Mainlay, J. and Tan, S. (2012) Mainstreaming Gender and Climate Change in Nepal. International Institute for Environment and Development (IIED), London.

Nepal Ministry of Environment (2011) Climate Change Policy 2011. Government of Nepal, Kathmandu.

Office of the United Nations High Commissioner for Human Rights (undated) Bangladesh. Available at: www.docstore.ohchr.org/SelfServices/FilesHandler.ashx?enc=6QkG1d%2FPPRiCAqhKb7yhsmPYo5NfAsNvhO7uZb6iXOQMUolEzAt%2BNbvA6Ys6yBTKrvLAGl9QbV2i57%2FT5rxnpnUkQ6hdbeDihWrUJ9%2BTwpA7XynabqKH3miyqqIqCtXrLgUkm6aa3M9w7jOd7zgjDw%3D%3D (accessed 3 February 2021).

Ojha, S.K. (2011) Nepal's climate change policies and plans: local communities' perspective. HELVETAS Swiss Intercooperation Nepal, Lalitpur, Nepal.

Olson, J. (2014) Whose voices matter? Gender inequality in the United Nations Framework Convention on Climate Change. *Agenda* (Durban, South Africa) 28, 184–187.

Raney, T., Anriquez, G., Croppenstedt, A., Gerosa, S., Lowder, S. *et al.* (2011) The Role of Women in Agriculture. Food and Agriculture Organization of the United Nations, Rome.

RUWON Nepal (undated) Microfinance. Available at: www.ruwonnepal.org.np/projects/empowerment/microfinance/ (accessed 3 February 2021).

Shabib, D. and Khan, S. (2014) Gender-sensitive adaptation policy-making in Bangladesh: status and ways forward for improved mainstreaming. *Climate and Development* 6, 329–335.

Sørvald, M. (2005) Evaluation of the 'Strategy for Women and Gender Equality in Development Cooperation (1997–2005)'. Norwegian Institute for Urban and Regional Research (NIBR) and the Norwegian Agency for Development Cooperation (NORAD), Dhaka.

Sugden, F., Maskey, N., Clement, F., Ramesh, V., Philip, A. and Rai, A. (2014) Agrarian stress and climate change in the eastern Gangetic plains: gendered vulnerability in a stratified social formation. *Global Environmental Change* 29, 258–269.

Sustainable Development Institute (2017) Gendered impacts of climate variability, particularly in relation to slow-onset change and climate events in Bangladesh. Oxfam-Monash Partnership, Monash University, Melbourne, Australia.

Transparency International (2018) Corruption Perceptions Index 2017. Transparency International e.V, Berlin.

UN (undated_a) Bangladesh, UNData. United Nations, New York. Available at: www.data.un.org/en/iso/bd.html (accessed 3 February 2021).

UN (undated_b) Nepal, UNData. United Nations, New York. Available at: www.data.un.org/en/iso/np.html (accessed 3 February 2021).

UN Refugee Agency (2007) Bangladesh: women's rights organizations. United Nations Refugee Agency, Geneva, Switzerland.

UN Women (2001) Gender mainstreaming: strategy for improving gender equality. Available at: www.un.org/womenwatch/osagi/pdf/factsheet1.pdf (accessed 3 February 2021).

UNFCCC (2014) National Adaptation Programmes of Action (NAPAs). United Nations Framework Convention on Climate Change, United Nations, New York. Available at: www.unfccc.int/national_reports/napa/items/2719.php (accessed 3 February 2021).

USAID (2017) Climate Risk Profile, July: Nepal. United States Agency for International Development (USAID), Washington, DC. Available at: www.climatelinks.org/resources/climate-risk-profile-nepal (accessed 1 February 2021).

USAID (2018) Climate Risk Profile, March: Bangladesh. United States Agency for International Development (USAID), Washington, DC. Available at: www.climatelinks.org/resources/climate-risk-profile-bangladesh (accessed 3 February 2021).

17 Conclusion: Final Thoughts and Future Directions

K. Dupuy and J. Eastin

What Have We Learned? Where Do We Need More Research?

The contributions in this book highlight several key points about the relationship between gender, climate change, and livelihoods. First is perhaps the most obvious: climate change is a gender issue. Gender matters in recognizing and mitigating climate change vulnerabilities, in facilitating disaster recovery, and in supporting climate adaptation. A gender lens also enables us to better evaluate the implications of climate change for livelihoods because livelihoods themselves are gendered and, as Perkins and Osman note in Chapter 15 (this volume), because the justness and effectiveness of policy interventions are linked to their impacts on societies' most vulnerable members. The persistence of gender-specific vulnerabilities and barriers to adaptation make it difficult to understand, much less address, the risks of climate change without approaching the problem from a gender-based perspective. As gender disparities often result from a society's structural inequalities and inequities, addressing the ways in which gender discrimination interacts with vulnerability and adaptation capabilities is also vital to enhancing societal resilience. Necessary first steps in integrating gender into research on climate change include collecting gender disaggregated data, both qualitative and quantitative; evaluating how climate change differentially affects women and men (or girls and boys); and identifying the ways in which gender intersects with other identity categories and demographic variables such as race, ethnicity, class, age, and geography. In the absence of these actions, understanding will remain thin and public policies uninformed and potentially misguided.

The second key point illustrated in this book is that the gender implications of climate change are not universal; different social, cultural, and ecological contexts can require distinct diagnoses and prescriptions. This book presents research on a diverse array of countries – Bangladesh, Honduras, Indonesia, Kenya, Lao PDR, Lesotho, Morocco, Nepal, Tunisia, Uganda, and Australia – distributed across five continents and all with distinct cultures and ecologies. While southern Asia and well-researched countries in Africa such as Kenya have received a greater share of attention in the past, solid evidence from other, often neglected countries and regions is also required to increase understanding of the full extent of the climate threat. In line with the contributions in this book, such research should take the form of both cross-country and within-country comparisons, as culture and the effects of climate change vary without regard to national borders and in ways that we do not fully understand. Moreover, since climate change is an ongoing and progressively more dangerous phenomenon, longitudinal research is required

 Gender, Climate Change and Livelihoods: Vulnerabilities and Adaptations (eds J. Eastin and K. Dupuy)
DOI: 10.1079/9781789247053.0017

to understand evolving and dynamic relationships between gender, livelihoods, and climate change over time.

Third, climate change generates a broad range of direct and indirect outcomes for women and their livelihoods. Those examined in this book address issues ranging from food security (Mwesigye, Chapter 4), to outmigration (Alston, Chapter 13; Hummel, Chapter 12; Najjar *et al.*, Chapter 14), women's health and well-being (Onyango, Chapter 5; Rezwana, Chapter 3; Eastin, Chapter 8), family and community care work (Garner, Chapter 7; Onyango, Chapter 5), agricultural practices (Babugura, Chapter 9; Bryan *et al.*, Chapter 10), infrastructure (Braun, Chapter 6), and natural disaster impacts (Momtaz *et al.*, Chapter 2; Rezwana, Chapter 3), among others. Yet, more research is required to better understand how these and other outcomes intersect with livelihoods in specific ways. For example, physical and mental health, patterns of asset ownership, access to financial capital, and household relationships are all affected by climate change, and all have implications for women's ability to manage or adapt their livelihood strategies. Greater awareness of these linkages should enhance the ability of policy makers to target interventions more efficiently and effectively.

Fourth, for governments, a laissez-faire approach to dealing with the gender dimensions of climate change is not only insufficient, it is counterproductive. As Shehwar notes in Chapter 16: 'Gender mainstreaming is an important component of countries' national adaptation plans (NAPs) and actions because it ensures that both women's and men's concerns are integral to the design, implementation, monitoring, and evaluation of the plan.' We agree, and would add that a failure to incorporate such concerns, and to increase women's representation in climate policy decision making, will not only diminish the effectiveness of any adaptation or vulnerability reduction scheme, it will also reinforce existing discriminatory practices and further undermine gender equality.

Fifth, while governments must be active in their approach to mitigating climate vulnerabilities and fostering adaptations, the content, design, and implementation of public policy is important. Here, more work is required to fully appreciate the gendered impacts of the various types of policy interventions currently in use, as well as those under consideration. For example, as Bryan *et al.* argue in their chapter on climate-smart agriculture (Chapter 10), it is necessary to better understand the gender-specific constraints of adopting particular interventions over others, as well as how these constraints are influenced by cultural, political, economic, and ecological contexts.

Finally, women have agency and are using it, as grassroots organizers and activists, and as mothers, sisters, daughters, wives, and community members fighting for their lives and livelihoods. Several of the chapters in this book highlight the importance of women's agency in taking charge of their livelihoods, adapting to climate change, and in pushing for greater inclusion in climate governance and decision making. Rezwana (Chapter 3) calls women 'agents of survival,' and points out that: 'despite gendered socio-cultural and economic barriers, and persistent discrimination, women adopt distinct strategies to survive and carry on.' Unfortunately, women's capabilities and leadership are not always recognized, she argues, because '... they are often presented as victims in the literature and in media, and their survival and savings capacity remain unclear or underrated.' Similarly, Perkins and Osman (Chapter 15) note that 'women have shown and continue to show a clear sense of the capacities needed to adapt to climate change,' and that 'most women demonstrate diverse adaptation and coping strategies and mechanisms, including moving to safer places, saving their assets, adopting strict dietary adaptations, energy saving, adapting agricultural practices, earning income from alternative sources, alternative healthcare, organization, and collective action.' Women are also increasingly taking the initiative in organizing and mobilizing for policy changes that address gendered vulnerabilities and that enhance societal resilience. The reason, Perkins and Osman argue, is because 'women, and marginalized people in general, place a stronger priority on addressing climate change than other groups' primarily because they are more affected by it. In the future, more research is required to systematically identify women's coping strategies and the conditions under which they are successful in taking on leadership roles in their communities.

Gender, Livelihoods, Climate Change and COVID-19

While work on this book began before the COVID-19 pandemic, much of the research and writing took place during the crisis, as societies around the world cycled through various stages of lockdown, quarantine, reopening, and more lockdown. This situation provided an intimate view of the pandemic's impact on academic livelihoods, especially women's, as several of our authors were compelled to withdraw their contributions to attend to the needs of their families and households. While our initial author list is far from a representative sample of those experiencing pandemic-related hardships, the unanticipated withdrawals indicate just how pervasive the effects have been. While it is not possible for us to identify whether or how disease transmission might be linked to climate change, the pandemic does offer some instructive parallels with the climate crisis that can add to our understanding of what to expect if it continues unabated.

One such parallel is that like climate change, the COVID-19 pandemic is a global disaster that requires global cooperation to solve. Just as the success of climate change mitigation hinges on choices made by individual states, so too does the control of disease transmission (UN, 2020a). Unfortunately, and also in line with the climate crisis, meaningful international cooperation has not been forthcoming. A key reason is that both a clean atmosphere and the international public health efforts required to fight the virus are 'global public goods,' non-rival, non-excludable resources that, in absence of market pressure, require cooperative governmental intervention to supply (Brown and Susskind, 2020). Sadly, the Olsonian (1965) collective action challenges that hinder interstate cooperation in the best of times are even more burdensome when they require policy makers to implement measures – lockdowns for COVID-19, and limitations on greenhouse gas (GHG) emissions for climate change – that generate economic hardships within their respective jurisdictions. Also similar to climate change is that the solution to COVID-19 hinges, at least in part, on the development of new technology that must be made cheap and widely available to be effective – in this case a vaccine rather than a sustainable energy source. We might have more optimism regarding COVID-19, given the tremendous resources devoted to vaccine development, though these efforts will also depend on the nature and extent of reinfection (Kissler *et al.*, 2020; Shaman and Galanti, 2020). The same optimism is less warranted in the case of climate change, as almost three decades of negotiations have produced no global agreements or technologies that will facilitate GHG emission reductions sufficient to keep global temperatures at or below pre-industrial levels, or even below those levels required to stave off major crises in vulnerable countries (IPCC, 2018).

The second, and more relevant (for our purposes) parallel between COVID-19 and climate change is that the impact of both crises falls disproportionately on the marginalized and vulnerable, especially women. A report published in 2020 by UN Women claims that the pandemic is likely to force nearly 100 million people into extreme poverty, with women and girls experiencing the brunt of these effects (Azcona *et al.*, 2020). Just as with climate change, a major reason why women's livelihoods have been so disproportionately affected during the pandemic is the persistence of pre-existing gender inequalities in the economy and society. Labor force concentrations offer one account, as feminized sectors such as healthcare, social work, food service, domestic work and childcare, and hospitality have taken substantial economic hits. Indeed, another report finds that women's rate of job loss has been 1.8 times greater than that of men, and that while women comprise only 39% of employees worldwide, they account for 54% of all lost jobs (Madgavkar *et al.*, 2020).

Yet, as we found in our experience putting this book together, gendered labor force concentrations do not tell the whole story. The pandemic is also forcing women to withdraw from the labor market in greater numbers than men in order to accommodate the rising demands of unpaid care work, including cooking, cleaning, caring for children, the elderly and sick, and the gathering of food, water, and biomass necessary to sustain a family and household. In India, for example, one survey finds that since the pandemic began 9 months ago, women's unpaid domestic work has increased by 30%, an amount which, at current minimum wage rates, is valued at US$300 billion (Purushothaman *et al.*, 2020).

These changes can strengthen the barriers women face in pursuing or sustaining formal employment, or in engaging in entrepreneurship, including the building of small businesses which have been so vital to women's economic participation across the developing world. The increased domestic burdens are also likely to hinder or reverse gains in gender equality across societies, especially if they lead to greater workforce automation, which could further depress women's employment, or most ominously, if they entrench or magnify regressive attitudes regarding a woman's place in society (Madgavkar *et al.*, 2020).

In addition to economic challenges, the pandemic, like climate change, has also imposed significant costs on women's minds and bodies. Home confinement has led to increased reports of domestic violence (Evans *et al.*, 2020). The relatively greater role women play in caregiving, whether in hospitals or the home, has exposed them to higher viral loads, and has put women at greater risk of suffering from mental health conditions such as depression and anxiety (CDC, 2020). The lockdowns and travel restrictions implemented to stem the pandemic have also reduced women's access to health facilities, especially those related to sexual and reproductive health, because in many places these services were not deemed to be essential, or the facilities have been adapted to treat COVID-19 cases (Cousins, 2020). Tragically, the impact of these restrictions is likely to be a significant increase in unintended pregnancies, unsafe abortions, and pregnancy-related fatalities, especially for impoverished women in developing countries.

As this discussion suggests, the impacts of the COVID-19 pandemic on women and women's livelihoods correspond closely with those imposed by the climate crisis, albeit at greater average rates and over a shorter period of time. In this sense, the COVID-19 crisis offers us a microcosm at the macroscale – a temporally condensed but globally distributed vision of the effects that climate change has and will continue to have absent concerted efforts to combat gendered disparities in climate vulnerability, adaptation capacity, and participation in climate policy and decision making. Discussion of COVID-19 is also useful because the gender disadvantages laid bare by the pandemic are likely to exacerbate the gendered impacts of climate change in the post-pandemic period. While the COVID-19 crisis is still ongoing as of this writing – the first vaccines have not yet hit the market – experience from past epidemics is instructive: the West African Ebola crisis, for example, had longer-lasting negative economic effects on women than men because male incomes recovered far more quickly than female incomes in the aftermath (UN, 2020b). Recovery from COVID-19 will also have a significant gender dimension given women's lower access to educational resources, assets such as land and credit, and opportunities in the formal labor market. To thwart these negative outcomes, gender must be mainstreamed into COVID-19 recovery policies in order to reduce the gender inequalities enhanced by the pandemic. On the other side, it would also be helpful to integrate environmental concerns into COVID-19 recovery to ensure that the reductions in global GHG emissions fostered by the pandemic are sustained when economic activity again reaches pre-pandemic levels.

References

Azcona, G., Bhatt, A., Encarnacion, J., Plazaola-Castaño, J., Seck, P., Staab, S. and Turquet, L. (2020) From insight to action: gender equality in the wake of Covid-19. UN Women, New York. Available at: www.unwomen.org/en/digital-library/publications/2020/09/gender-equality-in-the-wake-of-covid-19#view (accessed 3 February 2021).

Brown, G. and Susskind, D. (2020) International cooperation during the COVID-19 pandemic. *Oxford Review of Economic Policy* 36(Suppl. 1), S64–S76.

CDC (2020) Women, caregiving, and COVID-19. Centers for Disease Control and Prevention, Atlanta, GA. Available at: www.cdc.gov/women/caregivers-covid-19/index.html (accessed 3 February 2021).

Cousins, S. (2020) COVID-19 has 'devastating' effect on women and girls. *Lancet* 396, 301–302.

Evans, M.L., Lindauer, M. and Farrell, M.E. (2020) A pandemic within a pandemic – intimate partner violence during Covid-19. *The New England Journal of Medicine.* Available at: www.nejm.org/doi/full/10.1056/NEJMp2024046 (accessed 3 February 2021).

IPCC (2018) Global warming of 1.5°C: an IPCC Special Report on the impacts of global warming of 1.5°C above pre-industrial levels and related global greenhouse gas emission pathways, in the context of strengthening the global response to the threat of climate change, sustainable development, and efforts to eradicate poverty. Intergovernmental Panel on Climate Change, United Nations, Geneva, Switzerland. Available at: www.ipcc.ch/sr15/ (accessed 3 February 2021).

Kissler, S.M., Tedijanto, C., Goldstein, E., Grad, Y.H. and Lipsitch, M. (2020) Projecting the transmission dynamics of SARS-CoV-2 through the post-pandemic period. *Science* 368, 860–868.

Madgavkar, A., White, O., Krishnan, M., Mahajan, D. and Azcue, X. (2020) COVID-19 and gender equality: countering the regressive effects. McKinsey Global Institute. Available at: www.mckinsey.com/featured-insights/future-of-work/covid-19-and-gender-equality-countering-the-regressive-effects (accessed 3 February 2021).

Olson, M. (1965) *The Logic of Collective Action: Public Goods and the Theory of Groups*. Harvard University Press, Cambridge, MA.

Purushothaman, R., Madgavkar, A. and Pandit, V. (2020) Opening with care. *Hindustan Times, 19 June 2020*. Available at: www.hindustantimes.com/analysis/prioritise-care-work-to-integrate-women-working-from-home-into-the-economy/story-QCCWBFL2j5Qm6G6NuULm1O.html (accessed 3 February 2021).

Shaman, J. and Galanti, M. (2020) Will SARS-CoV-2 become endemic? *Science* 370, 527–529.

UN (2020a) COVID-19 pandemic demonstrates multilateral cooperation key to overcoming global challenges, President stresses as General Assembly concludes annual debate. United Nations, New York, 29 September 2020. Available at: www.un.org/press/en/2020/ga12273.doc.htm (accessed 3 February 2021).

UN (2020b) The impact of COVID-19 on women. *United Nations Secretary General Policy Brief, 9 April 2020*. Available at: www.unwomen.org/en/digital-library/publications/2020/04/policy-brief-the-impact-of-covid-19-on-women#view (accessed 3 February 2021).

Index

Note: The page numbers in bold and italics represents tables and figures respectively.

access levels 21
activism/emissions reductions 198–199
Addressing the gendered impacts of climate change on livelihoods 7
Africa
  agricultural systems 107
  CASU 117
  CSA (*see* climate-smart agriculture (CSA), Africa)
  female farmers 107, 108
  GCCASP 115–117
  gender 107–109
  PROMARA project 117–118
African National Congress (ANC) 201
African Union Development Agency (AUDA) 115
agricultural feminization 144
agricultural livelihood strategy 7
agriculture
  care work 83–85
  corn and beans 84
  Uganda
    annual crops 39
    livelihoods 39
    occupation 53
    production 39
    seeds and fertilizers 39
    women's vulnerabilities 40–41
agroforestry 59
Akaike Information Criterion (AIC) 100
Alianza Internacional de Reforestación (AIRES) 200
Anti-Privatisation Forum (APF) 201
anxiety 75–76
asset ownership 145
Australia climate-induced disasters 179–180
Australian Centre for International Agricultural Research (ACIAR) 177

Badan Pusat Statistik (BPS) 99
Bangladesh
  achievement 215
  adverse impacts 123
  agroecological zones 123
  appreciable rate 215
  awareness and adoption 125–126
  civil society sector 217
  climate change policies 212–213
  climate-induced disasters 178–179
  climatic events 11
  community-based approaches 216
  coping strategies (*see* coping strategies)
  CSA (*see* climate-smart agriculture (CSA), Bangladesh)
  cyclones 11
  data, methods and study sites 27–29
  family farming 124
  gender differences and information types 128–129
  gender mainstreaming 26, 212–213
  political instabilities 216
  recommendations 23, 34–35
  sex-disaggregated data 124
  site description and sample descriptives 127–128
  socio-political structures 216
  study setting and procedure 12–14
  survey and data collection approach 126–127

Bangladesh (*Continued*)
theoretical underpinning 26–27
victims, women 215
vulnerability (*see* vulnerability)
weather events 123
women's access 20–23, *22*
Bangladesh Bureau of Statistics (BBS) 14
Bangladesh Climate Change Strategy and Action Plan (BCCSAP) 212
Bangladesh National Web Portal 14
Bayesian Information Criterion (BIC) 100
bimodal cycle 39

capability/aspirations approach 162, 166
capital assets 162
caregivers 65
care work
advantage 90–91
climate change and agriculture 83–85
definition 83
framework 82
subsistence agriculture 83, 86–90
theory 85–86
Central Intelligence Agency (CIA) *72*
Char Livelihoods Program (CLP) 216
citizen science 201–202
civic engagement programs 84
civil society organizations 217
climate adaptation 220
climate adaptation planning 215
climate change; *See also* gender; livelihoods
Bangladesh (*see* Bangladesh)
care work 83–85, 87–90
developing countries 3–4
double injustice 1, 2
food security and nutrition 38
gender implications 220
gender issue 1, 220
goals 5
laissez-faire approach 221
policy engagement 2–3
threat distribution 1
Tunisia 143–144
Uganda
climate events and weather shocks 39
international frameworks 40
mean annual temperature 39
socio-economic impacts 40
variability 39
vulnerabilities and fostering adaptations 221
water 75
women and livelihoods 221
women's agency 221
Climate change, agriculture and food security (CCAFS) 125
climate disruption 1
climate governance 196
climate-induced disasters
agricultural communities 174
Australia 179–180
Bangladesh 178–179
catastrophic dimensions 174
critical points 175
developing policies 174
frequency and intensity 174
gender and livelihood strategies 175–176
gendered analysis 181–182
gendered livelihood adaptations 174
Lao PDR 177–178
methodology 177
outmigration 176–177
reserve army of labor 180
climate-induced migration 159, 160
climate refugees 160
climate-sensitive industries 6
climate-smart agriculture (CSA)
Africa
best practices 118–119
concept 109–110, **111–112**
gender consideration (*see* gender consideration)
gender inequalities 108
gender-responsive 115
gender-sensitive actions 114–115
GHG emissions 107
interventions 108, 119
principles **111–112**
Bangladesh
adaptive capacity and resilience 123
awareness and adoption 125–126, 129–138, **130–133, 136–137**
policy implications 138–139
practices 123
women's livelihoods 7
Common Market for Eastern and Southern Africa (COMESA) 110
communication infrastructure 21
community-based education 199–201
Comprehensive Disaster Management Program (CDMP) 216
Comprehensive Food Security and Vulnerability Assessment (CFSVA) 41
Conference of the Parties (COP) 175
Conservation Agriculture Scaling-Up (CASU) 117
Co-Operative and Policy Alternative Centre (COPAC) 201
coping strategies
cyclone experiences 29–30
food preparation without money 31
gender
commonalities 33
family displacement 31–32
migration 32
violence 32–33

household dynamics 26, 27
indirect effects 29
jobless men and earning women 29
male outmigration 27
post-disaster periods 29
resilience 33–34
riverbank erosion and floods 30–31
women against climatic events **19,** 19–20
corporeal–material–socio-cultural framework 83, 86
corruption perception 216
Corruption Perceptions Index 216
COVID-19 pandemic 222–223
COVID-19 recovery policies 223
crop yields 82
cultural differences 82

decision making power 145
devaluation framework 89
disaster crunch model (DCM) 6, 11, 12
disaster loans 19
disaster-prone regions 26
disease transmission 222
displacement
challenges 57
ethics and economies 62–64
forced movement 56
identity with land and sustenance 64–65
land dispossession 59–60
disproportional climate change 196
domestic violence
climate shocks 103
climatic instability 94
control variables 99
dependent variable 97
economics and livelihoods 95–96
independent variable 97–99
methodology and results 99–102
prevalence 94
rice-planting seasons 95
statistics **98**
drought adaptation mechanisms 157

economic development 40, 78, 79, 103, 213
economic dynamism 187
economics 95–96
ecosystem services 15
education 99, 102, 169
Empowering Women for Improved Food and Nutrition Security (EWIFNS) 116
energy transition 198
enumeration areas (EA) 41
environmental conservation efforts 59
environmental degradation 201
environmental refugees 160
epitomes of distress 65–66
erosion-hit communities 34
ethnocentric governance 62
evapotranspiration 96

family displacement 31–32
family lineage 64–65
female-headed households
food insecurity 45
hypothesis 47
percentage *42*
South Africa 196
statistics 42
wealth quintiles *44*
Feminist Dalit Organization (FEDO) 217
feminist movement 199
feminist political ecology (FPE) 70–71
feminization
labor 156, 186
managerial 156, 186
victimisation 181
financial stress 96
focus group discussions (FGD) 13, 17, 28, 57
fodder production 157
Food, Agriculture and Natural Resources Policy Analysis Network (FANRPAN) 113
Food and Agriculture Organization (FAO) 3–4, 38, 109, 211
food domains 86
food insecurity
causes *45*
coping strategies 41
determinants 49
education level 53
family size 53
global challenges 38
household disaggregated, gender 43–45, **44**
national- and urban-based households 39
weather shocks 49
women 17
food security
agricultural production 5, 38
dimensions 17
education rises 169
floods and irregular rains 49
household dataset 38
weather shocks 46–48
forest land management 63

gender; *See also* climate change; livelihoods
African context 108–109
characteristics 1
climate-induced disasters 175–176
commonalities 33, **34**
defined 107
developing countries 3–4

gender (*Continued*)
 equality 2
 family displacement 31–32
 ideologies 78
 impacts 4–5
 inequities and inequalities 40
 lens 161, 170–171
 linkages 2
 local narratives 75–76
 migration 32
 norms and disparities 40
 policy engagement 2–3
 responsive 2
 social construction 1–2
 Tunisia 143–144
 Uganda household characteristics disaggregated 42–43, **43**
gender-based violence (GBV) 27, 32–33
Gender, climate change and agriculture support program (GCCASP) 115
gender–climate gap 198
gender consideration
 agricultural productivity 113
 cash crops 113
 disparity 113
 domains 114
 female farmers 113
 formal and non-formal sphere structures 114
 gender-transformative 114
 income constraints 113
 land law and policy reforms 113
 land rights 112
 livestock farming systems 112
 paying attention 112
 rural communities 112–113
 structures 110
gender-disaggregated vulnerability analysis 201
gender disparities 220
Gendered Livelihood Vulnerabilities to Climate Change 6
gendered migration patterns 166–169, *167*, *168*
gender empowerment measure (GEM) 99
gender mainstreaming 108, 116–117, 211
gender matters
 disaster and weather-related events 164–165
 heterogeneous knowledge 163
 natural resources and natural capital 163–164
 pathways 163
gender-responsive 115
gender-sensitive actions 114–115
Global Climate Rate Index 212
globalized river basin development 69
government reforestation scheme 62
Greenbelt Movement 203
greenhouse gasses (GHG) 1, 107, 175
Green Morocco Plan (GMP) 187
gross domestic product (GDP) 26, 212
gross value added (GVA) 212
group-level variations 4

Heckman selectivity model results 129–138
Heckman selectivity regression model 125
home-based energy conservation practices 198
Homestead orchard development (HOD) 116
Honduras case study 83–85
Honduras Secretary of Agriculture and Husbandry 84
household dynamics change 27
human mobility 159

IMPACTLite 127
Inclusive Climate Change Adaptation for a Sustainable Africa (ICCASA) 115
income expenditures 190–192
in-depth interviews (IDI) 28, 57
Indonesia
 domestic violence (*see* domestic violence)
 economics 95–96
 hypothesis 96–97
 information and descriptive statistics **100**
 livelihoods 95–96
 subsistence resources 94
 vulnerable targets 94
Indonesian rice-planting season 102
informal economy 84
infrastructure development 6
infrastructure projects 70, 79
integrated pest management (IPM) 129, 134
Intergovernmental Panel on Climate Change (IPCC) 5, 11, 160
internal displacement 56
Internal Displacement Monitoring Centre (iDMC) 56
internally displaced persons (IDP) 56
International Association for Feminist Economics (IAFFE) 201
International Labor Organization 204
international river basin development 79
intersectionality 70–71
irrigation technologies 155

Joint Permanent Technical Commission 73

Kenya
 climate change 60–61
 displacement (*see* displacement)
 economic situations 67
 epitomes of distress 65–66
 family lineage 64–65
 land dispossession 59–60
 land overview and land tenure 57–59
 landlessness 60–62

livelihoods 60–61
materials and methods 57
post-independence 56
senior women's health 61–62
well-being 60–62
women's health 60–61
Kenya National Bureau of Statistics (KNBS) 63
key informant interviews (KII) 28
koukh 192

labor availability 135
labor feminization 156, 186
labor force 222
Land Consolidation and Registration (LCR) 58
land dispossession 57, 59–60
land ownership systems 56
land reform movements 85
land tenure system
history 57–59
ill-conceived forest conservation strategies 62
Lao PDR 177–178
large dams 69
large-scale infrastructural development
consequences 70
feminist political ecology and intersectionality 70–71
LHWP 73–75, *74*
research context 71–73
water (*see* water)
Lesotho Highlands Development Authority (LHDA) 73
Lesotho Highlands Water Commission 73
Lesotho Highlands Water Project (LHWP) 69, 73–75, *74*
livelihoods; *See also* climate change; gender
adaptation and aspirations 169–170
care perspectives 199, 203–205
climate-induced disasters 175–176
climate mitigation and adaptation 2
climate-sensitive 3
developing countries 3–4
diversification 1
economics and domestic violence 95–96
household dynamics 29–33
intersectional, gendered impacts 196–198
Kenya 60–61
local narratives 75–76
preservation 196
security 40
Uganda 39
water 75
women's involvement in livestock 144–145
livestock 143–144
Local Adaptation Program of Action (LAPA) 214
local food systems 90
Local Self-Governance Act (LSGA) 215
low-value agriculture 89, 90

Maasai Wilderness Conservation Trust (MWCT) 200
male-headed households
food insecurity 45
statistics 42
wealth quintiles *44*
male-dominated political system 217
male outmigration 27
managerial feminization 156, 186
mangrove plantations 3
material domain 87
microcredit loans 85
microcredit schemes 217
migration 32
monocausal explanatory approaches 160–161
multilateral development banks (MDB) 217
multipurpose cyclone center (MPCS) 18

national adaptation plans (NAPs) 211, 221
National Adaptation Program of Action (NAPA) 212–214
National Agriculture Marketing Council (NAMC) 113
national development plan (NPA) 54
nationally determined contributions (NDC) 5
National Social Protection Policy 65
National Violence Monitoring System (NVMS) 94
natural disasters 5, 26
negative binomial regressions 99–100, **101, 102**
Nepal
adaptation and mitigation strategies 214
bottom-up and top-down approaches 215
climate change policies 213–214
gender mainstreaming 213–214
hazards 212
male-dominated political system 217
political instabilities 216
political participation and representation 217
socio-political structures 216
victims, women 214
networking 199–201
nexus, climate and migration
adaptation/livelihood strategy and aspirations 169–170, *170*
climate, environment and migration 159–161
gender
education and age 169
lens 161, 170–171
matters 163–165
migration patterns 166–169, *167, 168*
integrating gender 161–163
public debate 159
West African Sahel 165–166
non-governmental organizations (NGO) 17

Older Persons Cash Transfer Program (OPCTP) 65
ordinary least squares (OLS) regression 47

Orphans and Vulnerable Children – Cash Transfer (OVC-CT) 65
outmigration 144, 176–178, 181, 182

Paris Climate Agreement (PCA) 2, 3
pastoralist communities 59
Plantation Establishment and Livelihood Schemes (PELIS) 63
Poisson regressions 99
policy document reviews 57
policy-table approach 204
Population Reference Bureau (PRB) 14
potential evapotranspiration (PET) 98
poverty 94, 99, 100, 102, 160, 213, 215
poverty eradication 40
probability proportional to size (PPS) 41
Prodifil project 150
PROMARA project 117–118
public engagement 113

recognize, reduce/redistribute care 86
REDD Preparedness Plan (RPP) 214
Reducing Emissions from Deforestation and Degradation (REDD) 214
Reducing Vulnerability through Climate Change (RVCC) 216
regression analysis 48–53
reserve army of labor 180
resilience 33–34, 83, 202
'the river shook the earth' 76–77
risk management strategy 90
rural–urban migration 71
Rural Women's Network (RUWON) 217

Saiss, Morocco
- adaptation measures 185
- case study *186*, 187
- decision making, migration and gender 186–187
- extension programs 193
- gendered decision making
  - asset-related income 192–193
  - assets 192
  - income expenditures 190–192
  - research communities *189*, **191**, *191*
  - work participation 188–190
- informal settlements 185
- methods and data analysis 187–188
- social and political transformation 185
- water sources 193

Sathi's experience 30
securing women's land tenure rights 34
self-exclusion 202
sex disaggregation 171
sexual and reproductive health 223
sexually transmitted infections (STI) 72
Social Afforestation Programs 215
social contexts 197
social–ecological systems (SES) 162–163, 166
social exclusion 64
social inequality 161
social injustice 56
social marginalization 161
social protection policies 83
social vulnerability 82, 162
societal resilience 220
socio-cultural systems 33, 87
solar technologies 3
Standardized Precipitation and Evapotranspiration Index (SPEI) 94, 98, *99*, *101*, *103*
Strategic Program for Climate Resilience (SPCR) 217
stratification 41
sub-Saharan Africa (SSA) 38
subsistence agriculture
- food domains 86
- material analysis 87–88
- material domain 87
- natural dispositions 87–88
- reframing 83
- socio-cultural realm 87
- value 88–90
- women's participation 83, 86

Sustainable Development Goals (SDG) 54
sustainable economic growth 40
sustainable livelihoods approach (SLA) 6, 161–162, 197
sustainable livelihoods research 171
Swynnerton Plan 58

techno-economic rationales 70
third national development plan (NDP III) 40
Trans Caledon Tunnel Authority (TCTA) 73
transformative approach 86
Tunisia
- asset ownership and decision making power 145
- climate, implications and coping mechanisms 148–150, **149–151**
- credit sources **152**
- feminization 156
- gender roles 145–147, **146–147**
- grazing responsibility **149**
- innovation availability and adoption in past 5 years 154–156, **155**, **156**
- land ownership **147**
- livestock
  - climate change and gender 143–144
  - control **148**
  - livelihoods 144–145
  - ownership **148**
- methodology 144
- reasons for not obtaining credit **152**
- research findings 144
- rural services and resilience 150–154
- training last attended **154**

training through extension programs **153**
visits by extension agents **153**

Uganda
agriculture, climate change and variability 39–40
data 41
gender
household characteristics disaggregated 42–43, **43**
household food insecurity disaggregated 43–45, **44**
policy recommendations 53–54
regression analysis 48–53
seasonal rainfall 39
weather shocks (*see* weather shocks)
women's vulnerabilities and agriculture 40–41
Uganda National Meteorological Authority (UNMA) 39
Uganda National Panel Surveys (UNPS) 42
unemployment 99, 102
unexploded ordinance (UXO) 178
unimodal cycle 39
United Nations Framework Convention on Climate Change (UNFCCC) 2, 40, 175, 212
Upazila Parishad Manual 14

vaccine development 222
Village Development Committees (VDC) 215
vulnerability
adaptation capacity 3, 124
assessment 12
climate change 2, 7
concept 11–12
DCM 12
education 17
male- and female-headed households 46
marital status 48
motivational/attitudinal 12
proxy 49
resilience 83
women (*see* women vulnerability)
vulnerable group development (VGD) 20

wage-labor migration 8
water
agent 76–77
anxieties 69
as commodity 78–79
as risk 77–78
climate change and livelihoods 75
commodification 69
gender, anxiety and livelihoods 75–76
purpose, delivery scheme 73
weather shocks
climate variability and welfare effects 38
differential effect 49
effects 38
empirical estimation, gendered effect 46–48
gender of household head (HH) and food insecurity **48, 50–52**
households 45, **46**
1 year preceding survey 46, *47*
survey captured information 41
West African Sahel 165–166
women
access to major facilities 20–23, *22*
bottom-up climate action 203–205
coping strategies 11, **19**, 19–20
economic empowerment 217
employment 2, 28–29
empowerment 85
engagement 181
involvement 144–145
livelihood perspectives 197–198
organizing and participation 198–199
socio-economic conditions 28
victim, survivor and rescuer 33–34
vulnerability (*see* women vulnerability)
women's collective activism
citizen science and data collection 201–202
commoning 203
community-based education and networking 199–201
innovative strategies 199
youth movements 202–203
women vulnerability
agriculture, Uganda 40–41
areas and extent *16*
categories 15
climatic events/hazards 15, 16
communication 18
decision making processes 15
disaster awareness and preparedness 19
ecosystem services 15
education system 17
FGD participants 17–18
food security 17
health-related issues 17
household assets 14
informant 17–18
land and security of tenure 14
limited education 15
physical assets **14**
prevailing/consequential conditions 14
rural sanitation facilities 18
survey, cyclone Aila 16
weather events 14
workforce automation 223
work participation 188–190
World Food Program (WFP) 41
World Health Organization (WHO) 15

youth movements 202–203